Earthquake Engineering

Earthquake Engineering

Edited by **Agnes Nolan**

R Callisto Reference

New York

Published by Callisto Reference,
106 Park Avenue, Suite 200,
New York, NY 10016, USA
www.callistoreference.com

Earthquake Engineering
Edited by Agnes Nolan

International Standard Book Number: 978-1-63239-156-8 (Hardback)

Printed in the United States of America.

Contents

Preface

This book has been a concerted effort by a group of academicians, researchers and scientists, who have contributed their research works for the realization of the book. This book has materialized in the wake of emerging advancements and innovations in this field. Therefore, the need of the hour was to compile all the required researches and disseminate the knowledge to a broad spectrum of people comprising of students, researchers and specialists of the field.

The book aims to educate the readers with information regarding the field of earthquake engineering. The mitigation of earthquake-related hazards is essential in the modern society. The mitigation of such kind of hazards extends from detailed studies on seismicity, assessment of sites affected and seismo-induced landslides. It also expands from analysis of disasters like tsunami along with the design and analysis of structures to refrain such actions. The study of earthquakes integrates science, technology and expertise in infrastructure and engineering in an endeavor to curtail human and material losses due to earthquakes. This book deals with a variety of topics aiming to mitigate geo-hazards like seismic hazard analysis, ground investigation for seismic design, seismic design, assessment and remediation, earthquake site response evaluation and soil-structure interaction assessment.

At the end of the preface, I would like to thank the authors for their brilliant chapters and the publisher for guiding us all-through the making of the book till its final stage. Also, I would like to thank my family for providing the support and encouragement throughout my academic career and research projects.

Editor

Geophysical Techniques

The Use of Source Scaling Relationships in the Simulation of a Seismic Scenario in Mexico

Alejandro Gaytán, Carlos I. Huerta Lopez,
Jorge Aguirre Gonzales and Miguel A. Jaimes

Additional information is available at the end of the chapter

1. Introduction

The general knowledge of seismicity of the Mexican Pacific Coast has been described in terms of its relationship within the regional context with the Middle America Trench, the convergence rate observed and predicted by plate tectonics processes, the use of seismic gap theory for forecast earthquakes on the region (Reyes [1]; Eissler and McNally [2]; Singh [3]; as well as Heaton and Kanamori [4]; among others) and prediction of strong ground motions using empirical relationship and seismic source scaling procedures (Somerville [5] and Somerville [6]).

The work here presented, mainly deals with the Colima-Jalisco region (ruptures areas of 1932, and 1973 earthquakes) as well as the remained gap of the Tecoman (2003) earthquake. The conduction of studies in this region to simulate seismic scenarios acquires special significance because of the hazard of an eventual large magnitude earthquake. For the large earthquakes recording in this region last and present century the rupture areas of the 1932, 1973 earthquakes and the small gap remainder from Tecoman earthquake the seismic convergence period have been exceeded or is in its limit.

Let us look into this with detail. Firstly, for the earthquakes of 3 and 18 June 1932, Nishenko and Singh [7] found that the average displacement on the fault of 1932 earthquakes as 155 cm. Pardo and Suarez [8] found that subduction rate of Rivera plate in this region is estimated to be from 2 to 5 cm/yr, take the small estimation of convergence rate of 2 cm/yr yields a recurrence period of about 77 years assuming that the convergence is entirely taken up by the seismic slip (Nishenko and Singh [7]). However considering a rate larger than 2 cm/yr the recurrence period could be much lower. The 79 years elapsed since 1932 indicate that the recurrence period has expired. Although the rupture area of Manzanillo 1995 Mw 8.0 earthquake invades the rupture area of 1932 earthquakes, but represent less than 15 % of rupture area. Secondly, for the January 30 1973 earthquake (Mw 7.3) Reyes [1] estimate an average slip of 144 cm. For this region Minster [9] estimate a convergence rate of 5.6 cm/yr, this suggest a recurrence time period of about 25 years Reyes [1]. This recurrence period consider the various uncertainties and is in very good agreement with the interval between this earthquake and the preceding 1941 earthquake (32 years). The 38 years trascurred since 1973 indicate that the recurrence period have expired Reyes [1].

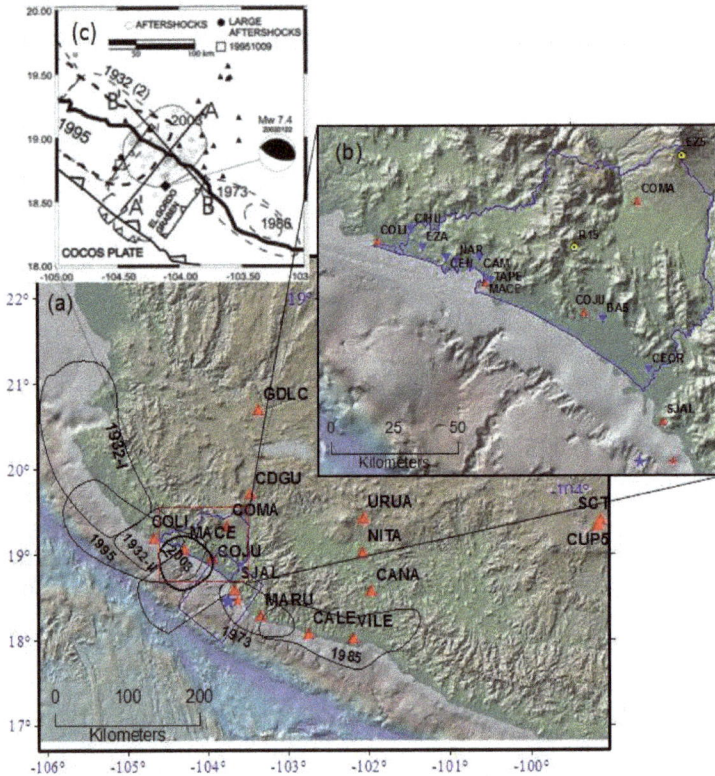

Figure 1. Geographic regional map. (a) Contours show historic earthquakes in region (modified from Singh [10], local and regional stations where the 13 August 2006 earthquake (small star) were recorded: red triangles, big star epicentral location of simulated earthquake. (b) Local and regional stations within Colima state where earthquake was recorded; blue triangles, IINGEN array and pentagons, RESCO array. (c) From Quintanar [11], map showing the best aftershock locations (empty circles). Large full square, shows the epicenter location of the Tecoman earthquake. Its fault plane solution is also shown. Large empty square shows the epicenter location of the Colima–Jalisco earthquake of 1995 (Mw = 8.0). Profiles AA' and BB' are cross sections onto which aftershocks were projected. Triangles show the locations of seismic stations.

Thirdly, is the existence of a seismic gap in the region located between the rupture areas of the 1973 and the Tecoman earthquakes. In figure 1 Quintanar [12] shows that the aftershocks location of the Tecoman earthquake lies north of El Gordo graben and the aftershock area encompass part of the rupture area of the 1932 and 1995 earthquakes. The area between the limits of the rupture areas of the 1995 and the1973 earthquakes is what has been called the Colima seismic gap. The northwest area of this gap ruptured with the Tecoman earthquake in 2003. The other half of the gap, roughly to the southeast, remains quiet Quintanar [12].

The later show the existence of 3 different zones where is necessary to make a simulation of strong ground motions to estimate the response spectra. From these 3 zones the more important (by magnitude and area) is the region broken by earthquakes of 1932; however, the possibility of conduct a simulation in this area is limited by the poor instrumentation in the past and therefore the absence of data the region.

In the case of small gap remainder from Tecoman and the rupture area of 1973 earthquakes, the Mw 5.3 earthquake of August 13, 2006 (element event) offered the opportunity to generate a seismic scenario for the studied area that constitutes a potential seismic source in this region in the near future. We apply: (i) the empirical Green's function method (EGFM) proposed by Irikura [13] to estimate the peak ground acceleration (PGA) and the acceleration time histories using the Somerville [6] relationship allowing us scaling from moderate to major earthquake. Traditionally the EGFM is well known technique commonly used to simulate an already occurred earthquake. The scientific community thinks that would be of great benefit if this technique can be applied to forecast an expected earthquake in Mexico. Under this premise, an immediate doubt arise because the absence of observed records (event we intent simulate) to compare with synthetics. Conduct a study to simulate strong ground motions under this situation makes difficult to constrain and validate the results, but at the same time is one of the more important challenge simulating a future earthquake. In order to overcome the absence of observed records two methodologies were applied: (i) The first one was Somerville [6] relationship together with the EGFM, and (ii) the second one was ground motions predictive equations (GMPE) relationships to make accurate estimation of strong motions. Aguirre and Irikura [14] used acceleration records of Mexican subduction earthquakes to validate the Somerville [5, 6] relationship for subduction earthquakes and found that the asperities size were well predicted by this relationship. The EGFM is a well established methodology used in the field of earth sciences to estimate the ground motions. What the EGFM requires to estimate the ground motions according to the tectonic conditions of the region, are the fault parameters determination. Works like the one done by Irikura [15] address the study of the factors involved with the inner and outer fault parameters of the source to do more accurate estimations of ground motions. Those outer and inner fault parameters are estimated from the inversion of the waveform in studies of rupture processes using strong ground records. Using the results of many kinematic inversions, Somerville [6] obtained some relationships that synthesize the main characteristics of the earthquakes, stated as follows: (i) the seismic moment, (ii) the rupture area, (iii) the slip average, (iv) the combined area of asperities, and (iv) the area of the largest asperity, among others. In this study we are using these relationships because of their great utility in the estimation of seismic ground motions. The above statement provides an efficient way to work when a limited number of parameters to be considered in numerical simulations is available.

Finally from both estimations; it is said the PGAs, and the curves obtained with the two GMPE here used: (Ordaz [16], and Young's [17]), an trial and error iterative process of residuals minimization was conducted to identify the result that in the statistical sense better matched with the two GMPE here used.

The main contribution of this method is that it reflects a model that considers the source, the path, and the site effects. Another important contribution of the method is that reliable estimations about the energy distribution can be achieved in the high frequency band (between 0.1- and up to 10-Hz). This frequency range is of engineering interest because of the following reasons: (i) Many structures, including tall buildings and long bridges have their natural frequencies in the above frequency range, and (ii) 8 of the 10 major cities of this state are located in the sedimentary basins of the Colima graben and could amplify the ground motions in the frequency range of 0.1 to 10 Hz. It is therefore important to investigate how ground motions up to 10 Hz are generated from great subduction-zone earthquakes. This kind of investigations play a vital role in the effort to propose an scenario of strong ground motions from future large subduction earthquakes in the area in study and to evaluate the performance of structures subject to ground motions.

2. Tectonic

The Colima state is located in Mexico's Pacific coast. The tectonic of the region is complex, in which the Rivera, the Cocos, and the North American plates converge. In addition to the above, the existence of a microplate has also been proposed by DeMets and Stein [18], and Bandy [19]. There are significant changes in the parameters of the subduction process along the subduction zone on the Pacific coast of Mexico, which has been divided in four sections by Pardo and Suárez [8]. Although the dip of the interplate contact geometry is constant to a depth of 30 km, lateral changes in the dip of the subducted plate are observed once it is decoupled from the overriding plate. In front of the Jalisco block, the Rivera plate has a dip of 45° and its subduction rate below the North American Plate is estimated to be from 2 to 5 cm/yr. The Cocos plate below Colima shows a similar dip to that of the Rivera but the subduction rate below the North American Plate is estimated to be from 4 to 6 cm/yr. To the south, the dip of the Cocos plate decreases gradually and is almost sub-horizontal at Guerrero (where it subducts with a velocity from 6 to 7 cm/yr) before increasing again farther south to the large values observed in Central America. Pardo and Suárez [8] explained the observed no parallelism between the volcanic belt and the subduction zone by these large lateral variations.

3. Data

To simulate acceleration time histories of the Tecomán earthquake, we used two sources of data. Firstly, records of the 13 August 2006 earthquake. These records were obtained from previous temporal campaign in that area and from the support of other institutions with permanent instrumentation in that zone. Its epicenter, focal mechanism, and seismic moment were obtained from Centroid Moment Tensor Project [20]. The location of instruments that recorded this event (figure 1) is next described. The instruments were from permanent seismic networks: 15 Etna episensor wideband accelerographs from d.c. to 200 Hz at 200 samples per second from the national accelerations network of Instituto de Ingeniería (IINGEN) of Universidad Nacional Autonoma de Mexico (UNAM); two Guralp CMG40T-DM24 flat response wideband velocity type seismographs from 0.5 to 100 Hz at 100 samples per second from the network Red Sismica del Estado de Colima (RESCO). Secondly, data from temporal networks installed in the region as part of this project as follow: (i) four Altus Etna wideband accelerographs from d.c. to 100 Hz at 100 samples per second, four (ii) Geosig strong-motion recorder model 18 with analogue-digital converter, wideband accelerometers from d.c. to 100 Hz recording at 100 samples per second. Because 2 of the 25 records used in this study were velocity records, it was necessary to transform them to acceleration. Also, it was necessary to remove the instrumental response of each of the different instruments.

4. Method

The method used to model the target event requires a small magnitude event (earthquake of 13 August 2006) called element event, with hypocenter in close proximity to the earthquake that we want to simulate. For this particular case, the magnitude, location, focal mechanism and source parameters of the element event was reported by Harvard CMT (Mw 5.3, 18.45°N latitude and -103.63°W longitude, depth 23.5 km, strike 38°, dip 23°, and slip 96°, seismic moment 1.12e24 dyne-cm). For the simulated event (target event) and taking in consideration that the area in study is the region between the limits of the rupture areas of the Tecoman (2003) and the 1973 earthquakes, the hipocentral location was proposed just inside of the area in study and near to element event (18.45°N latitude and -103.75°W longitude). Considering the rupture area of 1973 earthquake and area of remainder gap of Tecoman earthquake we proposed 70 km along strike of fault area. Along the dip, we propose 80 km considering an intermediate value of dip length of neighbors earthquakes. Based on the above considerations the proposed effective rupture area is of 5600 km². Using equation (1), Somervile [6] we estimate a seismic moment of 1.1091e27 dine-cm.

$$A = 5.2^{-15} * Mo^{2/3} \tag{1}$$

Where A is the rupture area and Mo is the seismic moment.

Using equation (2) by Kanamori [21], the maximum estimated M_W magnitude is 7.3.

$$M_w = (1/1.5)\log(M_o) - 10.73 \tag{2}$$

Where Mw is the moment magnitude and Mo is the seismic moment.

We use the relationships of Somerville [6] to characterize the source parameters as follows: (i) equation (3) to estimate the combined area of asperities ($A2$), (ii) equation (4) to estimate area of largest asperity ($A1$), (iii) equation (5) to estimate the hipocentral distance to center of closest asperity (CA), and (iv) equation (6) to estimate the rise time (Rt) that is related to seismic moment of a small earthquake. For the S-wave propagation velocity we used 3.4 km/s.

$$A2 = 1.21e - 15 * Mo^{2/3} \tag{3}$$

$$A1 = 8.87e - 16 * Mo^{2/3} \tag{4}$$

$$CA = 1.76e - 8 * M0^{1/3} \tag{5}$$

$$Rt = 1.79e - 9 * Mo^{1/3} \tag{6}$$

The fault plane was defined considering: an azimuth of 38°, a dip of 23° and a slip of 96°. These parameters were taken assuming that the mainshock will have the same focal mechanism as the element earthquake.

To estimate the number of sub-events, we applied the ω^{-2} spectral model, Aki [22], obtaining the number of sub-events necessary and estimate N^3 by using the relationship between seismic moments of the target event (M_0), and the element event (m_0) that is used as empirical Green's function. N^3 is equal to the number of sub-faults in direction of the strike (Nx), the dip (Nw) and the time (Nt).

The above description clearly states that it is necessary to find the parameter N, which will be used to scale the fault area for the event to simulate. Since it is divided into N x N subfaults, N^3 is obtained using the equation (7), and the relationship between these parameters is stated through equations (7), (8), and (9).

$$N^3 = N_x x\ N_w x\ N_t \tag{7}$$

$$\frac{\bar{U_0}}{u_0} = \frac{M_0}{m_0} = N^3, \tag{8}$$

where $\bar{U}_0$ and $\bar{u}_0$ is the flat level of the displacement Fourier spectrum for the target and element events respectively. On the other hand M_o and m_o are the seismic moments of the target and element events respectively. The relationship for high frequency is given by:

$$\frac{\overline{A_0}}{a_0} = \left(\frac{M_0}{m_0}\right)^{1/3} = N,$$ (9)

where $\overline{A}_0$ and $\overline{a}_0$ are the flat level of the acceleration Fourier spectrum of the target and element events respectively.

Then the synthetic motion of the target event $A(t)$, is given by the element event $a(t)$ using the equations 10 and 11.

$$A(t) = \sum_{i=1}^{Nx}\sum_{j=1}^{N_W}\left(\frac{r}{r_{ij}}\right)F(t - t_{ij}) * a(t)$$ (10)

$$F_{ij}(t - t_{ij}) = \delta(t - t_{ij}) + \frac{1}{n}\sum_{k=1}^{(N-1)n}\delta[t - t_{ij} - \frac{(k-1)\tau}{(N-1)n}]$$ (11)

where n' is an appropriate value to eliminate spurious periodicity, r is the distance from the station to the element event hypocenter, r_{ij} is the distance from the station to the sub-fault (i, j), t_{ij} is the sum of the delay times due to the rupture propagation and the differences of distances between the location of the element event and the location of the target event (i,j) at the observed site.

During the simulation process, we use the Somerville [6] relationships to assume and vary the inner and outer fault parameters in order to simulate acceleration records. Such parameters are: the rupture velocity, the rise time and the point where the rupture starts among others.

The PGAs were estimated for the three orthogonal components. On the other hand, earthquake magnitude, focal depth, hypocentral distance, and site characteristics (rock or soil) were the controlling parameters to estimate curves of ground motion by using the GMPE from Ordaz [16] and Young's [17] used in this study. To compare our results with respective GMPE the PGAs of two horizontal components was computed according with table 1.

GMPE	Horizontal components	Equation
Youngs et al. (1997)	Geometric mean *	$A_G = \sqrt{\max|a_1(t)|_{fort}\ \max|a_2(t)|_{fort}}$
Ordaz et al. (1989)	Cuadratic mean	$A_{max} = \sqrt{\dfrac{A^2_N + A^2_E}{2}}$

* Douglas (2003)

Table 1. Computation of two horizontal components in the two GMPEs.

After the above steps were completed, we proceeded to generate and compare the mean value of the residual between the PGAs and each one of the GMPEs by applying the definition of mean residuals as the weighted sum of the residuals of the logarithmic values between observed and estimated. The above step was applied to identify, in the statistical sense trough the estimation of the residuals, how realistic our PGA estimations are.

5. Results and discussion

We applied the EGFM to generate a lot of models using the Somerville [6] relations to characterize the parameters of the source. The total rupture area was 5,571.90 km^2 with an average slip of 3.2 m and a combined area of asperities of 1,269.11 km^2, in which the area of largest asperity was of 812.33 km^2. The hipocentral distance to the center of the closest asperity was estimate at 18 km. It should be pointed out that the Somerville [6] relationships do not define an azimuth to specify the location of asperity, this mean the largest asperity can be located within an azimuth of between 0° to 360°. Then in the modeled process the position of largest asperity was varied from 0 to 360°.

The main goal of this study was in the context of obtaining the most probable scenario of the ground response in the major cities of region upon the occurrence of a M_W 7.3 earthquake generated in the area in study. In this work, all the simulations were carried out to obtain a statistical sample that may represent the most probable scenarios of the ground response at the studied sites. For that purpose, the fault parameters and rupture process (i.e., (i) the azimuth of the closest asperity to hypocenter, (ii) the rise time, (iii) the rupture velocity, and (iv) the location of SMGA within the fault plane) were varied in an iterative process in order to generate a statistical sample of the most probable ground response scenario.

According to Somerville [6] an earthquake with a moment magnitude of 7.3 should have around 2.4 asperities. Based on the above, our procedure was divided in two stages. In the first stage, we used 2 SMGA and 3 SMGA in the second stage. These SMGA were positioned inside of different points into the dislocation area adopting the Somerville [6] criteria. In order to evaluate which scenario is the most probable we compare the PGA obtained from each scenario to those predicted by two GMPE, from Ordaz [16] and from Young's [17].

It should be pointed out that both GMPE were obtained considering earthquake data from subduction tectonic environments as follows: (i) Ordaz [16] used subduction earthquakes from the Mexican pacific coast, and (ii) Young's [17] subduction earthquakes from around the world. The process of finding the best fit of the simulated PGA with respect to Ordaz [16], and Young's [17], GMPEs was based on the smallest residual criteria between the simulated PGA and the estimated GMPEs. The above, allowed us to identify 2 different source models that provided PGA values that better matched with the used GMPE as follows: (i) the first one with 2 SMGA, and (ii) the second one with 3 SMGA.

In order to find a best residual in each interaction, we varied the position of SMGA, the rupture velocity, rise time, radiation pattern, and the size of SMGA changes in the strike or dip directions, according to the azimuth of the station or stations that had poor adjustment with GMPE. In the process of modeling we found little sensitivity of the synthetics to the rise time variations. On the other hand, we found high sensitivity of the synthetics to the rupture velocity variation, the size of the SMGA and its location inside of fault plane. The parameters with major weight in the modeled are the number, the size and the location of SMGA. The optimal model is a combination of all these parameters. The best model for each stage was determined by minimizing the residual between synthetic and observed PGA.

Figures 2a and 2b show the comparison between both GMPE versus our results for the lowest residual case of the source models (two SMGA). Figure 2a shows the Young's [17] GMPE curves for rock and soil.

Numbers of SMGA	SMGA	Length (km)	Width (km)	Area (km²)	Total area	Vr Km/s	Residual with Young's GMPE (Rock)	Residual with Young's GMPE (Soil)	Residual with Ordaz GMPE (Rock)
2	1	24.94	32.06	799.51	1256.38	3.0	0.009	0.024	0.245
	2	21.37	21.37	456.86		2.9			
3	1	35.62	24.94	888.35		2.3			
	2	17.81	10.69	190.36	1281.76	2.5	0.011	0.101	0.252
	3	14.25	14.25	203.05		2.4			

Table 2. Shows that the mean residual for the 25 stations decrease when using source models with 2 SMGA instead of when using source models with 3 SMGA. Table 2 shows the comparison of residuals between the theoretical values of each GMPE and the PGAs for all 25 stations.

Our data showed three clusters that according to their hypocentral distances are distributed as follows: (i) The first group was distributed within the distance from 35 to 60 km, in this group 5 of the simulated PGA were located nearby of both curves; (ii) The second group is defined for distances range from 60 to 120 km, on this case the PGA are distributed almost evenly below the GMPE curves; (iii) Finally, the third group is defined for distances range from 120 to 500 km, on this particular situation the PGA values are over-estimated by the GMPE and show a clear tendency to attenuate faster than the pattern showed on the GMPE. Figure 2b shows the Ordaz [16] GMPE, the author uses thrust subduction earthquakes from Mexico (such as event simulated in this study). In general the comparison shows that 90% our results are located above this curve.

For the GMPE of Ordaz [16], we compare only the 19 stations seated on rock. For the GMPE of Young's [17] we compare the 19 stations seated on rock, and 6 stations seated on soil sites (table 3), each of these groups with the respective curves for sites on rock and for sites on soil. From the modeling process of the target event, when three SMGA were used the lowest residuals we obtained between the PGA and the GMPE were: (i) 0.011 for Young's [17] GMPE

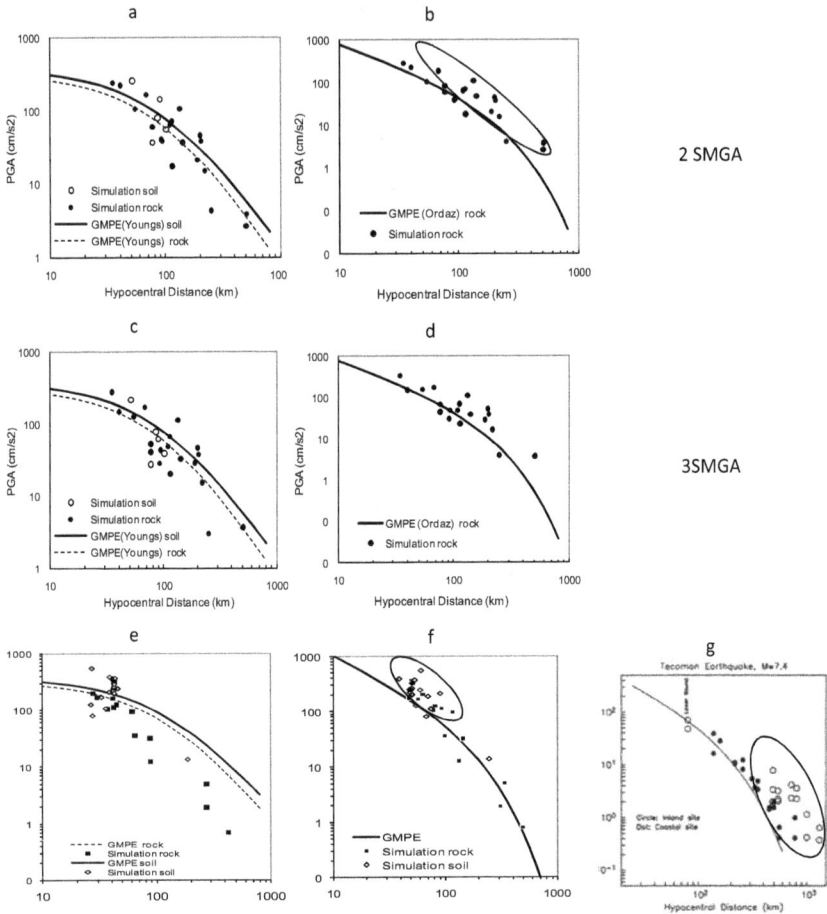

Figure 2. Comparison of synthetics PGA versus PGA from GMPE. In first row comparison of PGA from our simulated acceleration time histories using 2 SMGA versus PGA predicted using the following GMPE: (a) Young's [17], (b) Ordaz [16]. In second row comparison of PGA from our simulated acceleration time histories using 3 SMGA versus PGA predicted using the following GMPE: (c) Young's [17], (d) Ordaz [16]. In third row comparison of simulated PGA of Tecoman earthquake from Ramirez-Gaytán [23] versus PGA predicted using the following GMPE: (e) Young's [17], (f) Ordaz [16]. In (e) comparison of real PGA of the Tecoman earthquake using records of stations with distances larger than 50 km versus GMPE of Ordaz [16] from Singh [10].

(Rock), (ii) 0.101 Young's [17] GMPE (Soil), and (iii) 0.252 for Ordaz [16] GMPE for the sites shown in Figure 2c and 2d. The best fit between the estimated values of PGA and GMPE, estimated by means of the lowest residual, was obtained for the source model with two SMGA. The respective estimated residuals were: (i) 0.009 for Young's [17]GMPE (Rock), (ii) 0.024 for Young's [17] GMPE (Soil) and (iii) 0.245 for Ordaz [16].

Ramirez-Gaytán [23] simulate PGA and acceleration time histories for Tecoman earthquake located in the adjacent area at the event here simulated. The particularity and importance of this study is that the authors used observed strong motion records as a comparison reference to adjust the synthetics and found that from the comparison between PGA synthetics with respect to Young's [17] GMPE for rock and soil curves, and the Ordaz [16] GMPE curve (Figures 2e and 2f) behave very similar to the description previously provided for the same ranges of distances obtained in this study.

No	Station	Hipocentral distance	Soil type	PGA EW	Maximum acceleration (cm/s/s)					
					T 0.1	T 0.3	T 0.5	T 1	T 2	T 3
1	SJAL	34.54	Rock	457.66	819.32	1934.28	958.19	193.3	159.37	30.9
2	CEOR	40.07	Rock	125.38	218.08	265.65	328.99	178.02	273.65	57.36
3	BA5	51.22	Soil	157.32	230.26	402.48	428.66	150.87	325.69	54.23
4	COJU	54.02	Rock	201.97	248.95	484.43	190.27	65.25	39.12	8.63
5	MARU	68.11	Rock	194.29	419.83	559.86	627.33	261.75	180.06	29.4
7	TAPE	76.76	Rock	30	60.46	84.91	94.25	20.38	25.95	5.56
6	R15	76.78	Soil	27.33	45.06	47.41	51.59	16.83	7.49	2.14
8	MANZ	77.11	Rock	91.74	133.89	260.72	174.85	37.56	26.15	4.94
9	CAM	85.23	Soil	78.43	113.16	366.71	135.51	24.06	30.26	4.63
10	NAR	89.25	Soil	51.95	70.01	215.24	139.55	18.02	13.37	2.1
11	CEN	92.34	Rock	20.73	33.06	49.66	60.03	13.19	19.46	2.87
12	COMA	93.68	Rock	63.3	78.45	146.38	207.83	117.09	112.31	19.73
13	EZA	100.73	Soil	25.49	32.19	87.7	59.22	14.75	13.34	2.06
14	CIHU	108.51	Rock	42.84	47.23	91.04	213.08	26.49	12.49	2.56
15	EZ5	113.45	Rock	52.18	92.98	113.73	150.01	32.7	39.9	5.93
16	COLL	113.83	Rock	29.68	35	76.05	157.38	24.04	17.21	3.07
17	CALE	132.32	Soil	71.7	84.37	120.62	167.18	109.63	60.38	10.92
18	CDGU	140.21	Rock	50.83	57.72	177.79	172.63	43.18	15.1	3.34
19	VILE	188.48	Rock	31.74	32.76	44.1	98.98	60.66	48.56	8.31
20	NITA	199.77	Rock	68.46	75.31	188.1	193.54	120.04	80.13	13.41
21	CANA	203.51	Rock	31.33	43.4	85.63	77.88	60.92	46.25	8.62
22	URUA	217.28	Rock	20.51	21.37	34.56	69.81	38.08	56.54	9.33
23	GDLC	247.67	Rock	1.71	2.1	5.37	5.72	0.98	0.29	0.06
24	CUP	504.35	Rock	3.53	3.95	5.79	9.29	7.18	6.58	0.97
25	SCT	509.27	Soil	3.55	3.78	7.32	14.28	3.78	0.57	0.21

Table 3. Relation of 25 stations where acceleration time histories was simulated, soil type, PGA, spectral acceleration corresponding to structural period of 0.1, 0.2, 0.3, 1.0, 2.0 and 3.0 s, EW component.

We made an additional comparison, for Tecoman earthquake Singh [10] used records of stations located at distances larger than 50 km and comparing the PGA versus Ordaz [16]. In figure 2g it can be seen that the behavior is the same: observed PGAs lie above the curve. The comparison made are important because in the first case Ramirez-Gaytán [23] generate PGA based in a previous model Ramirez-Gaytán [24] when using observed records to compare with synthetics. In the second case Singh [10] use real data, in both cases the results are very similar with the obtained in this study. All mentioned studies correspond at the same tectonic environment. Figure 2g show than for Singh [10] major PGA locate at distances above 120 km this explained because he use data from regional stations. In the case of Ramirez-Gaytan [23] major PGA are located at distances from 10 to 100 km this explain why authors only use data from local networks.

In this study, we compare our PGA's with those obtained by the GMPE of Ordaz [16] and because they used data from Mexico. However, all of these studies considered all of the data available at the time of their analyses. This means that the data used were essentially strong motion data from the southern part of the country Tejeda- Jácome and Chávez-García [25]. None of these studies included data from western Mexico, along the northern section of the subduction zone. It is uncertain that ground-motion prediction equations developed for Guerrero in a very different tectonic setting can be applied to Colima Tejeda-Jácome and Chávez-García [25]. For this reason, it is important to compare our results with GMPE of other parts of the world, such as the GMPE of Young's [17]. They used some similar parameters to those of the event simulated in this study (tectonic environment or hypocentral distance magnitude, etc.). This seems to be justified when we observed that the minor residual is reached when compared with the GMPE of Young's [17] with minor residual (0.009) than Ordaz [16] with residual of 0.245 for the model with best adjust (model with two SMGA). However, we expect that any of the two GMPEs compared in this paper satisfy the detailed similarities with the event simulated in this study for the Colima region. For this reason, our intention is to use local and world parameters in order to validate our results and to give a degree of confidence when applying this methodology to future earthquakes for determination of acceleration time histories and PGA.

The main contribution of the process detailed above is not obtaining PGA, for this case is sufficient to consult GMPE. The purpose of this paper is to prepare acceleration time histories to be used by structural engineers on the analysis and design of structures. In modern seismic design approaches the quality of a structural solution frequently depends on the detailed knowledge that designers have on the characteristics of the seismic ground motion that the structure will suffer at a site if an earthquake of a given magnitude occur at a given location. Figures 3a-3d show the acceleration time histories for the model with minor residual after applying the process previously described. The largest acceleration was obtained in the near-to-source station SJAL (rock site) with 0.47 g. It is important to point out that stations CUP and SCT (figure 3c) located in Mexico City with hipocentral distances larger than 470 km still produced considerable values of peak accelerations (3.53 and 3.55 gal) similar of those of the GDLC station (with 1.71 and 5.3 gal) located to 273.48 km from the epicenter. This is due to the fact that seismic waves are enormously amplified at lake-bed sites respect to hill-zone sites in Mexico City, although it has been suggested that even hill-zone sites suffer amplification and Singh [26].

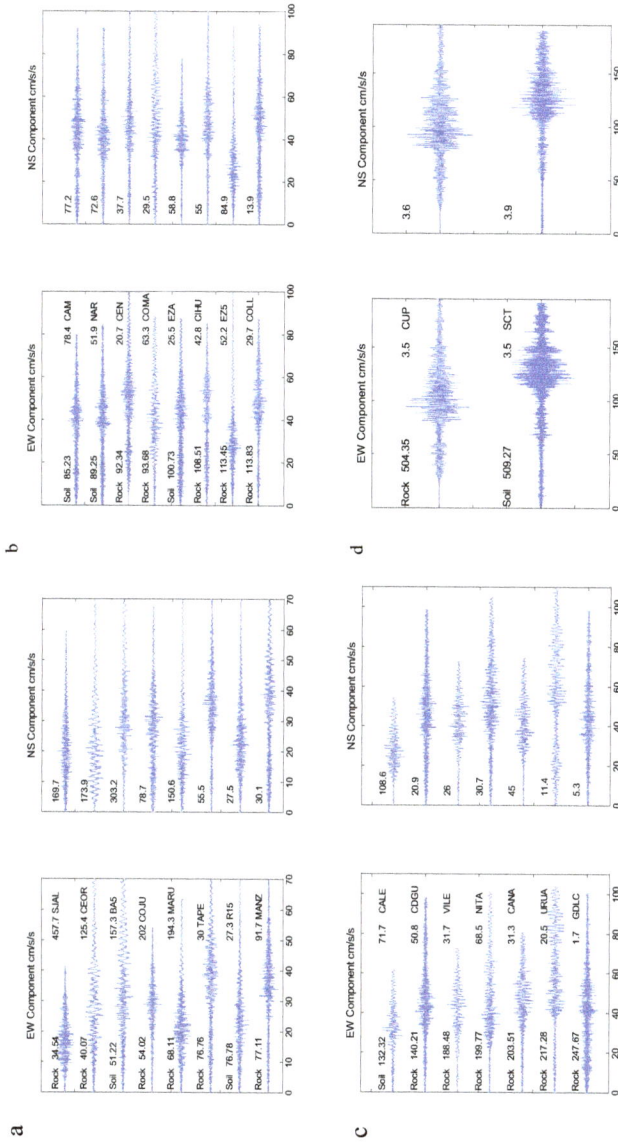

Figure 3. Simulated synthetic acceleration time histories (intense period) with next distances from hypocenter: (a) 34 to 78 km, (b) 85 to 114 km, (c) 132 to 248 km, (d) major to 500 km. In left column type of soil and epicentral distance are denoted above and left of each trace. PGA and code of station is denoted above and right of trace. In right column PGA is denote above and left of trace.

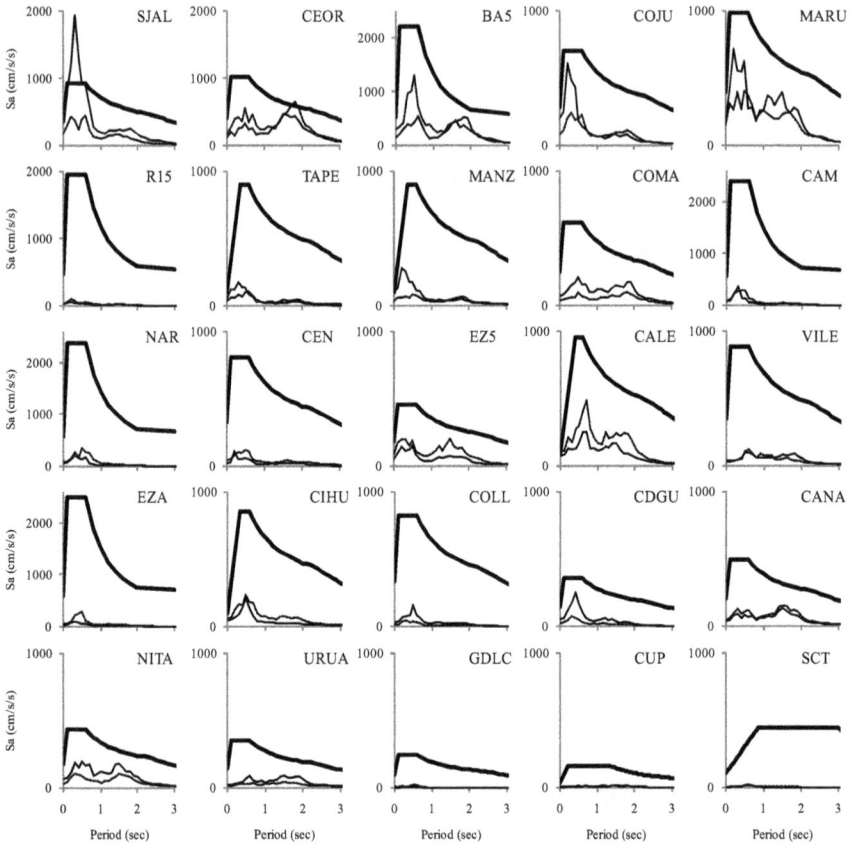

Figure 4. Comparison of the 25 response spectra ordered for hipocentral distance of the simulated event for east–west component (thin continuous line) and north–south component (discontinuous line) with the design spectra (dark continuous line). It can be observed that the response spectra obtained with EGFM is realistic.

On the other hand, with acceleration time histories it is possible to generate a response spectrum, which considers forces related to parameters of maximum response like spectral acceleration. Response spectra are essential for the seismic design, any effort to accurately predict these should be done. The synthetic acceleration response spectra for an equivalent viscous damping of 5 percent were calculated and compared with the elastic acceleration design spectra for structures of group B (standard occupancy) of the Manual of Civil Structures MOC-2008 [27], a model design code in Mexico and seismic provisions for current Mexico's Federal District Code NTCS-2004 [28], Sites CUP and SCT. In the MOC-2008 [27] code, seismic hazard in Mexico is defined as a continuum function where peak accelerations in rock are associated with return periods that were obtained using an optimization design criterion to define the seismic coefficients for the plateaus of the elastic design spectra for standard occupancy structures Tena-Colunga [29].

It can be also seen that only in 1 of 25 stations with distances comparable to the source dimensions the seismic ordinates are underestimated respect to the elastic design spectra. This station is SJAL located in rock site having seismic ordinate of 1.97 g in structural period of T= 0.3 sec. This is possibly because radiation patterns and source heterogeneity, which are taken into account in this work, causes spatial variations in ground motion around the fault Somerville [30]. These changes are very clear in the EGFM, but may be unnoticed in the GMPE. Therefore, obtaining future acceleration time histories and response spectra is a matter of essential interest for the present seismic design to achieve more efficient and safer structures.

6. Conclusions

We use the 25 records of August 13 2006 earthquake Mw 5.3 as element event to simulate strong ground motions for an eventual earthquake Mw 7.3 in the studied area. To reach this objective we integrate the advantages of three methodologies (EGFM, Somerville [6] relations and GMPE) to estimate the possible PGA, acceleration time histories and response spectra for an eventual earthquake in the studied area. We apply the empirical Green's function method (EGFM) whose main contribution is to reflect a model that considers the source, the path, and the site effects. In Mexico this method has been used to simulate an event that already occurred. In this study we applied it to predict some probable earthquake which may be expected in the region. To overcome the absence of observed records we made use of Somerville [6] relations to be able to make more accurate predictions of strong motions and two GMPE adequate for region to compare our results. The process of finding the best adjustment generated 2 different models (2 and 3 SMGA). This process of minimizing the residual between synthetics and observed PGA clearly shows that the mean residual for 25 stations is obtained when comparing with GMPE of Young's [17] and modeled with 2 SMGA. Ramirez-Gaytan [23] simulate PGA for Tecoman earthquake, whit difference that in this case the earthquake had occurred and exist observed records to compare and adjust synthetics, results are similar to those obtain in this study. Singh [10] comparing real PGA for Tecoman earthquake versus GMPE of Ordaz [16], results are similar to those obtain in this study. The purpose of this paper is to rescue the acceleration time histories of simulated event prepared to be used by structural engineers to analyze and design structures. Response spectrum show that for 1 of 25 stations (this station is near the source or with distance comparable with the source dimensions) the seismic ordinates are underestimated with the design spectra of the MOC-2008 [27] due possibly to radiation patterns and source heterogeneity, which is still to be confirmed by future records. For any practical evaluation of the seismic hazard in terms of response spectra is possible to integrate the advantages of three methodologies aboard in this study to estimate the possible PGA, acceleration time histories and response spectra for an eventual future earthquake.

Acknowledgements

We are grateful to Leonardo Alcantara Nolasco for providing acceleration records from the network of the Instituto de Ingeniería of UNAM, and Tonatiuh Dominguez for providing records of Red Sismica del Estado de Colima (RESCO).

Author details

Alejandro Gaytán[1], Carlos I. Huerta Lopez[2], Jorge Aguirre Gonzales[3] and Miguel A. Jaimes[3*]

*Address all correspondence to: gramirez@sciences.sdsu.edu.; huerta@cicese.edu.mx; joagg@pumas.iingen.unam.mx

1 Universidad de Guadalajara, Departamento de Ciencias Computacionales, Centro Universitario de Ciencias Exactas e Ingeniería, Mexico

2 University of Puerto Rico at Mayaguez Civil Engineering and Surveying Department, Puerto Rico Strong Motion Program Home Institution: Research Center and Higher Education at Ensenada, CICESE Seismology Department, Earth Sciences Division, Puerto Rico

3 Instituto de Ingeniería, Universidad Nacional Autónoma de México, Circuito Interior, Ciudad Universitaria, Mexico

References

[1] Reyes A, Brune J, Lomnitz C, Source Mechanism and Aftershock Study of the Colima, Mexico Earthquake of January 30, 1973, Bulletin of Seismological Society of America, 69 (1979) 1819-1840.

[2] Eissler H K, McNally K C, Seismicity and Tectonics of the Rivera plate and implications for the 1932 Jalisco Mexico earthquake, Journal of Geophysical Research, 89 (1984).

[3] Singh S K, Astiz L, Havskov J, Seismic gaps and recurrence periods of large earthquakes along Mexican subduction zone: a reexamination, Bulletin of the Seismological Society of America, 71 (1981) 827 - 843.

[4] Heaton T H, Kanamori H, Seismic potential associated with subduction in the northwestern United States, Bulletin of the Seismological Society of America, 74 (1984) 933 - 942.

[5] Somerville P, Irikura K, Graves R, Sawada S, Wald D, Abrahamson N, Iwasaki W, Kagawa T, Smith N, Kowada A, Characterizing crustal earthquake slip models for the prediction of strong motion, Seismic Research Letters, 70 (1999).

[6] Somerville P, Sato T, Toru I, Collins N, Dan K, Hiroyuki F, Characterizing subduction earthquake slip models for the prediction of strong motion, Proceedings of the Eleventh Symposium on Earthquake Engineering, 1 (2002) 163-166.

[7] Nishenko SP, Singh KS, Conditional probabilities for the recurrence of large and great interplate earthquakes along the Mexican subduction zone, Bulletin of Seismological Society of America, 77 (1987) 2095-2114.

[8] Pardo M, Suarez G, Shape of the subducted Rivera and Cocos plate in southern Mexico: Seismic and tectonics implications, Journal of Geophysical Research, 100 (1995) 357-374.

[9] Minster J, Jordan T H, Molnar P, Haines E, Numerical modeling of instantaneous plate motions, Geophysics Journal, 36 (1974) 541-576.

[10] Singh KS, Pacheco F, Alcantara L, Reyes G, Ordaz M, Iglesias A, Alcocer S, Gutierrez C, Valdez C, Kostoglodov V, Reyes C, Mikumo T, Quass R, Anderson R, A preliminary report on the Tecomán, México earthquake of 22 January 2003 (Mw 7.4) and its effects, Seismological Research Letters, 74 (2003) 279-289.

[11] Quintanar L, Rodriguiez-Lozoya H, Ortega R, Gomez Gonzalez J, Dominguez T, Javier C, Alcantara L, Rebollar T, Source Characteristics of the 22 January 2003 Mw =7.5 Tecoman, Mexico, Earthquake:New Insights, in: Pure and Aplplied Geophysics, 2010.

[12] Quintanar L, Rodriguiez-Lozoya H, Ortega R, Gomez Gonzalez J, Dominguez T, Javier C, Alcantara L, Rebollar T, Source Characteristics of the 22 January 2003 Mw =7.5 Tecoman, Mexico, Earthquake:New Insights, Pure and Aplplied Geophysics, 168 (201) 1339 - 1353.

[13] Irikura Kojiro, Prediction of strong accelerations motions using empirical Green's function, Procedures of 7 th Japan Earthquake Engineering Symposium, (1986).

[14] Aguirre J, Irikura K, Source Characterization of mexican Subduction Earthquakes from Acceleration Source Spectra for the Prediction of Strong Ground Motions, Bulletin of Seismological Society of America, 97 (2007) 1960-1969.

[15] Irikura K, Miyake T, Kamae K, Kawabe H, Dalguer LA, Recipe for predicting strong ground motion from future earthquakes, in: Proceedings of the 13th World Conference on Earthquake Engineering, available on CD-ROM, 2004.

[16] Ordaz M, Jara J, Singh KS, Riesgo sísmico y espectros de diseño en el estado de Guerrero, Memorias del VIII Congreso Nacional de Ingenieria Sismica, (1989) D40-D56.

[17] Youngs R, Chiou J, Silva J, Humphrey R, Strong ground motion attenuation relationships for subduction zone earthquakes, Bulletin of Seismological Society of America, 68 (1997) 58-73.

[18] DeMets C, Stein S, Present-day kinematics of the Rivera plate and implications for tectonics of southwestern Mexico, Journal of Geophysical Research, 95 (1990) 21931-21948.

[19] Bandy W, Mortera C, Urrutia J, Hilde T, The subducted rivera Cocos plate boundary: where is it, what is it, and is its relationship to the Colima rift?, Geophysical Research Letters, 22 (1995) 3075-3078.

[20] Centroid Moment Tensor (CMT) Catalog, www.seismology.harvard.edu/ CMTsearch.html (last access August 25 of 2008), in.

[21] Kanamori H, The energy released in great earthquakes, Journal of Geophysical Research, 82 (1997) 2981-2987.

[22] Aki K, Scaling law of seismic spectra, Journal of Geophysical Research, 72 (1967) 5359-5376.

[23] Ramírez-Gaytán A, Aguirre GJ, Huerta CI, Simulation of Accelerograms, Peak Ground Accelerations, and MMI for the Tecomán earthquake of 21 January 2003, Bulletin of Seismological Society of America, 100 (2010).

[24] Ramírez-Gaytán A, Aguirre GJ, Huerta CI, Tecomán Earthquake: Physical Implications of Seismic Source Modeling, Applying the Empirical Greens Function Method, and Evidence of non-Linear Behavior of Ground, ISET Journal of Earthquake Technology, 47 (2011) 1-23.

[25] Jacome T, Chavez-Garcia FJ, Empirical ground motions estimations in Colima from weak motions records, ISET Journal of Earthquake Technology, 44 (2007) 409-419.

[26] Ordaz M, Singh KS, Source spectra and spectral attenuation of seismic waves from Mexican earthquakes, and evidence of amplification in the hill zone of Mexico City, Bulletin of Seismological Society of America, 82 (1992) 24-43.

[27] Comision Federal de Electricidad, Manual de Diseno de Obras Civiles 2008, Diseño por sismo, in: Instituto de Investigaciones Electricas (Ed.), 2008.

[28] Reglamento de construcciones del Gobierno del Distrito Federal, NTCS-2004 Normas tecnicas complementarias para diseño por sismo, in: Gaceta Oficial del Departamento del Distrito Federal (Ed.), 2004.

[29] Tena Colunga A, Mena Hernandez U, Perez -Rocha LE, Aviles J, Ordaz M, Vilar J, Updated Seismic Design Guidelines for Model Building Code of Mexico, Earthquake Spectra, 25 (2009) 869-898.

[30] Somerville P, Smith N, Graves R, Modification of empirical strong ground motion attenuation relations to include the amplitude and duration effects of rupture directivity, Seismological Research Letters, 68 (1997) 199-222.

Chapter 2

Seismic Hazard Analysis for Archaeological Structures — A Case Study for EL Sakakini Palace Cairo, Egypt

Sayed Hemeda

Additional information is available at the end of the chapter

1. Introduction

The modern architectural heritage of Egypt is rich, and extensively variable. It covers all kinds of monumental structures from palaces, public buildings, residential and industrial buildings, to bridges, springs, gardens and any other modern structure, which falls within the definition of a monument and belongs to the Egyptian cultural heritage. We present herein a comprhensive geophysical survey and seismic hazard assesment for the rehabilitation and strengthening of Habib Sakakini's Palace in Cairo, which is considered one of the most significant architectural heritage sites in Egypt. The palace located on an ancient water pond at the eastern side of Egyptian gulf close to Sultan Bebris Al-Bondoqdary mosque, a place also called "Prince Qraja al-Turkumany pond". That pond had been filled down by Habib Sakakini at 1892 to construct his famous palace in 1897.

Various survey campaigns have been performed comprising geotechnical and geophysical field and laboratory tests, aiming to define the physical, mechanical and dynamic properties of the building and the soil materials of the site where the palace is founded. All these results together with the seismic hazard analysis will be used for the seismic analysis of the palace response in the framework of the rehabilitation and strengthening works foreseen in a second stage. We present herein the most important results of the field campaign and the definition of the design input motion.

The seismic hazard analysis for El Sakakini Palace has been performed based on historical earthquakes, and maximum intensity.PGA with 10% probability of exceedance in 50 and 100 years is found equal to 0.15g and 0.19g respectively. P-wave and S-wave seismic refraction indicated a rather low velocity soil above the seismic bedrock found at depths higher than 20m. Ambient noise measurements have been used to determine the natural vibration frequency of soil and structure of El-Sakakini Palace. The fundamental frequency of El-

Sakakini palace is 3.0Hz very close to the fundamental frequency of the underlying soil, which makes the resonance effect highly prominent.

Some floors are considered dangerous since it show several resonance peaks and high amplification factors (4th and 5th floors) these floors are made of wood so, warnings to decision makers are given for the importance of such valuable structures.

The seismic design and risk assessment of El Sakakini palace is performed in two steps. In the first one we perform all necessary geotechnical and geophysical investigation together with seismic surveys and seismic hazard analysis in order to evaluate the foundation soil properties, the fundamental frequency of the site and the structure, and to determine the design input motion according to Egyptian regulations. The second phase comprises the detailed analysis of the palace and the design of the necessary remediation measures. IN the present pare we present the results of the first phase.

2. Seismic hazard

2.1. Historical seismicity

Egypt possesses a rich earthquake catalogue that goes back to the ancient Egyptian times. Some earthquakes are reported almost 4000 years ago. Figure 1 shows the most important historical events affecting ElSakakini palace. We can see that the Faiyum area as well as the Gulf of Suez is the most important earthquake zones affecting the place.

2.2. Maximum intensity

Historical seismicity and maximum reported intensity is a good preliminary index of the expected severity of a damaging earthquake. Available isoseismal maps in the time period 2200 B.C. to 1995 were digitized and re-contoured to determine the maximum intensity affecting the place. This was done using a cells value of equal area 0.1 lat. × 0.1 long. Figure 2 present the produced IMM intensity showing that a maximum IMM of VII is good design value.

2.3. Probabilistic hazard assessment

An improved earthquake catalogue for Egypt and surrounding areas affecting El Sakakini Palace has been prepared for the purposed of this study partially based on recent work of Gamal and Noufal, 2006. The catalogue is using the following sources:

- For the period 2200 B.C to1900: Maamoun,1979; Maamoun et al., 1984 ; Ben-Menahem 1979 and Woodward-Clyde consultants, 1985.

- For the period 1900 to 2006: Makropoulos and Burton, 1981; Maamoun et al., 1984 ; Ben-Menahem 1979; Woodward-Clyde consultants, 1985; Riad and Meyers, 1985; Shapira, 1994 and NEIC, 2006; Jordan seismological observatory 1998-2000.

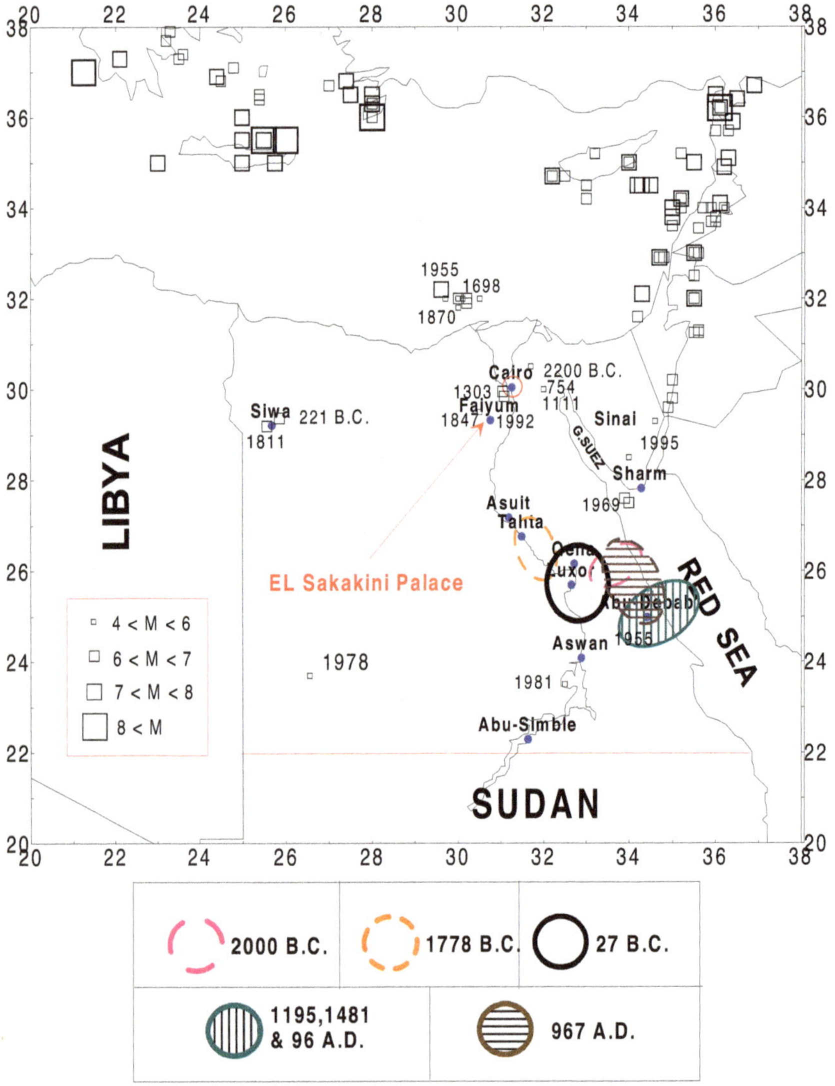

Figure 1. Important and historical earthquakes occurred in and around El Sakakini Palace area in the period 2200 B.C to1995.

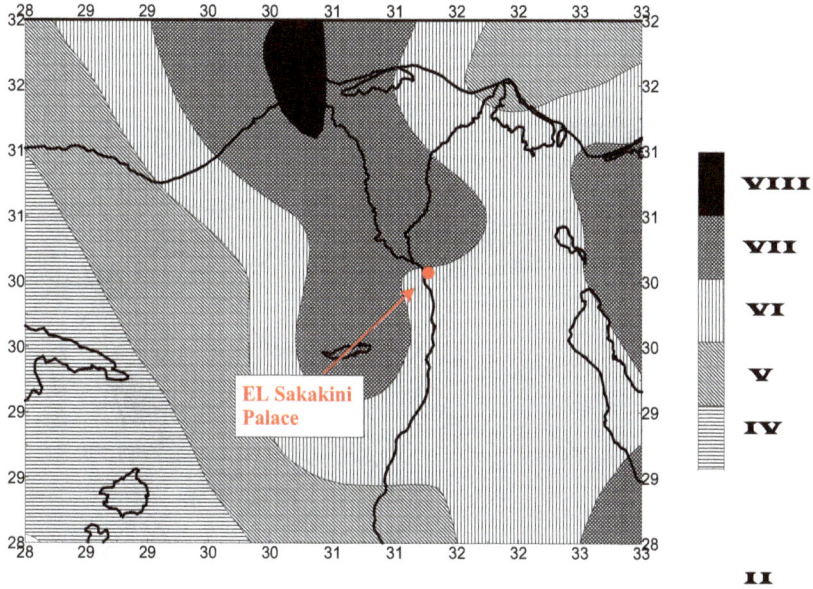

Figure 2. Maximum intensity zonation map based on the historical seismicity reported in the time period 2200 BC to1995

The horizontal peak ground acceleration over the bedrock of El Sakakini area was estimated using Mcguire program 1993. 37 seismic source zones were used to determine the horizontal PGA over the bedrock (Figure 3), while PGA attenuation formula of Joyner and Boore, 1981 was used because of its good fitting to real earthquake data in Egypt. A complete analysis for the input parameters to estimate the PGA values over the bedrock can be found in Gamal and Noufal, 2006.

$$\left(\text{PGA}\right)= 2.14 \; e1.13^{M}D^{-1}e^{-0.00590} \quad D=\left(R^{2}+4.0^{2}\right)^{0.5} \tag{1}$$

The probabilistic analysis provided the following results: The peak horizontal acceleration in gals with 10 % probability of exceedance over 50 years is 144cm/sec^2(or 0.147g) For 10% probability in 100 years the estimated PGA for rock conditions is 186 (cm/sec^2) (or 0.19g) (Figures 3 and 4). These values are quite high and considering the local amplification they may affect seriously the seismic design and stability of El. Sakakini Palace.

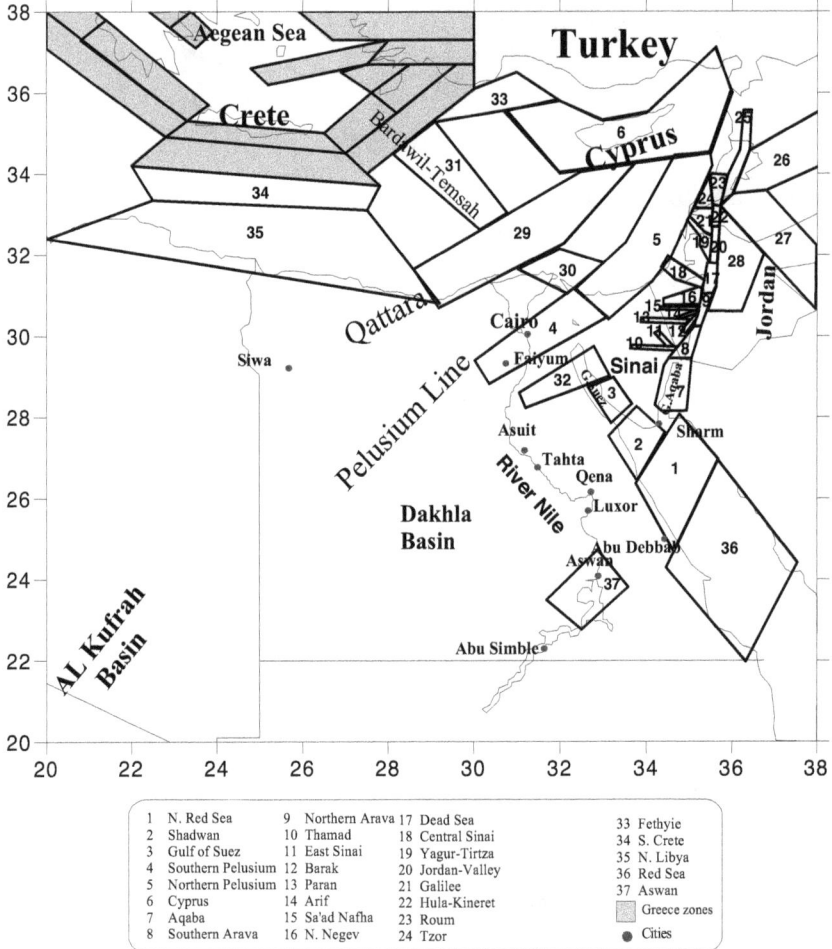

Figure 3. Seismic source regionalization using 37 seismic source zone (except greece zones) adopted for Egypt and surrounding areas (Gamal and Noufal, 2006).

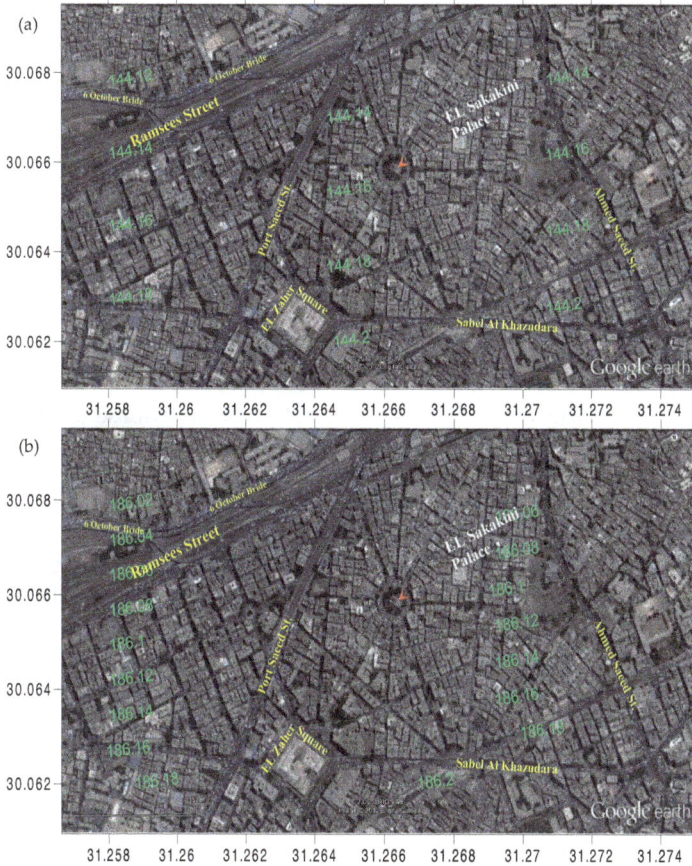

Figure 4. a: Peak Horizontal Acceleration in gals (cm/sec²) for the seismic bedrock with10 % probability of exceedance in 50 years. b: Peak Horizontal Acceleration in gals (cm/sec²) for the seismic bedrock with 10 % probability of exceedance over 100years

3. Geotechnical investigation

Core drilling is among the routine methods for subsurface exploration. Most commonly, NX-size core drill is used, representing a hole diameter of 76 mm (3″) and a core diameter of 54 mm (2 1/8″). The drilling often has multiplier purposes, of which the following are in most cases the most important:

Verification of the geological interpretation. Detailed engineering geological description of rock strata. To obtain more information on rock type boundaries and degree of weathering. To supplement information on orientation and character of weakness zones. To provide samples for laboratory analyses. Hydro geological and geophysical testing. Input data for engineering classification of rock masses.

The geotechnical investigation, six geotechnical boreholes with Standard Penetration Test (SPT) measurements have been carried out in the archaeological site included the drilling of three geotechnical boreholes with integral sampling to a depth 20 meters, one borehole to depth 15 meters and two boreholes to depth 10 meters at six locations in the site. The geotechnical data also indicated the ground water level at the archaeological site. We did all the boreholes inside the site with hand boring machine.

The results of laboratory tests which have been carried out on the extracted soil samples from the boreholes, which include specific gravity (Gs), water content (Wn), saturated unit weight (γsat), unsaturated unit weight (γunsat), Atterberg limits and uniaxial compressive strength (UCS), in addition to the ground water table (GWT), are shown in the figures (7a,7b).

The shear wave profile obtained by using ReMi compared very well to geotechnical boreholes and geophysical survey data. In addition, the shear wave profile obtained by using ReMi Performed much better than commonly used surface shear-wave velocity measurements.

Geotechnical boreholes (1) through (3) indicated that:

Filling of Fill (silty clay and limestone fragments, calc, dark brown) From ground surface 0.00m to 3.50m depth. Sand Fill (silty clay, medium, traces of limestone& red brick fragments, calc, dark brown) From 3.50m to 5.00m depth. Silty clay, stiff, calc, dark brown From 5.00m to 6.50m depth. Clayey silt, traces of fine sand & mica, yellowish dark brown From 6.50m to 8.50m depth. Silty sand, fine, traces of clay & mica. Dark brown. From 8.50m to 11.00m depth. Sand, fine, some silt, traces of mica, yellowish dark brown. From 11.00m to 14.00m depth. Sand, fine to medium, traces of silt& mica, tracesof fine to medium gravel, traces of marine shells, yellowish dark brown. From 14.00m to 16.00m depth. Sand, fine, traces of silt & mica, yellowish dark brown. From 16.00m to 18.00m depth. Sand & Gravel, medium sand, graded gravel, traces of silt, yellow darkbrown. From 18.00m to 20.00m depth. End of drilling at 20.00m.

Geotechnical boreholes (4) through (6) indicated that:

Fill (silt, clay and fragments of limestone and crushed brick, from ground surface 0.00m to 4m depth. Fill (silty cal with medium pottery and brick fragments, calc dark brown) from 4 to 5 m depth. Brown stiff silty clay and traces of limestone gravels, from 5.00m to 7.50m depth. silt, traces of brown fine sand & traces of clay from 12.00m to 14.00m depth. Dark brown clay silt with traces of fine sand. from 14.00m to 15.00m depth.

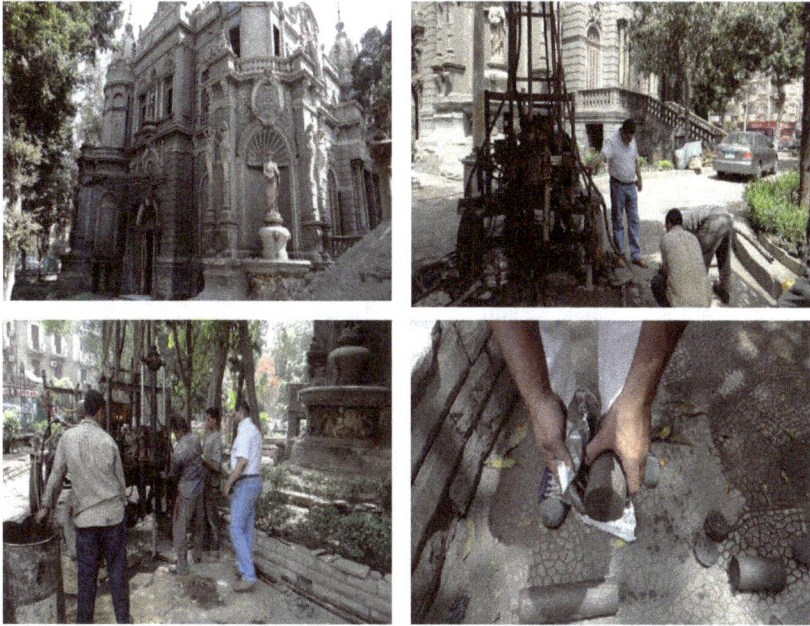

Figure 5. El- Sakakini palace and the Geotechnical investigations.

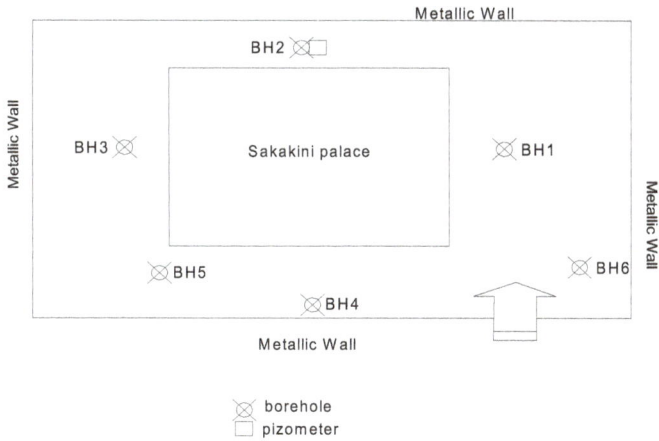

Figure 6. General layout & boreholes locations.

project	: existing . habib pasha elsakakeeny palace	file no : sakakeeny feb10
location	:eldaher-cairo	date commenced : jan . 15- 2012
Boring. no	: 1	datecompleted : jan . 18- 2012
DRILL METHOD	: MANUAL DRILLING	weather : cold
drill fluid	:none	ground level :
driller	: alaa amin drilling co	initial / final gwd : 2.30/1.20

depth (m)	legend	classification	end of layer (m)	qu (kg/ cm2)	spt n/30 cm	yb (um3)	f.s.	wL (%)	wP (%)	RECOVERY (%)	R.Q.D. (%)
1		fill(silty clay and limestone fragments . calc dark brown									
2											
3			3.40	1.10							
4		fill(silty clay .medum u of limestone &red bock fragments .calc dark browen)		1.00							
5			4.80								
6		silty clay stiff calc dark browen		1.80							
7		clayey silt . traces of fine sand & mica yellowsh dark brown	6.30	1.80							
8			8.00								
9		more sand / more silt	8.50	0.80				39.82	21	42	
10		silty sand fine traces of clay &mica dark brown	9.50								
11			11.00								
12		sand fine some silt traces of mica yellowish dark brown			21						
13											
14		sand fine to med u of silty clay dark brown	13.00 / 14.00								
15		sand fine to medum traces of silt & mica of fine to medum gravel u of marine shells yellowish dark brown									
16			16.00								
17		sand fine traces of silt & mica yellowish dark brown			28						
18			18.00								
19		sand & gravel medum sand graded gravel u of silt yell dark brown			33						
20			20.00								
		end of drilling at 20.00m									

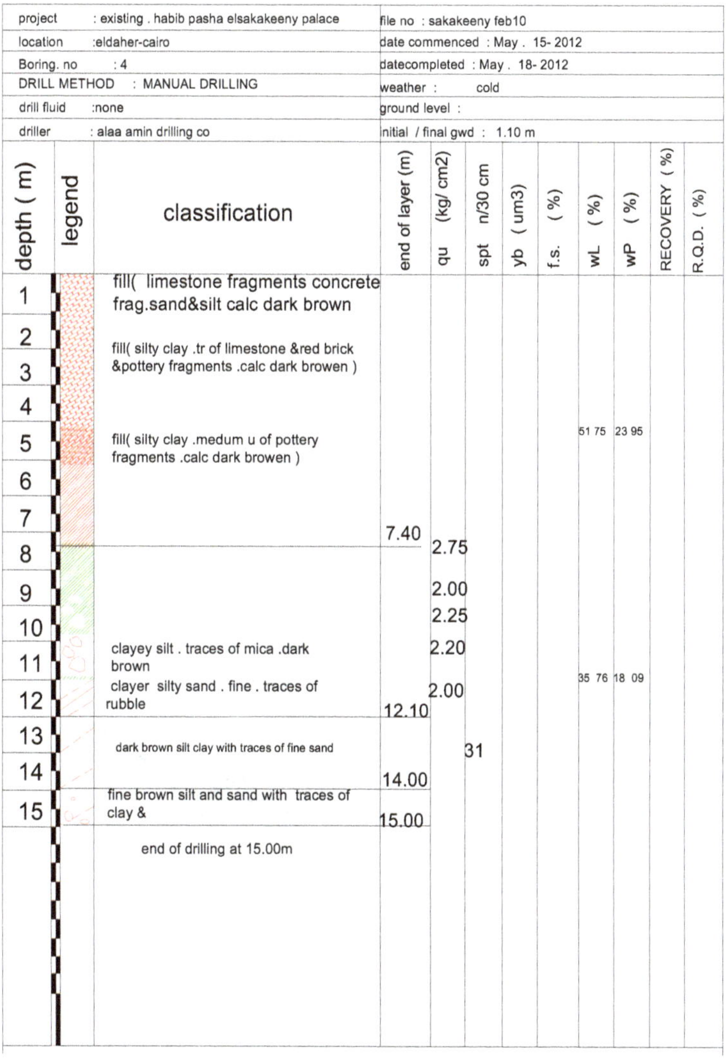

project	: existing . habib pasha elsakakeeny palace	file no : sakakeeny feb10
location	:eldaher-cairo	date commenced : May . 15- 2012
Boring. no	: 4	datecompleted : May . 18- 2012
DRILL METHOD	: MANUAL DRILLING	weather : cold
drill fluid	:none	ground level :
driller	: alaa amin drilling co	initial / final gwd : 1.10 m

depth (m)	legend	classification	end of layer (m)	qu (kg/ cm2)	spt n/30 cm	yb (um3)	f.s. (%)	wL (%)	wP (%)	RECOVERY (%)	R.Q.D. (%)
1		fill(limestone fragments concrete frag.sand&silt calc dark brown									
2											
3		fill(silty clay .tr of limestone &red brick &pottery fragments .calc dark browen)									
4											
5		fill(silty clay .medum u of pottery fragments .calc dark browen)						51 75	23 95		
6											
7			7.40								
8				2.75							
9				2.00							
10				2.25							
11		clayey silt . traces of mica .dark brown		2.20							
12		clayer silty sand . fine . traces of rubble	12.10	2.00				35 76	18 09		
13		dark brown silt clay with traces of fine sand		31							
14			14.00								
15		fine brown silt and sand with traces of clay &	15.00								
		end of drilling at 15.00m									

Figure 7. a. Geotechnical Borehole_1, El Sakakini Palace. b. Geotechnical Borehole_4, El Sakakini Palace.

4. Geophysical campaign

4.1. P-wave refraction

A total of 10 seismic profiles are conducted at El Sakakini palace area (Figure 8). All profiles are carried out using 12 receivers, P-type geophones with 5m intervals and 2 shots. The forward and reverse shots were carried at a distance of 1 m at both ends. The seismic shots layouts are described in Table 1.

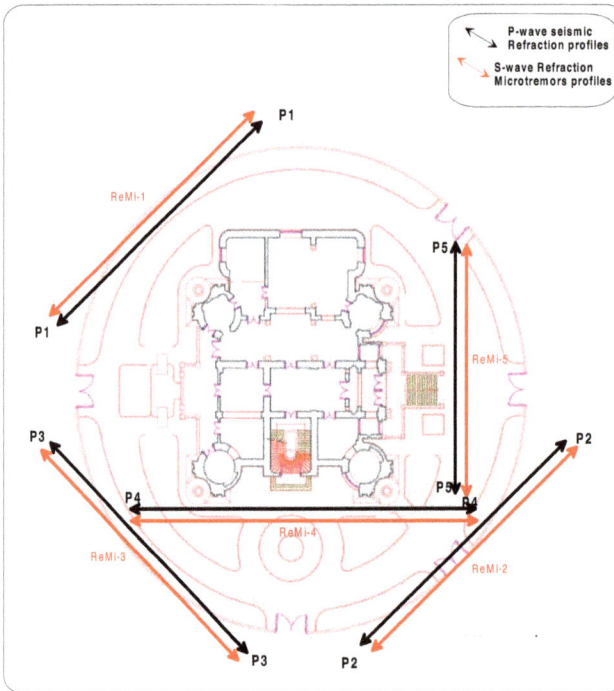

Figure 8. Location of the P-wave seismic refraction, S-wave refraction and ReMi profiles conducted at El Sakakini Palace.

Shot #	Name	Offset X (m) (relative to R1)
S2	Forward	-1
S4	Reverse	56

Table 1. Seismic shots.

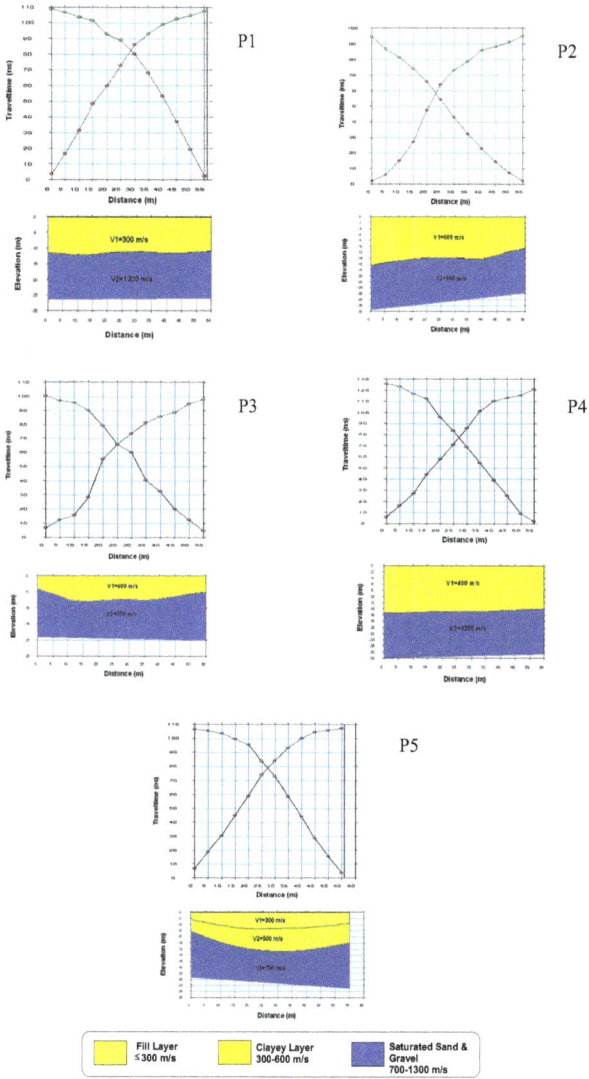

Figure 9. P-wave travel time distance curve and its corresponding geoseismic model for profiles # P1-P5 (Figure 8).

The conducted profiles are interpreted using time-term inversion method; an example of the conducted profiles and corresponding geoseismic model is shown in Figure 9. Table 2 summarizes the measured Vs values and the corresponding soil thicknesses. The soil stratification is not uniform and horizontal, as it should be expected for a filled area. However it is possible to distinguish the following three main layers: the soil layering can be summarized in the following (table 2).

Soil A- Fill (<300 m/s): A surface highly heterogeneous material (mainly man-made fill) with an average thickness of 10 m and an average velocity Vs lower than 300m/s. It is composed of very loose and low strength sediments such as silt, clay and limestone fragments. It is not found in all locations.

Soil B-Clayey soil (400-600 m/s): Below the surface layer (soil A) there is a clayey or silty clay layer with an average thickness of 10 m meters and Vs velocity 400-600 m/s.

Soil C-Saturated Sand & Gravel (700-1300 m/s): Below soil B there is a stiff soil layer with various thicknesses. it shows a considerable increase of Vs seismic velocity reaching sometimes values as high as 1300m/s. The soil is composed of compacted stiff saturated sand and gravel with an average Vs velocity equal or higher than 700m/s. It may be considered as the "seismic bedrock" for the local site amplification analyses.

	Layer A	Layer B		Layer C	
Profile N°	Velocity in m/s	Velocity in m/s	Depth (m)	Velocity in m/s	Depth in (m)
1		300	10	1300	25
2		600	16	900	32
3		400	9	700	20
4		400	14	1200	30
5	< 300	300	5	500	10

Table 2. P-wave refraction geophysical campaign conducted at El-Sakakini palace area.

4.2. Refraction- microtremor (ReMi method)

We have used the ReMi (refraction microtremors) method to determine the S-wave seismic velocity with depth. The method is based on two fundamental ideas. The first is that common seismic-refraction recording equipment, set out in a way almost identical to shallow P-wave refraction surveys, can effectively record surface waves at frequencies as low as 2 Hz (even lower if low frequency phones are used). The second idea is that a simple, two-dimensional slowness-frequency (P-f) transform of a microtremors record can separate Rayleigh waves from other seismic arrivals, and allow recognition of true phase velocity against apparent velocities. Two essential factors that allow exploration equipment to record surface-wave velocity dispersion, with a minimum of field effort, are the use of a single geophone sensor at each channel, rather than a geophone "group array", and the use of a linear spread of 12 or more geophone sensor channels. Single geophones are the most commonly available type, and are typically used for refraction rather than reflection surveying. There are certain advantages of ReMi method: it requires only standard refraction equipment, widely available, there is no need for a triggering source of energy and it works well in a seismically noisy urban setting. (Louie, 2001, Pullammanappallil et al. 2003).

A 12 channel ES-3000 seismograph was used to measure background 'noise' enhanced at quiet sites by inducing background noise with 14Hz geophones in a straight line spacing 5m Figure 5 shows the map were ReMi measurements were made. Almost all the sites were noisy. In particular big hammer used to break some rocks generated noisy background at El Sakakini Palace.30 files of 30sec records (unfiltered) of 'noise' were collected at each site. Five profiles were taken inside the Palace (Figure 8). Figure 10-11 shows an example of the dispersion curves and its P-F image (Remi Spectral ratio of surface waves) for refraction microtremors profile ReMi-1. The estimated average Vs for all profiles are shown in Figure 12.

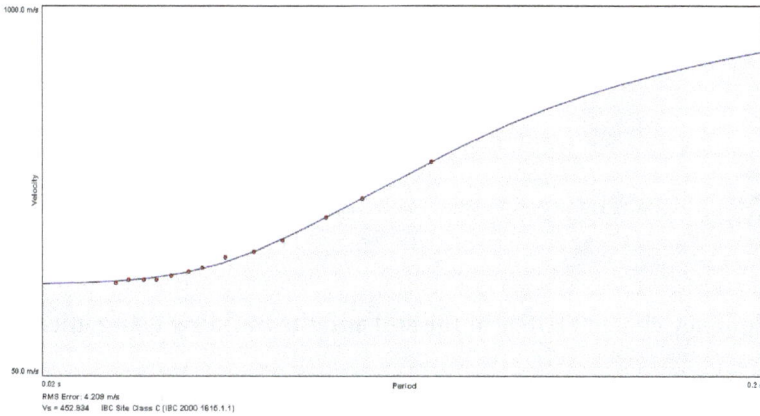

Figure 10. Dispersion curve showing picks and fit for Profile ReMi-1

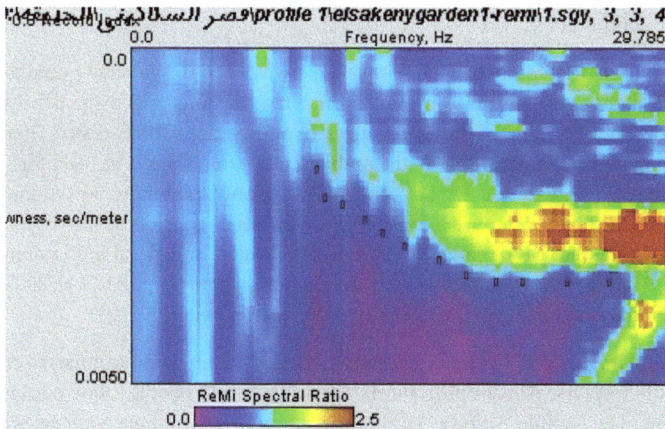

Figure 11. P-F image with dispersion modeling picks for Profile ReMi-1

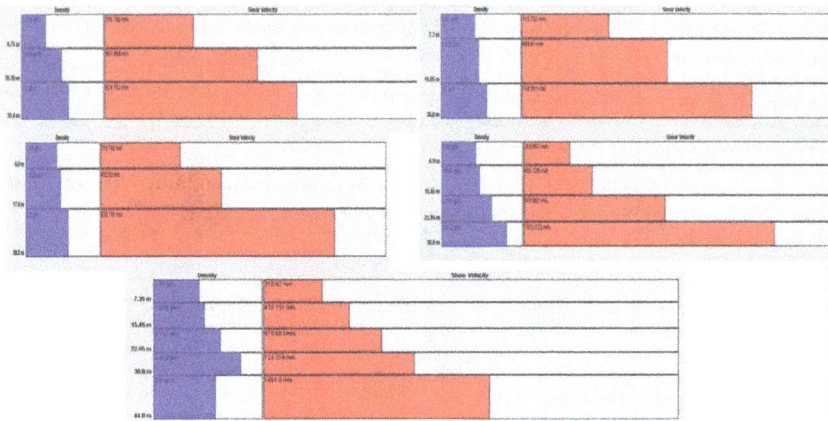

Figure 12. Shear wave velocity model calculated for refraction microtremors profiles ReMi-1 To ReMi-5 (Figure 9).

5. Frequency charactersitics of the soil and the building using microtremrs

Microtremors are omnipresent low amplitude oscillations (1-10 microns) that arise predominantly from oceanic, atmospheric, and urban or anthropogenic actions and disturbances. The
implicit assumption of early studies was that microtremors spectra are flat and broadband
before they enter the region of interest (soil or building). When microtremors enter preferable
body it changes and resonate depending on the nature of the material, shape, and any other
characteristics of this body.

It may be considered to compose of any of seismic wave types. We have two main types of
microtremors, Local ambient noise coming from urban actions and disturbances and long
period microtremors originated from distances (e.g. oceanic disturbances). There is still a
debateongoing on the characteristics of the ambient noise that should be used for site characterization and ground response. While some are using only the longer period microtremors
originated from farther distances (e.g. Field et al, 1990), others considered that traffic and other
urban noise sources are producing equally reliable results. In general low amplitude noise
measurements comparable results give with strong motion data (Raptakis et al, 2005., Pitilakis,
2011., Apostolidis et al., 2004., Manakou et al, 2010., Mucciarelli, 1998).

Kanai 1957, first introduced the use of microtremors, or ambient seismic noise, to estimate the
earthquake site response (soil amplification). After that lots of people followed this work but
from the point of soil amplification of earthquake energy for different frequencies (e.g. Kanai
and Tanaka 1961 and Kanai 1962, Kagami et al, 1982 and 1986; Rogers et al., 1984; Lermo et al.,
1988; Celebi et al. 1987).

5.1. Instrumentation and data acquisition

A high dynamic range Seismograph (Geometrics ES-3000 see Figure 13) mobile station with triaxial force balance accelerometer (3 channels), orthogonally oriented was used. The station was used with 4Hz sensors to record the horizontal components in longitudinal and transverse directions in addition to the vertical components. For the data acquisition and processing we followed the following steps:

• Recording 10-min of ambient noise data using a mobile station moving among variable soil stations or El Sakakini building floors/

• Zero correction to the total 10-minnoise at time domain

• Subdivision of each 10-minsignal into fifteen 1-min sub-windows,

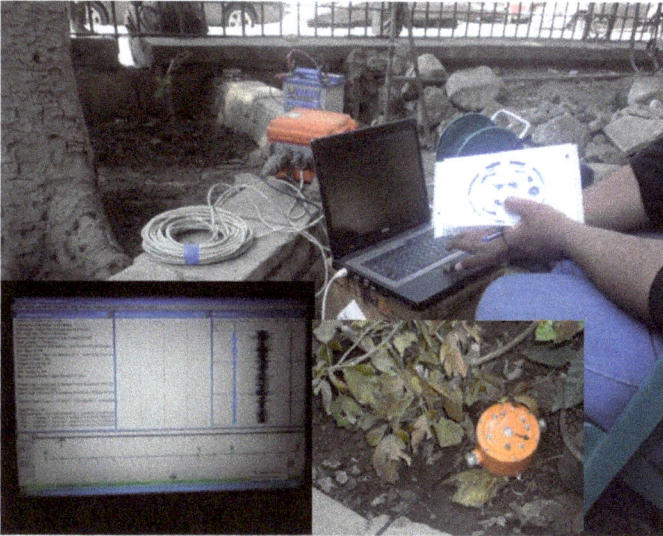

Figure 13. High dynamic range ES-3000 Geometrics mobile station and triaxial geophone used 4 Hz to drive soil response of El-Sakakini Palace.

Each of these series was tapered with a 3-sec hanning taper and converted to the frequency domain using a Fast Fourier transform,

• Smoothing the amplitude spectrum by convolution with 0.2-Hz boxcar window,

• Site response spectrum for a given soil site (or certain floor) is given by dividing the average spectrum of this site over the spectrum of the reference site. The reference site is choose carefully in the site as deepest and calmest station in the basement floor with least soil response (usually we choose a certain basement floor location with least soil response to be used as reference site).

- Smoothing the final response curves by running average filter for better viewing. A complete description of the methodology can be found in Gamal and Ghoneim, (2004).

5.2. Ground response

Figure 14 shows the locations of microtremors stations used to determine the ground response at EL Sakakini Palace area. The predominant frequency of the ground at EL Sakakini Palace is about 3 Hz (see Figure 15 & Table 1), a value almost identical to the theoretical estimation according to Kennett and Kerry (1979) (Figure 16 & Table 4). The amplification factor is about 2, which is relatively low.

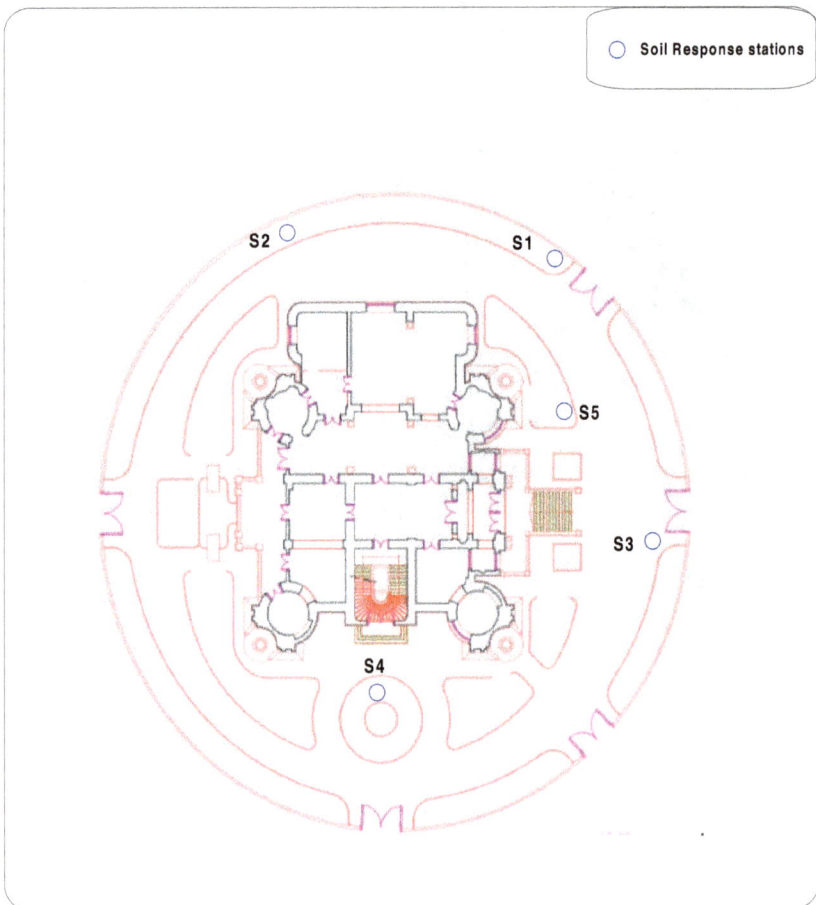

Figure 14. Ambient noise measurement locations

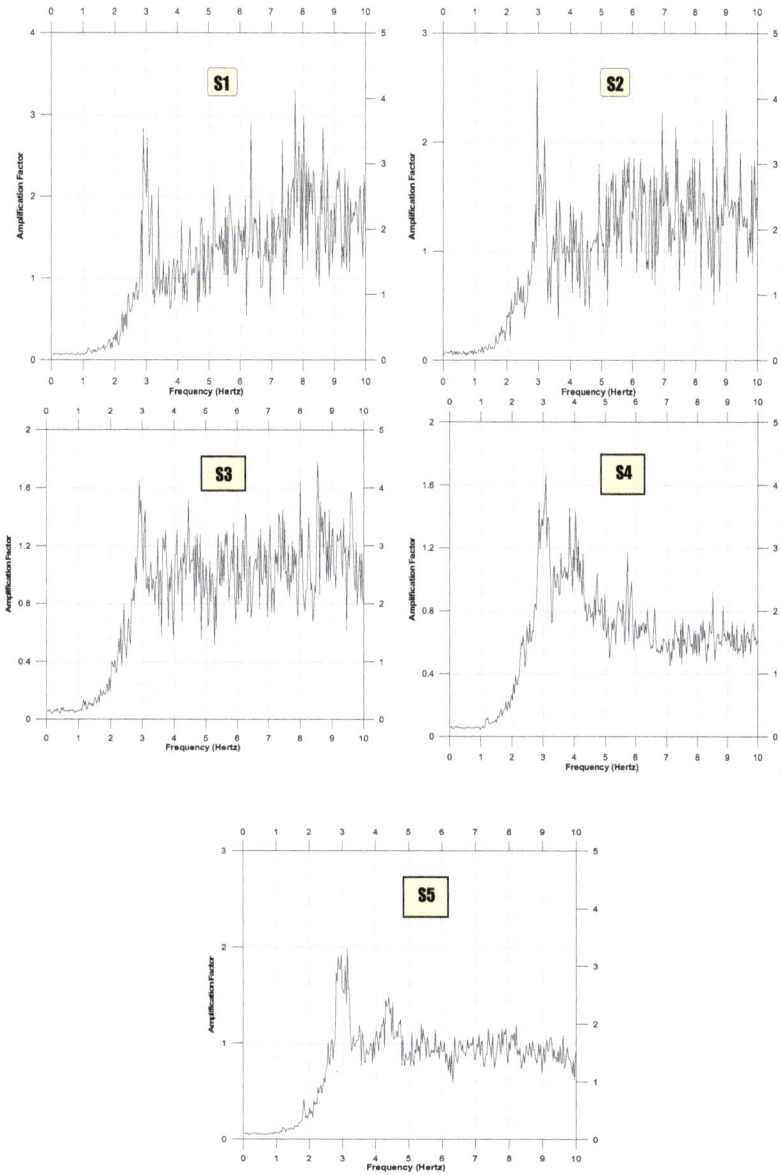

Figure 15. Microtremors soil response for El Sakakini Palace Sites S1 to S5.

Site	Fundamental frequency (Hz)	Amplification Factor
S1	3	2.5
S2	3.2	2
S3	3	1.6
S4	3	1.6
S5	3	2

Table 3. Fundamental frequencies and amplification factors at five locations

Thickness	P-wave velocity (m/s)	S-wave velocity (m/s)	Dry Density (gm/c.c)	Quality factor Qs
10	1300	315 (350?)	1.6	7
25	1300	500	1.7	15
>10.5	2000	700	2	100
	3000	1200	2.5	200

Table 4. Parameters used for the Kennett and Kerry method (1979)

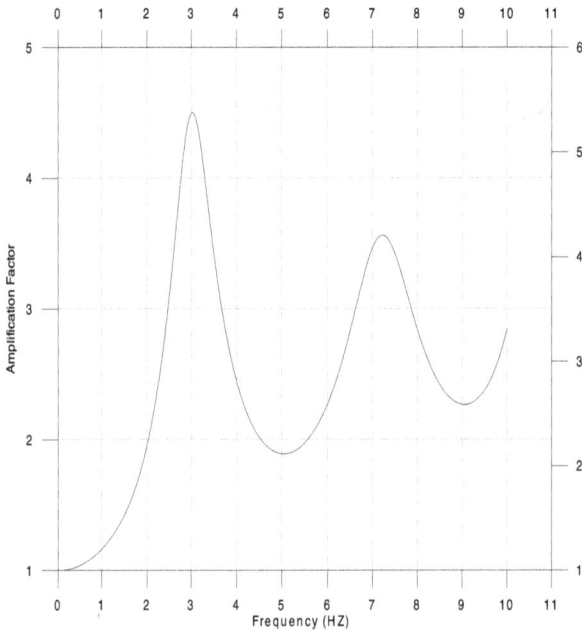

Figure 16. Theoretical ground response analysis at EL Sakakini Palace using Kennett at al. (1979) method.

5.3. Building response

The El Sakakini building is composed of a basement and five floors the upper two being wooden. Figures 17 to 19 show the locations of recording stations used to drive El Sakakini building response. Figures 21 to 26 and Table 5 show the recorded natural frequency of vibration for each floor. All floors show nearly the same resonance frequency with the soil (3-4 HZ). The wooden floors (Figure 25 & 26) show very high amplification and multi peak as fundamental and other harmonics. The fundamental natural frequency of vibration is always the most important frequency that insert the maximum earthquake vibration energy into structure. However when we find other mode of vibrations with big amplification factors we consider this as a warning that this structure may suffer from vibration. This could be very good warning for its unstable performance during vibration.

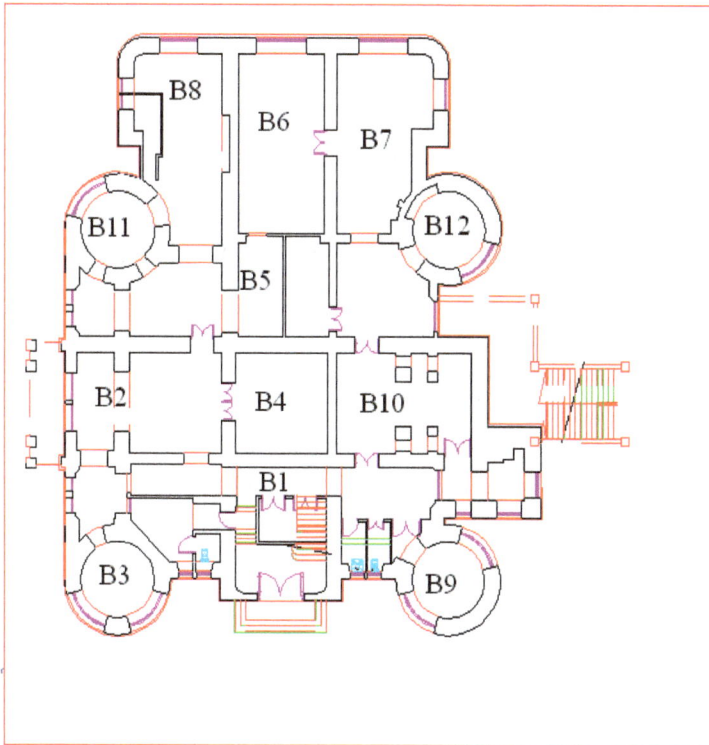

Figure 17. Location of stations at the basement of EL-Sakakini Palace.

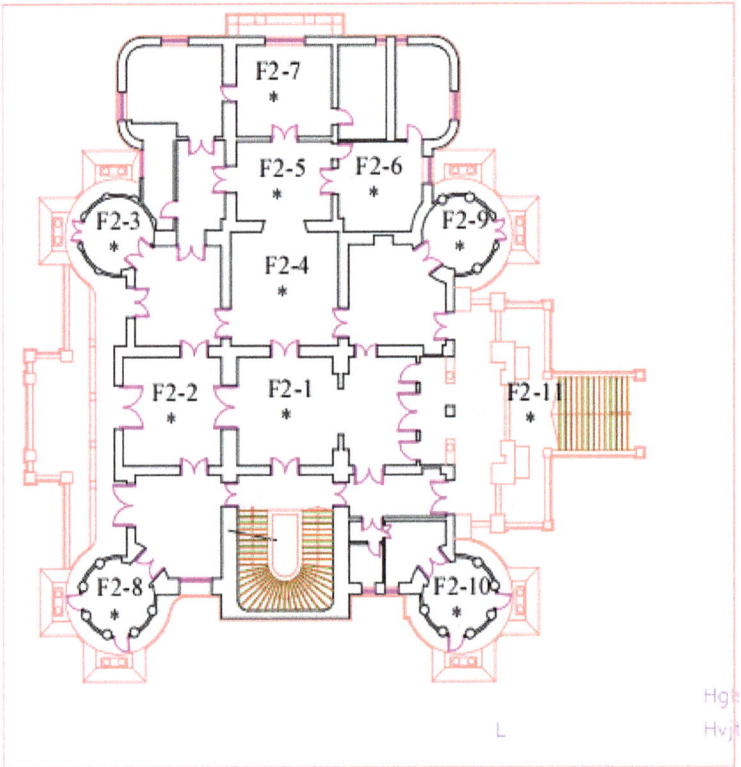

Figure 18. Location of stations at 2nd floor of El-Sakakini Palace.

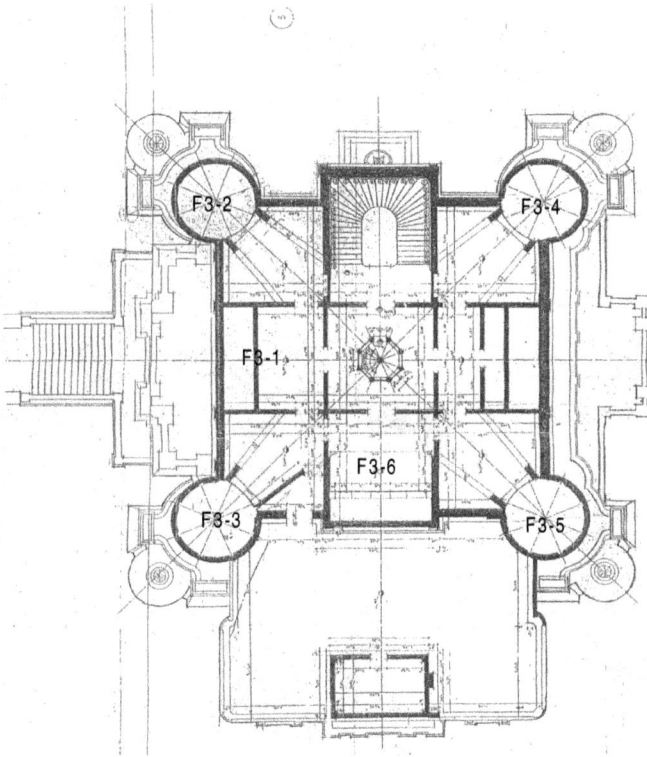

Figure 19. Location of stations at the 3rd floor.

Figure 20. High dynamic range ES-3000 Geometrics mobile station and triaxial geophone used 4 Hz to drive structure response of El-Sakakini Palace.

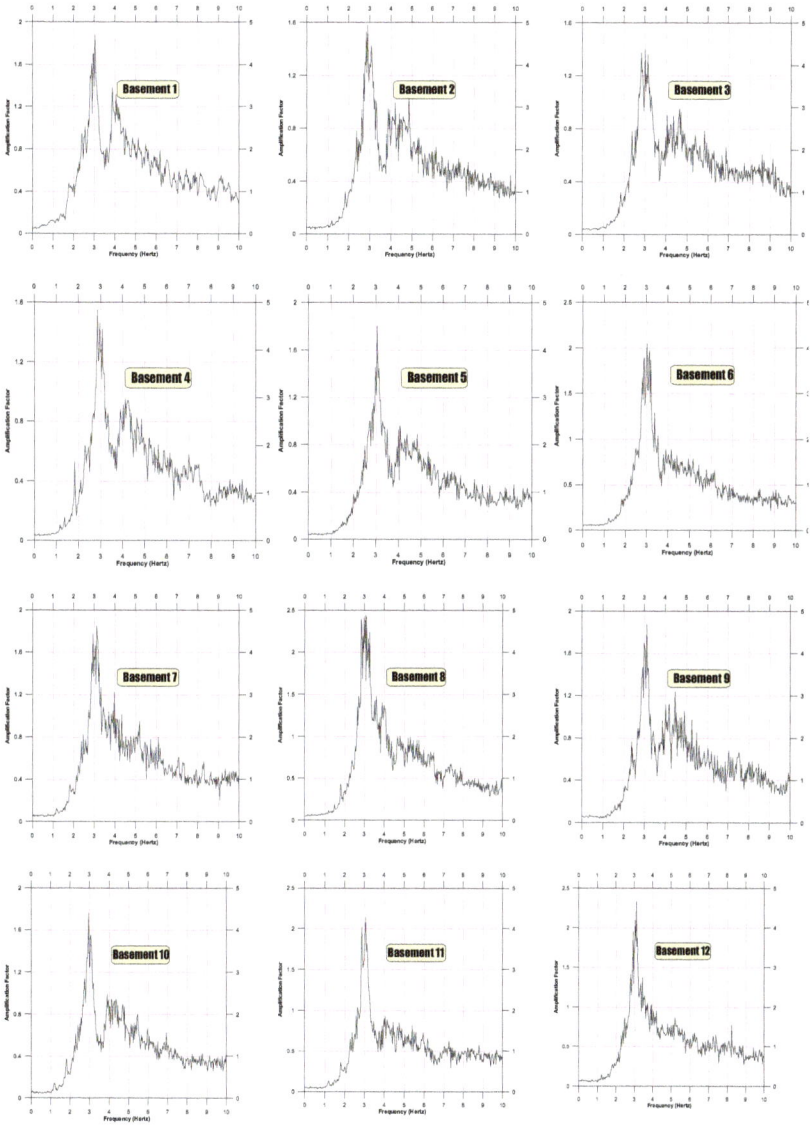

Figure 21. Natural frequency of vibration for basement floor.

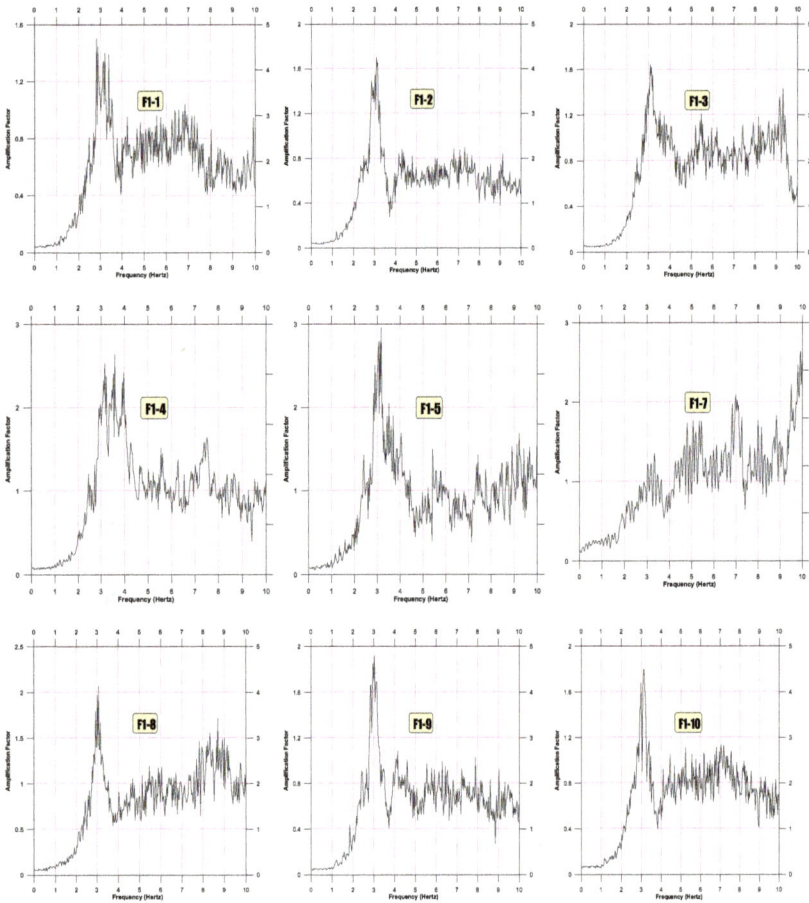

Figure 22. Natural frequency of vibration for the 1st floor.

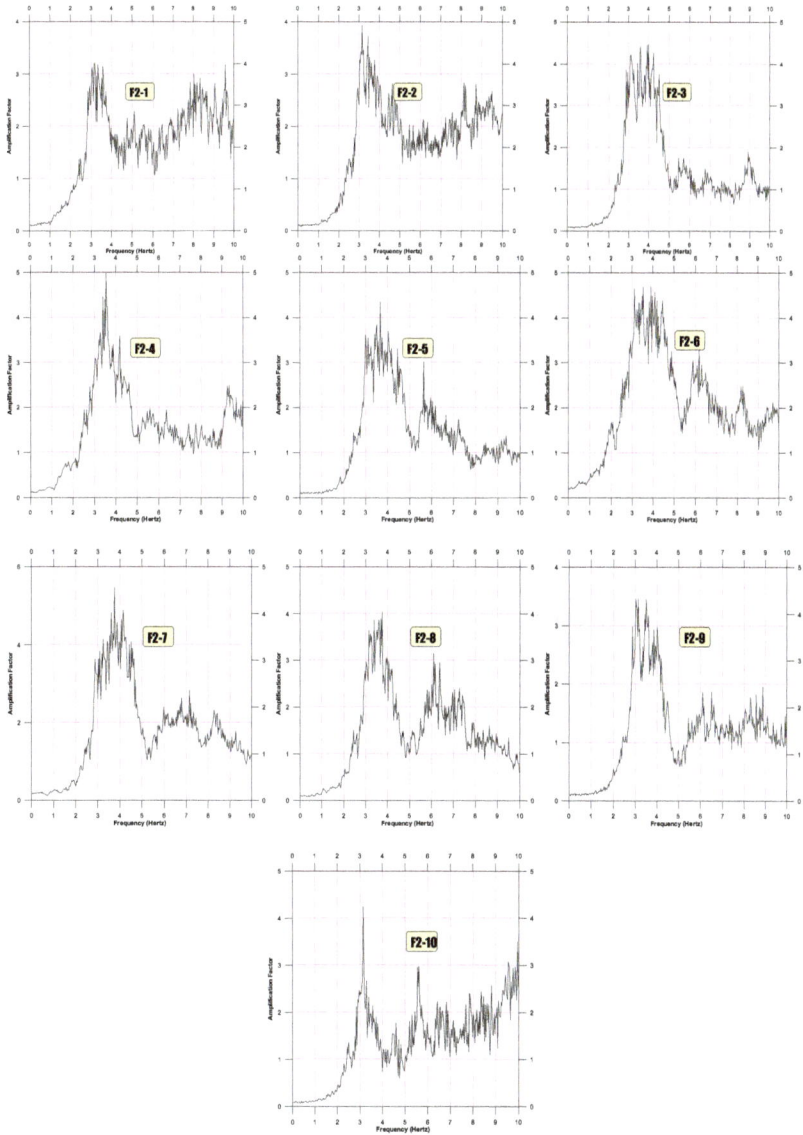

Figure 23. Natural frequency of vibration for the 2nd floor.

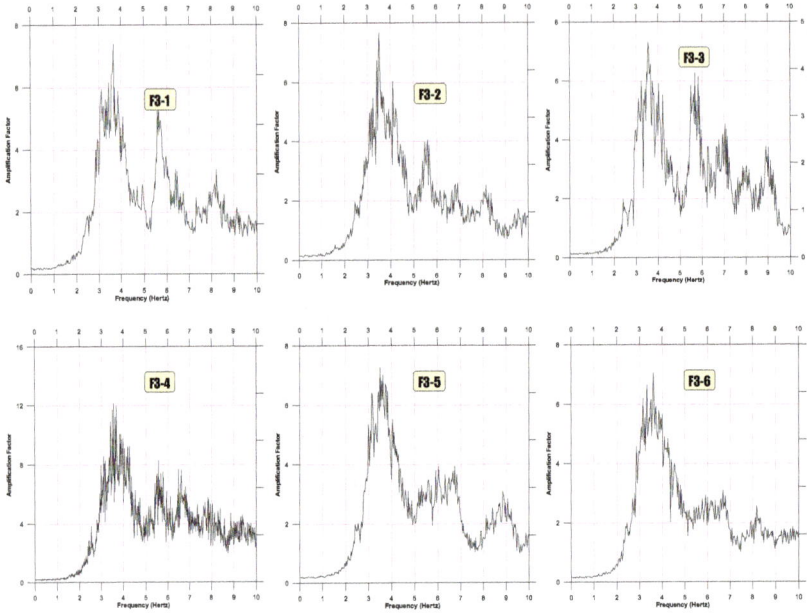

Figure 24. Natural frequency of vibration for the 3rd floor.

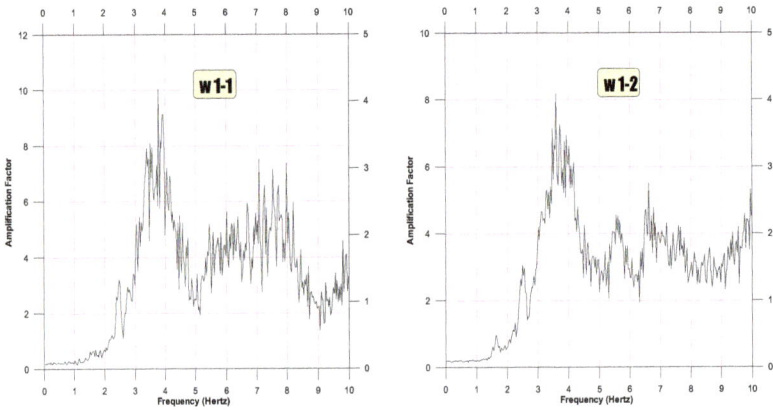

Figure 25. Natural frequency of vibration of the 4th floor (wooden).

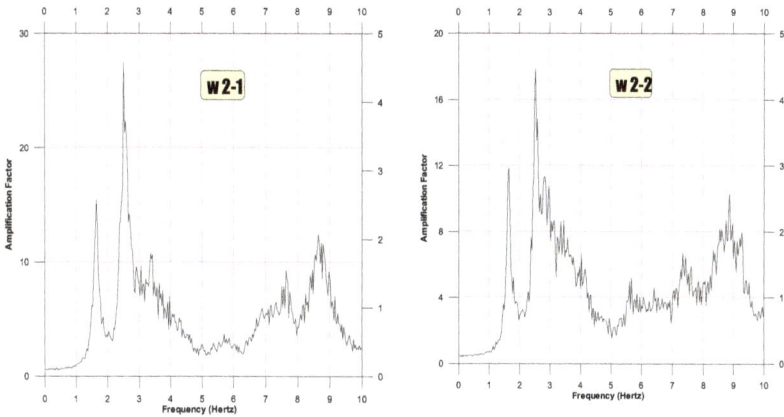

Figure 26. Natural frequency of vibration of the 5th floor (wooden).

Floor	Fundamental resonance frequency (HZ)	Amplification Factor
Basement	3	1.3-2
1st Floor	3-3.5	1.4-2.8
2nd Floor	3-4	3-4.5
3rd Floor	3-4.3	5.5-8
4th wood Floor	3-4.5	7.5-8
5th wood Floor	2.5	12-20

Table 5. Natural frequencies of vibration of El Sakakini Palace.

6. Conclusions

ElSakakini Palace is an important monument in Egypt. We presented the main results of the seismic hazard analysis and the geophysical campaign to estimate the main characteristics of the ground response and the structure. Based on the available maximum intensity maps for historical earthquakes (>2200BC) the maximum Mercalli Intensity expected at ElSakakini Palace site is VII.

The peak horizontal acceleration at the seismic rock basement found at -35m approximately, and for 10% probability of exceedance in 50 years is 144 cm/sec² (0.147g), while for 100 years is 186 (cm/sec²) (0.19 g). We determined the average soil profile using different geophysical campaigns. It is found that the upper layer has an average shear wave velocity lower than

300m/s and a thickness of 5 to 10meters. It is a man made fill material in rather loose conditions. Below there is a clayey material with average Vs velocity equal to 400-600m/s. At -35m in average we found saturated compacted sand and gravels with Vs velocity exceeding 700m/s. It is considered as the seismic bedrock for the foreseen detailed site-specific analysis of the ground response.

Based on the ambient noise campaign the fundamental frequency of the ground is of the order of 3.0 to 3.5sec very close to the fundamental frequency of the palace. Resonance phenomena should be expected and considered seriously in the detailed analysis of the structure. There are strong evidences that the upper two stories with wooden floors, which are presenting high amplification factors, are subjected to several damages and degradation of their bearing capacity.

Author details

Sayed Hemeda*

Address all correspondence to: hemeda@civil.auth.gr

Conservation Department, Faculty of Archaeology, Cairo University, Egypt

References

[1] Apostolidis P., Raptakis D., Roumelioti Z., Pitilakis K., (2004), Determination of S-Wave velocity structure using microtremors and SPAC method applied inThessaloniki (Greece). Soil Dynamics & Earthquake Engineering, 24, 1, 49-67.

[2] Ben-Menahem, A., (1979). Earthquake catalogue for the Middle East (92 B. C. - 1980 A. D.): Bollettinodi Geofisica Teoricaed Applicata, 21, No. 84, 245 - 310

[3] Castro, R. R., F. Pacor, A. Sala, and C. Petrungaro (1996). S Wave attenuation and site effects in the region of Friull, Italy, J. Geophysics. Res. 101, 22355-22369.

[4] Celebi, M., C. Dietel, J. Prince, M. Onate, and G. Chavez (1987). Site amplification in Mexico City (determined from 19 September 1985 strong-motion records and from records of weak motion), in Ground Motion and Engineering Seismology, A. S. Cakmak (Editor), Alsevier, Amsterdam, 141-152.

[5] Chopra, A. K. (1981). Dynamics of structures, a primer Earthquake engineering research institute.

[6] Chopra, A. K., (1995). "Dynamic of structures – Theory and Application to Earthquake Engineering, Englewood Cliffs, New Jersey.

[7] Clayton, R. W., and McMechan, G. A., 1981, Inversion of refraction data by wavefield continuation: Geophysics, v. 46, p. 860-868.

[8] Egyptian Code. Egyptian code of Practice (ECP-1993) for estimating loads and forces, 1993.

[9] F. Nateghi-A, (1994). Seismic strengthening of a ten story steel framed hospital: Earthquake resistant Construction & design, Savidis (ed.), Balkema, Rotterdam, ISBN 90 5410 392 2, P 849-856.

[10] Field E. H., S. E. Hough, and. H. Jacob (1990). Using microtremors to assess potential earthquake site response, a case study in Flushing Meadows, New york city. Bulletin of the Society of America, Vol. 80. No. 6, pp. 1456-1480, December, 1990.

[11] Gamal M. A. and Noufal A., A Neotectonics model AND seismic hazard assessment for Egypt, 8 th international conference on geology of the Arab world, Cairo university Egypt 13-16 2006.

[12] Jordan Sesimoloical observatory (1998-2000), national resources authority geophysics and technical services, bulletin (1998-2000).

[13] Joyner, W. B., and D. M. Boore, 1981, Peak horizontal acceleration and velocity from strong motion record including records from the 1979 Imperial Valley, California, earthquake, Bull. Seism. Soc. Am., 17, 2011-2038.

[14] Kagami, H., C. M. Duke, G. C. Liang, and Y. Ohta (1982). Observation of 1-5 second microtremors and their application to earthquake engineering. Part II. Evaluation of site effect uppon seismic wave amplification due to extremely deep soils., Bull. Seism. Soc Am. 72, 987-998.

[15] Kagami, H., S. Okada, K. Shiono, M. Oner, M. Dravinski, and A.K. Mal (1986). Obser- vation of 1-5 seconds microtremors and their application to earthquake engineering. Part III. A two-dimensional of the site effect in San Fernando valley, Bull. Seism. Soc. Am. 76, 1801-1812.

[16] Kanai, K. (1957). The requisite conditions for predominant vibration of ground, Bull. Earthq. Res. Inst., Tokyo University, 31, 457.

[17] Kanai, K. (1962). On the spectrum of strong earthquake motions, Primeras J. Argenti- nasIng.Antisismica 24, 1.

[18] Kanai, K. and Tanaka (1961). On microtremors. VII, Bull. Earthq. Res. Inst., Tokyo Univ., 39, 97-115.

[19] Langston, C. A. (1977). Corvallis, Oregon, Crustal and upper mantle receiver structure from teleseismic P and S waves, Bull. Seism. Soc. Am. 67, 713-724.

[20] Lermo, J. and F. J. Chavez-Garcia, (1993). Site effect evaluation using spectral ratios with only one station, Bull. Seism. Soc. Am. 83, 1574-1594.

[21] Lermo, J., M. Rodriguez, and S. K. Singh (1988). The Mexico Earthquake of September 19, 1985: Natural periods of sites in the valley of mexico from microtremorsmeasurments and from strong motion data, Earthquake Spectra 4, 805-814.

[22] Louie, J, N., 2001, Faster, Better: Shear-wave velocity to 100 meters depth from refraction microtremors arrays: Bulletin of the Seismological Society of America, v. 91, p. 347-364.

[23] Maamoun, M., 1979:Macroseismic observations of principal earthquake in Egypt, HelwanInstitue of Astronomy and Geophysics, Bulletin 183, 120.

[24] Maamoun, M., A. Megehed, and A. Allam, (1984).Seismicity of Egypt, Institute of Astronomy and Geophysics 4.

[25] Makropoulos, K. C., and P. W. Burton, (1981).A catalogue of Seismicity in Greece and adjacent areas.Geophys. J. R. Astr. Soc. 65, 741 - 762.

[26] Manakou M., Raptakis D., Apostolidis P., Chávez-García F. J., Pitilakis K., (2010), 3d soil structure of the Mygdonian basin for site response analysis, Soil Dynamics and Earthquake Engineering, vol. 30, pp. 1198-1211

[27] McGuire, R.K., (1993), FORTRAN computer program for seismic risk analysis. U.S. Geol. Surv.

[28] Miller, R. D., Park, C. B., Ivanov, J. M., Xia, J., Laflen, D. R., and Gratton, C., 2000, MASW to investigate anomalous near-surface materials at the Indian Refinery in Lawrenceville, Illinois: Kansas Geol. Surv. Open-File Rept. 2000-4, Lawrence, Kansas, 48 pp.

[29] Mohamed A. Gamal, A. Noufal and E. Ghoneim, ANeotectonics model and seismic hazard assessment for Egypt (2006): (Accepted for publication) 8th international conference of geology of the Arab world, 13-16 Feb. 2006 Cairo university Giza, Egypt.

[30] Mucciarelli M., (1998). Reliability and applicability of Nakamura's technique using microtremors: an experimental approach, Journal of Earthquake Engineering, Vol.2, No.4 625-638 Imperial college press.

[31] Nakamura, Y. (1989). A method for dynamic characteristic estimation of subsurface using microtremors on the ground surface. Q. R. Railway tech. Res. Inst. Rept. 30, 25-33.

[32] NEIC, (1973-2006). Preliminary determination of Epicenters, monthly listing, U.S. Department of the interior/ Geological survey, National Earthquake information center:.

[33] Nogoshi, M. and T. Igarashi (1970).On the propagation characteristics of microtremors, J. Seism. Soc. Jpn. 23, 264-280.

[34] Pitilakis K., Anastasiadis A., Kakderi K., Manakou M., Manou D., Alexoudi M., Fotopoulou S., Argyroudis S., Senetakis K., 2011, "Development of comprehensive earthquake loss scenarios for a Greek and a Turkish city: Seismic hazard, Geotechnical and Lifeline Aspects", Earthquakes and Structures,Vol. 2, No 3, September 2011.

[35] Pullammanappallil, S., Honjas, B., Louie, J., Siemens, J. A., and Miura, H., 2003, Comparative study of the refraction microtremors method: Using seismic noise and standard P-wave refraction equipment for deriving 1D shear-wave profiles, Proceedings of the 6th SEGJ International Symposium (January 2003, Tokyo), 192-197.

[36] Raptakis D., Manakou M., Chavez-Garcia F., Makra K., Pitilakis K., (2005), 3D configuration of Mygdonian basin and preliminary estimate of its site response,Soil Dynamics and Earthquake Engineering, 25, 871-887.

[37] Riad, S., and H. Meyers, (1985).Earthquake catalog for the Middle East countries 1900 - 1983. World Data Center.

[38] Rogers, A. M., R. D. Borcherdt, P. A. Covington, and D. M. Perkins (1984). A comparative ground response study near Los Angeles using recordings of Nevada nuclear tests and the 1971 San Fernando earthquake, Bull. Seism. Soc. Am. 74, 1925-1949.

[39] Shapira, A., (1994).Seimological Bulletin of Israel. 1900 - 1994.

[40] Thorson, J. R., and Claerbout, J. F., 1985, Velocity-stack and slant-stack stochastic inversion: Geophysics, v. 50, p. 2727-2741.

[41] Uniform Building Code (UBC,1985). International Conference of building officials whittier, California.

[42] Wilson, E.L, and Habiballah, A (1990)."SAP2000, Structural Analysis Programe" computer and structures Inc., Berkeley, California.

[43] Woodword Clyde Consultant WWCC, (1985).Earthquake activity and dam stability for the Aswan High Dam, Egypt.High and Aswan Dames Ministry of Trregation, Cairo 2.

Chapter 3

V$_S$ Crustal Models and Spectral Amplification Effects in the L'Aquila Basin (Italy)

M.R. Costanzo, C. Nunziata and V. Gambale

Additional information is available at the end of the chapter

1. Introduction

On April 6, 2009 a strong earthquake (M$_L$ 5.9, M$_W$ 6.3), hereafter called main shock, struck the Aterno Valley in the Abruzzo region (central Italy) causing heavy damage in L'Aquila and in several nearby villages and killing more than 300 people. The event had a pure normal faulting mechanism, with a rupturing fault plane NW striking and 45°SW dipping; hypocentral location was at 9.5 km depth and epicenter at a distance of about 2 km WSW from L'Aquila center [1]. Few days later, the aftershock activity involved also the area NE of L'Aquila toward Arischia and Campotosto. The overall distribution of the aftershocks defined a complex, 40 km long and 10-12 km wide, NW trending extensional structure. The largest damage was mainly distributed in a NW-SE direction [2], according to the orientation of the Aterno river valley. The area has a high seismic hazard level in Italy [3] and has experienced in the past destructive earthquakes such as the 1349, I=IX–X; the 1461, l'Aquila, I=X and the 1703, I=X [4]. Many active faults are recognized in the area and several of them are indicated as potential sources for future moderate and large earthquakes by several authors (see for a review [5]).

The Aterno river basin is a complex geological structure with a carbonate basement outcropping along the valley flanks and elsewhere buried below alluvial and lacustrine deposits with variable thickness. The surface geology is even more complicated by the presence at L'Aquila of breccias consisting of limestone clasts in a marly matrix. Such complex geological scenario reflected in a large spatial variability of amplitude and frequency content of the ground motion (e.g. [6]). Among several studies performed on the recorded events, there is not a note dedicated to the modeling of the shear seismic velocities of the crust structures, yet it is useful for the geological reconstruction and fundamental for computing seismograms. Simulation of the ground motion has been performed at L'Aquila based on literature data [7] by using the Neo-Deterministic Seismic Hazard Analysis (NDSHA) [8,9], an innovative modeling technique that takes into account source, propagation and local site effects. In order to estimate realistic ground motion we need physical parameters of rocks from surface to depths greater than the earthquake hypocenter. At engineering scale, microzoning activities promoted by the Italian Civil Defense Department [10] have performed V_S measurements at depths around 25 m, in gravelly soils with different degree of cementation, alternating to thin layers of finer deposits (sands and/or silts) that often include carbonate boulders (www.cerfis.it). The investigated depths are too shallow to define the vertical and lateral passage from soft sediments to rock basement (V_S at least of 800 m/s) which was sporadically found. At regional scale, a physical model is available extending to depths of about 300 km [11].

Aim of this paper is to retrieve V_S models of the shallow crust in the Aterno river valley from the non-linear inversion of the group velocity dispersion curves of the fundamental mode extracted with the FTAN method (e.g. [12,13,14]) from recordings of earthquakes with $M_L \geq$ 2.9 (Table 1) between April 5 and November 10, 2009, in the selected coordinate window of 42.4 ± 0.2 N and 13.4 ± 0.2 E. In addition, V_S of the superficial 30 m of Aterno alluvial soils are defined by an active seismic experiment in the Coppito area, and compared with nearby cross-hole measurements. The V_S profiles vs. depth are then attributed to lithotypes along a geological cross section from the epicenter to a seismic station at L'Aquila. Simulation of the main shock is performed with the NDSHA approach and the computed response spectra and the H/V spectral ratios are compared with those recorded.

2. Geological and geophysical setting

The epicentral area of the L'Aquila seismic sequence mainly corresponds to the upper and middle Aterno river valley which is bounded by predominantly NW-SE-striking and SW-dipping active normal faults (e.g. [15] and references therein) and characterised by the high variability of the geologic and geomorphologic patterns (Fig. 1). The valley is superimposed on a Quaternary lacustrine basin of tectonic origin and surrounded by carbonates. The thickness of the Quaternary deposits is variable, from about 60 m in the upper Aterno river valley to more than 200 m in the middle valley.

The upper and middle crustal structure of the Abruzzo region has been reconstructed from V_P and V_P/V_S images on a 3D grid with a node vertical spacing of 4 km, by using local earthquake tomography and receiver function modeling [16]. Low V_P velocities (smaller than 4.8 km/s) were found in correspondence with the main Plio-Quaternary basins (Fucino, l'Aquila and Sulmona basins) and high V_P velocities (V_P larger than 5.0 km/s) were mostly correspondent with the outcropping Mesozoic carbonates. Low V_P and high V_P/V_S anomalies were found beneath the L'Aquila and Fucino Quaternary basins between 4 and 12 km depth, suggesting the existence of fluid-saturated rock volumes. Very high P-wave (6.7–7.0 km/s) and S-wave (3.6–3.8 km/s) velocity bodies were observed below 8–12 km depth, and interpreted as deep crustal or mantle rocks exhumed before the sedimentation of the Mesozoic cover. This layer has a regional character as it is found at ~15 km of depth below the 1°x1° cell containing L'Aquila epicentral area, and lying on the mantle detected at ~35 km [11].

The analysis of strong and weak motion recordings in 1996-98 put in evidence amplification effect at low frequencies (0.6 Hz) in the town of L'Aquila and 2D numerical modeling allowed to fit it along a SW-NE section [17]. A sedimentary basin was inferred, filled by lacustrine sediments, with a maximum thickness of about 250 m, below the breccias formation about 50 m thick.

3. V_S models

3.1. Data analysis

The 2009 seismic sequence was recorded by Rete Accelerometrica Nazionale (RAN) network, managed by the Italian Civil Defense Department, some of which located at L'Aquila (AQK station) or in the NW of it (AQG, AQA, AQM, AQV), and by the station AQU operating since 1988 as part of the Mediterranean Network (MedNet), managed by the Italian Istituto Nazionale di Geofisica e Vulcanologia (INGV). They are equipped with three-component accelerometers set to 1 or 2 g full-scale, coupled with high resolution digitizers, while AQU is also equipped with a very broadband Streckeisen STS-1.

In order to obtain V_S models for the L'Aquila basin shallow crust, we have analysed about forty earthquakes ($M_L \geq 2.9$) recorded at the RAN and AQU stations, and rotated to get the radial and transverse component of motion. Rayleigh wave group velocities of the fundamental mode have been measured from vertical and radial components of 17 events (Fig. 1, Table 1).

As regards the shallow 30 m subsoil, the same analysis has been applied to recordings of an active seismic experiment performed in the Coppito area, about 500 m far from the AQV station (Fig. 1).

The group velocity is measured as function of period by the Frequency Time Analysis (FTAN) on single waveforms (e.g. [12,13,14]). The FTAN method allows to isolate the different phases in a seismogram, in particular the fundamental mode of surface waves. A system of narrow-band Gaussian filters is employed, with varying central frequency, that do not introduce phase distortion and give the necessary resolution in the time-frequency domain. The source-receiver distance is commonly assumed to be the epicenter distance when it is much greater than the event depth. When this assumption is not valid, in order to extract the correct dispersion curve of Rayleigh waves we have to add a time delay to seismograms as $\Delta t = h/V_P$, h being the source depth, V_P the average P-wave velocity from surface to the hypocenter, and then analyze the seismograms by FTAN, considering the hypocenter distance (e.g. [19] and references therein).

The dispersion curves obtained in such a way can be inverted to determine S-wave velocity profiles versus depth. A non-linear inversion is made with the Hedgehog method ([20,14] and references therein) that is an optimized Monte Carlo non-linear search of velocity-depth distributions. In the inversion, the unknown Earth model is replaced by a set of parameters and the definition of the structure is reduced to the determination of the numerical values of these parameters. In the elastic approximation, the structure is modeled as a stack of N homogeneous isotropic layers, each one defined by four parameters: V_P (dependent parameter), density (fixed parameter), V_S and thickness (independent parameters).

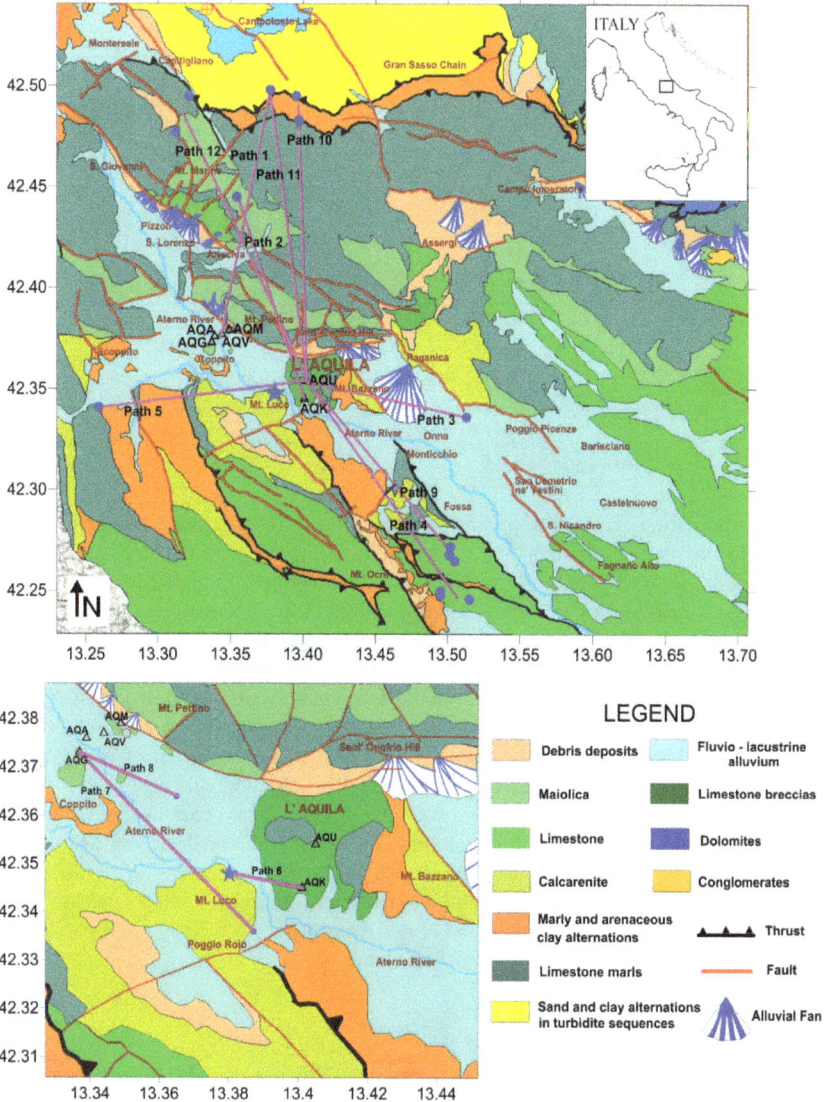

Figure 1. Simplified geological map of L'Aquila basin (modified after [18]) with location of the V$_s$ investigated paths connecting the analyzed events (blue circles) and the recording INGV stations (triangles), and of the deep drilling at Campotosto (red circled cross). April 2009 mainshock (M$_w$ = 6.3) epicenter is represented by star. Detail of the geological map at L'Aquila is shown in the bottom with the location of the V$_s$ investigated paths and the site of the active seismic experiment (red circle).

Events	Time origin (UTC)	M_L	Depth (km)	Lat. (°N) ; Long. (°E)	Path
04/06/09	01:32:41	5.9	90	42.348 ; 13.380	6
04/07/09	09:26:28	4.8	9.6	42.336 ; 13.387	7
04/07/09	21:34:29	4.3	9.6	42.364 ; 13.365	8
04/09/09	13:19:33	4.1	9.7	42.341 ; 13.259	5
04/09/09	20:47:01	3.3	10.8	42.495 ; 13.321	12
04/09/09	20:40:06	3.8	11.1	42.477 ; 13.312	12
04/11/09	19:53:53	3.0	90	42.336 ; 13013	3
04/13/09	21:14:24	5.0	9.0	42.498 ; 13.377	1-11
04/14/09	07:36:44	2.9	9.6	42.495 ; 13.395	10
04/20/09	11:43:06	3	9.7	42.272 ; 13002	9
04/24/09	22:51:29	3	10.6	42.265 ; 13005	9
05/05/09	18:03:41	3.2	9.7	42.268 ; 13001	9
05/14/09	06:30:22	30	90	42.483 ; 13.397	10
06/22/09	20:58:40	4.6	13.8	42.445 ; 13.354	2
07/23/09	22:37:33	3.1	10.1	42.250 ; 13.495	4
07/31/09	11:05:40	3.8	9.6	42.248 ; 13.495	4
08/31/09	14:09:10	3.1	9.5	42.246 ; 13.515	4

Table 1. List of the studied events and paths. The source parameters are from INGV (http://bollettinosismico.rm.ingv.it) and for the 04/06/2009 main shock from [1].

Given the error of the experimental phase and/or group velocity data, it is possible to compute the resolution of the parameters (parameter step), computing partial derivatives of the dispersion curve with respect to the parameters to be inverted ([14] and references therein). The theoretical phase and/or group velocities computed during the inversion with normal-mode summation are then compared with the corresponding experimental ones and the models are accepted as solutions if their difference, at each period, is less than the measurement errors and if the r.m.s. (root mean square) of the differences, at all considered periods, is less than a chosen quantity (usually 60–70% of the average of the measurement errors). Being the parameter step indicative of the parameter resolution, all the solutions of the Hedgehog inversion differ by no more than ±1 step from each other. A good rule of thumb is that the number of solutions is comparable with the number of the inverted parameters. From the set of solutions, we accept as representative solution the one with r.m.s error closest to the average r.m.s error of the solution set, and hence reduce, at the cost of loosing in resolution, the projection of possible systematic errors [20] into the structural model. Other selection criteria of the representative solution are discussed in detail by [21].

3.2. Results

Dispersion curves of Rayleigh wave fundamental mode have been extracted from the vertical and/or radial components of recordings of 17 events (Table 1, Fig. 1). Average dispersion curves have been computed along 12 paths and inverted with Hedgehog method [20,14] to get V_S models. A V_P/V_S ratio equal to 1.8 turned out to be a suitable value after a set of tests made varying it between 1.8 and 2.1. In other words, keeping all other values of the parameterization unchanged, the number of solutions maximizes for $V_P/V_S=1.8$. Moreover, the analysis of group velocity derivatives with respect to elastic parameters versus depth has allowed to define the sensitivity of the investigated periods on the S-wave velocity structure. Time corrections have been applied to the recordings by assuming the regional average V_P computed from the V_S model relative to the cell 1°x1° containing L'Aquila [11], assuming a V_P/V_S ratio of 1.8. Such value is in very good agreement with those used by INGV (5 km/s) and [1] (V_P=5.5 km/s) for earthquake location and attributed to the shallow 11.1 km of crust.

In the following, the results (dispersion data and Hedgehog V_S solutions) are presented by grouping the paths as northern paths (1-2-10-11-12), middle Aterno river valley paths (6-7-8), middle paths (3-5), and southern paths (4-9). The events are listed in Table 1 and the paths are located in Fig. 1. Moreover, the results obtained with the same methods from an active seismic experiment in the Coppito area, about 500 m far from the AQV station, are presented.

3.2.1. Northern paths

All the results obtained for the northern paths (1-2-10-11-12) are shown in Fig. 2. The dispersion curves relative to path 1 have been extracted from the recordings of the 4/13/2009 event, located nearby Campotosto, at AQG, AQV and AQM stations. Rayleigh data, sampled at periods of 1-3.8 s, have been inverted in V_S models of the shallow 6 km of crust. The representative solution is characterized by velocities increasing from 0.9 to 3.0 km/s at 3.9 km of depth.

The path 2 is relative to the recording of the 6/22/2009 event at AQK station. Dispersion curves have been extracted at periods of 0.5-1.6 s and V_S models of the shallow 2 km have been retrieved. The representative solution is characterized by velocities increasing from 0.9 to 2.7 km/s at 1.3 km of depth.

The dispersion curves relative to path 10 have been extracted from the recordings of 2 events at AQU station. Rayleigh data have been sampled in the 0.4-1.2 s period range and the V_S models are relative to the shallow 2 km. The representative solution is characterized by shear velocities increasing from 0.9 km/s to 2.2 km/s at 0.8 km of depth.

The dispersion curves along the path 11 have been extracted from the radial and vertical components of the 04/13/09 event recording at AQU station. The average dispersion curve is defined in the 1.0-2.5 s period range and V_S models of the shallow 3 km have been retrieved from it. The representative solution shows velocities increasing from 0.9 km/s to 2.9 km/s at 1.8 km of depth.

The average dispersion curve along the path 12 has been computed from the dispersion curves extracted from the vertical components of 2 earthquakes on 04/09/09 and is defined at periods between 0.7 and 1.2 s. The retrieved V_S models are relative to the shallow 2 km and the representative solution is characterized by velocities increasing from 1.1 to 2.8 km/s at 1 km of depth.

The interpretation of the obtained V_S models, attributed beneath the middle points of the paths (bottom of Fig. 2), is performed by taking into account available geological data [18]. Outcropping rocks along the cross section A through the paths 12-1-11-10 consist of limestone and marls, which are found with a thickness of ~1 km in the Campotosto drilling (located in Fig. 1), below a 1.2 km thick layer of clays and sandstones. The Mesozoic carbonate horizon is found at the well bottom (2.45 km). V_S range between 0.9 and 1.5 km/s in the shallowest 100-200 m of weathered rocks, and reach the value of 2.5 km/s at depths between 0.6 and 1.7 km. The velocity of 2.5 km/s can be reasonably attributed to fractured limestone rocks and, based on the Campotosto drilling stratigraphy, to the top of the Mesozoic limestone horizon. Such horizon has V_S of ~2.8 km/s (V_P ~5 km/s) at depths of 1-2.5 km and of ~3 km/s (V_P ~5.5 km/s) at 4 km of depth (detected only below the path 1).

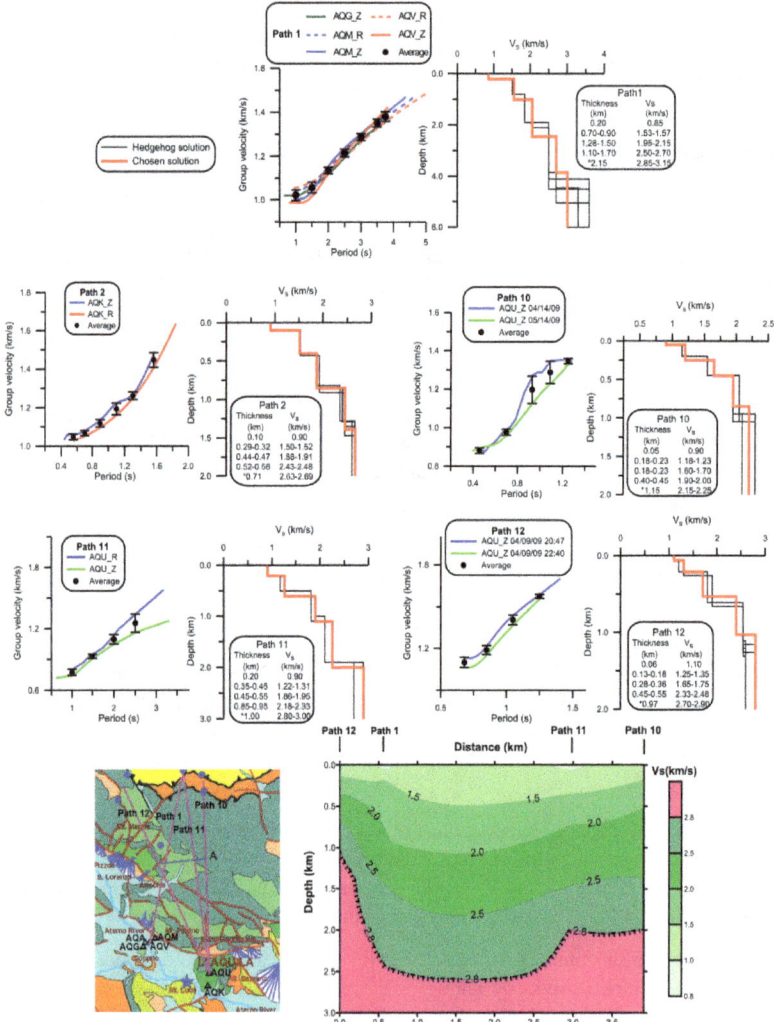

Figure 2. Top: dispersion curves and average with error bars (on the left) and V_S models (Hedgehog solutions) for northern paths 1-2-10-11-12. Z is the vertical component and R is the radial component of the rotated recorded events (Table 1). For each path, the chosen solution is represented by the solid bold red line and is shown in the table with the uncertainty for the inverted parameters. The thickness marked by * is not a truly inverted parameter, but it satisfies the condition that the total thickness from the free surface to the top of the first fixed layer is equal to a predefined quantity. Bottom: V_S pattern below the cross section A through the middle points of the paths 12-1-11-10 (located on the left).

3.2.2. *Middle Aterno river valley paths*

All the results obtained for the middle Aterno river valley paths (6-7-8) are shown in Fig. 3. The paths are relative to the recording of the 4/6/2009 mainshock event at AQK station (path 6), and of the 4/7/2009 events at the AQG station (paths 7 and 8).

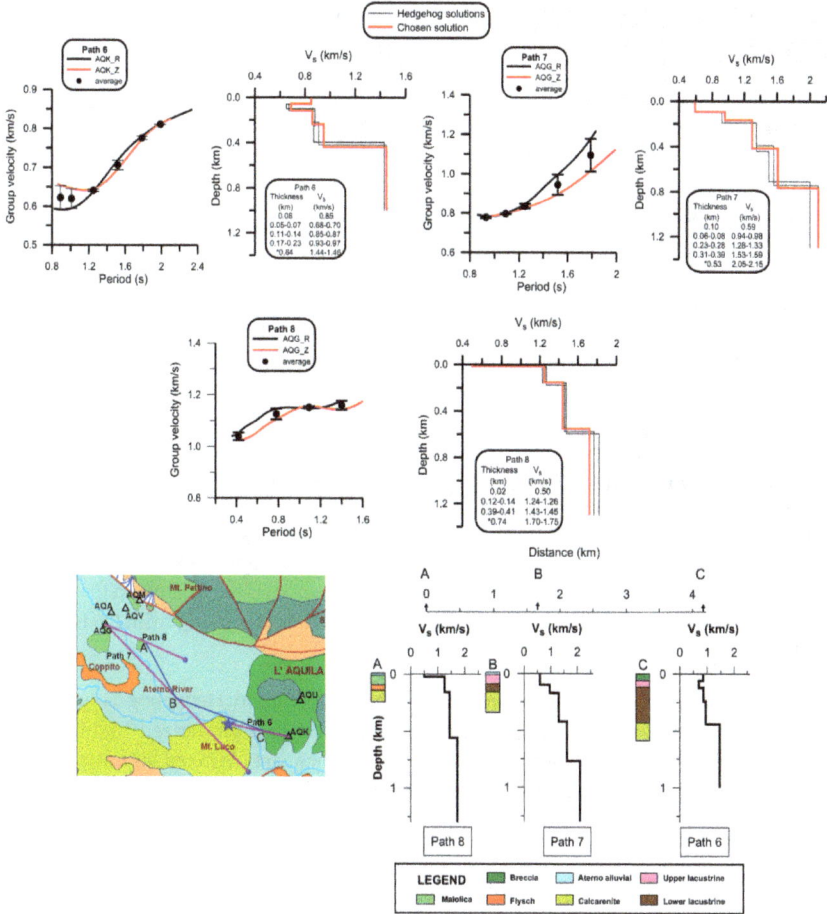

Figure 3. Top: dispersion curves and average with error bars (on the left) and V$_S$ models (Hedgehog solutions) for middle Aterno river valley paths 6-7-8. For detailed caption see Fig. 2. Bottom: V$_S$ profiles vs depth (chosen Hedgehog solutions) obtained along the paths 6, 7 and 8 (located on the left) and attributed at the respective middle points A, B and C. Stratigraphies are based on the geological studies [10].

As regards the path 6, dispersion curves have been extracted at periods of 0.9-2 s and V_S models of the shallow 1 km have been retrieved. The representative solution is characterized by velocities decreasing from 0.85 to 0.7 km/s in the shallow 0.1 km, and deeper increasing from 0.9 to 1.45 km/s at 0.4 km of depth. Dispersion curves relative to path 7 are defined in the period range of 0.9-1.8 s and V_S models of the shallow 1.3 km have been retrieved from their average curve. The representative solution presents velocities increasing from 0.6 to 2.1 km/s at 0.8 km of depth. The average dispersion curve relative to path 8, is defined in the period range of 0.4-1.4 s and V_S models of the shallow 1.3 km have been obtained. The representative solution presents velocities increasing from 0.5 to 1.7 km/s at 0.5 km of depth. Taking into account the geological data relative to the shallowest 0.2-0.3 km [10], stratigraphies may be attributed to the V_S profiles.

Lacustrine soils have a thickness increasing from about 0.2 km in the center of the valley (path 7) to about 0.4 km towards L'Aquila (path 6) with V_S increasing from 0.5 km/s to ~0.9 km/s. Maiolica and flysch layers have an average V_S of ~1.2 km/s while breccia has a V_S of ~0.9 km/s. The calcarenites with V_S of ~1.4 km/s deepen from 0.15 km to 0.45 km towards S (path 6).

3.2.3. Middle paths

All the results obtained for the middle paths (3-5) are shown in Fig. 4. The path 3 is relative to the recording of the 4/11/2009 event at AQU station. It crosses the Aterno valley delimited on the eastern side by the Paganica fault. Dispersion curves have been extracted at periods of 0.7-1.9 s and V_S models of the shallow 2 km have been retrieved. The representative solution is characterized by velocities increasing from 0.7 to 2.0 km/s at 0.8 km of depth.

The path 5 is relative to the recording of the 4/9/2009 event at AQU station and crosses the lower border of the middle Aterno valley, on the west of L'Aquila. Dispersion curves have been extracted at periods of 0.8-1.4 s and the retrieved V_S models are relative to the shallow 2 km. The representative solution is characterized by velocities increasing from 0.6 to 2.5 km/s at 0.9 km of depth.

From the comparison of the representative V_S profiles (chosen Hedgehog solutions) below paths 3, 5 and 6, it turns out that the western sector is characterized by high velocities (~1.8 km/s) at very shallow depth (~0.1 km) which are detected at 0.8 km in the eastern sector and are not found in the investigated shallow 1 km in the epicentral area of the L'Aquila main shock.

3.2.4. Southern paths

The paths 4 and 9 cross the south of L'Aquila consisting of outcropping alluvial deposits and carbonate rocks (Fig. 1). All the results are shown in Fig. 5. Both the paths are averaged on 3 events recorded at AQU station. Dispersion curves have been extracted at periods of 0.5-0.9 s, for the path 4, and of 0.6-1.2 s for the path 9. V_S models of the shallow 1 km have been retrieved which show homogeneous velocities increasing from 0.9 to 2 km/s at ~0.5 km of depth.

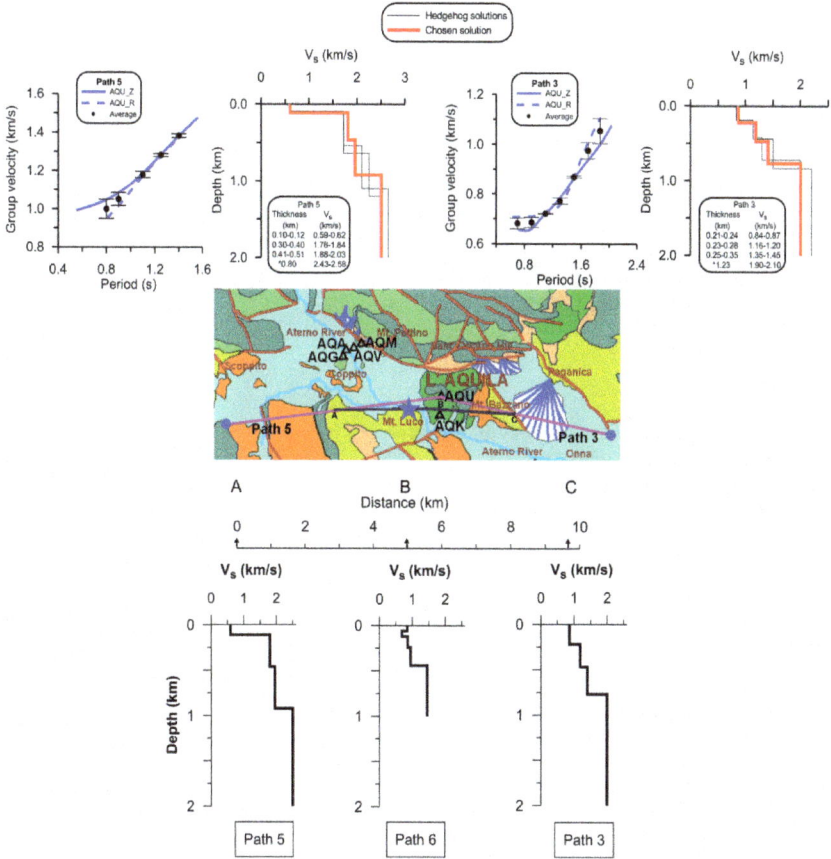

Figure 4. Dispersion curves and average with error bars (on the left) and V$_S$ models (Hedgehog solutions) for the middle paths 3-5. For detailed caption see Fig. 2. Bottom: V$_S$ profiles vs depth (chosen Hedgehog solutions) obtained along the paths 5, 6, and 3 and attributed at the respective middle points A, B and C.

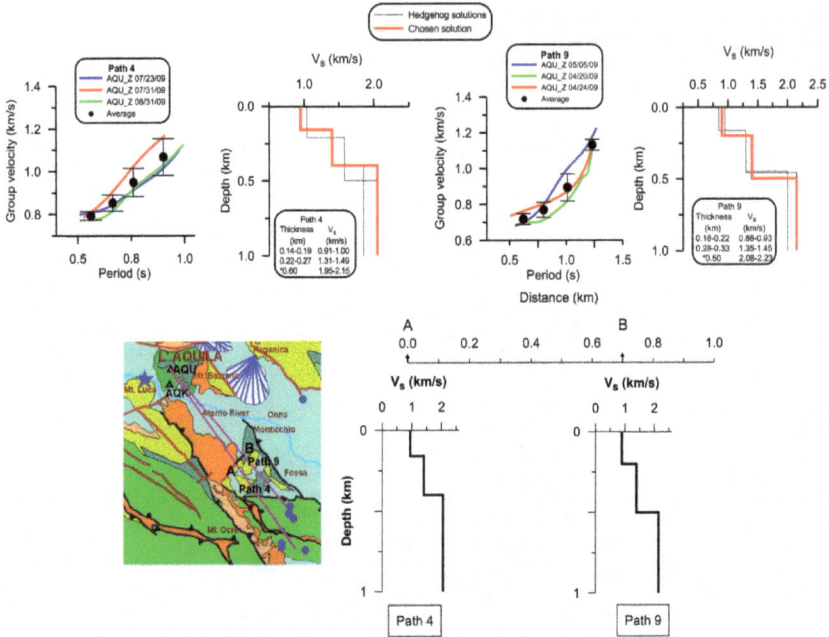

Figure 5. Top: dispersion curves and average with error bars (on the left) and V_S models (Hedgehog solutions) for southern paths 4-9. For detailed caption see Fig. 2. Bottom: chosen Hedgehog solutions obtained along the paths 4 and 9 (located on the left) and attributed at the respective middle points A and B.

3.2.5. Active seismic experiment

V_S models have been also retrieved from an active seismic experiment performed in the Coppito area (Fig. 1) with geophone offsets of 64 m and by using FTAN and Hedgehog methods. The models are characterized by an average velocity (V_{S30}) of 190 m/s in the shallow 30 m of alluvial soils (Fig. 6b). Instead a V_{S30} of 473 m/s is obtained from a cross-hole test, 500 m distant, performed close to the AQV station, in the same alluvial soils [22]. Such discrepancy has important consequences in the respect of the national building code as the soil classification changes from C to B [23].

The comparison of the frequency of the maximum peak of the H/V spectral ratio, relative to the main shock recorded at the AQV station, with the 1D spectral amplifications, computed with SHAKE program [24], by assuming the two different V_S data sets has evidenced the agreement with the V_S profiles relative to FTAN-Hedgehog measurements (Fig. 6c). Once more, such comparison evidences: 1) the strong lateral and vertical heterogeneities of such alluvial soils; 2) the cross-hole (and down-hole) point-like measurements, even though quite precise, may not be representative of the average seismic path (e.g. [25,26]). We remind that the surface measurements for the FTAN analysis do not need boreholes or conventional arrays, hence are particularly suitable in the urban areas.

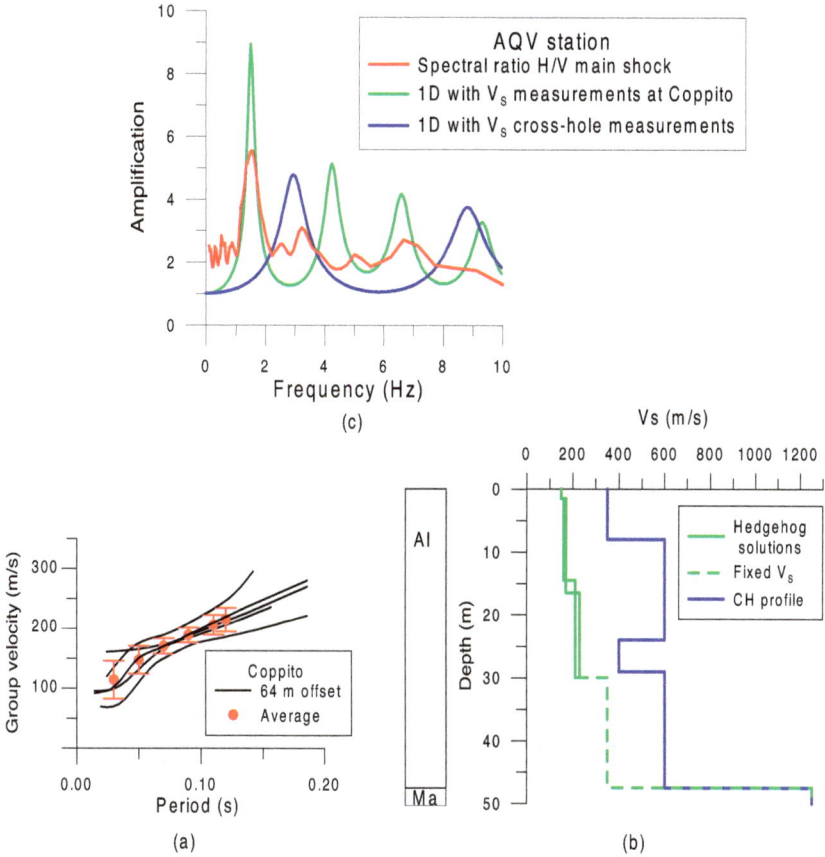

Figure 6. a) Rayleigh waves group velocity dispersion curves of fundamental mode extracted by FTAN method from signals of active seismic experiment at Coppito (location in Fig. 1); (b) V$_S$ velocities obtained from the non-linear inversion (Hedgehog method) of the average dispersion curve (a) as compared with Cross-Hole (CH) measurements at the AQV station site [22] (location in Fig. 1); (c) Comparison between the resonance frequency estimated from the spectral ratio H/V of the main shock recorded at the AQV station and the 1D amplifications [24] computed by assuming the V$_S$ profiles in (b). Legend: Al = Alluvial soils; Ma = Maiolica.

4. Ground motion modeling

Simulations of the 2009 L'Aquila main shock have been performed with the Neo-Deterministic Seismic Hazard Analysis (NDSHA), an innovative modeling technique that takes into account source, propagation and local site effects [8,9].

This approach uses a hybrid method consisting of modal summation and finite difference methods. The path from the source up to the region containing the 2-D heterogeneities is represented by a 1-D layered anelastic structure. The resulting wavefield for both SH- and P-SV- waves is then used to define the boundary conditions to be applied to the 2-D anelastic region where the finite difference technique is used. Synthetic seismograms of the vertical, transverse and radial components of ground motion are computed at a predefined set of points at the surface. Spectral amplifications are computed as response spectra ratios, RSR, i.e. the response spectra computed from the signals synthesized along the laterally varying section (2D) normalized by the response spectra computed from the corresponding signals, synthesized for the bedrock (1D). A scaled point-source approximation ([27] as reported in [28]) has been considered to scale the seismogram to the desired scalar seismic moment.

Modeling of the main shock has been done along a geological cross section at L'Aquila, from the epicentre through the AQK station (Fig. 7).

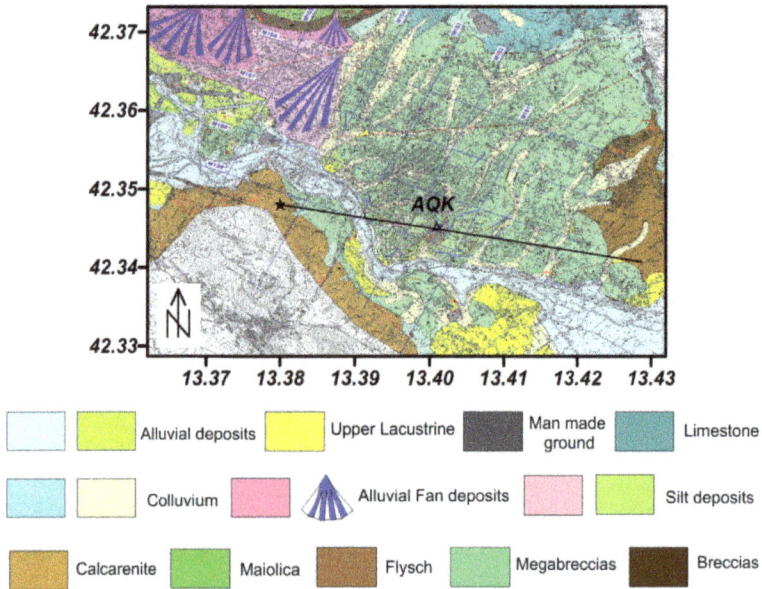

Figure 7. Location of the geological cross section from the main shock (star)[1] to L'Aquila town, passing through the AQK recording station. A new interpretation of gravity data has been performed along the blu lines (modified from [10]).

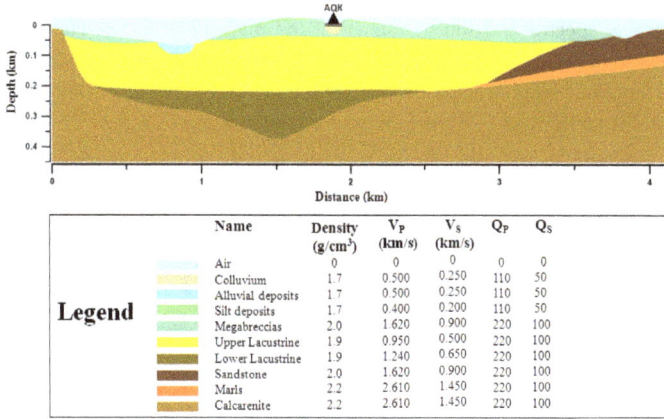

Legend	Name	Density (g/cm³)	V_P (km/s)	V_S (km/s)	Q_P	Q_S
	Air	0	0	0	0	0
	Colluvium	1.7	0.500	0.250	110	50
	Alluvial deposits	1.7	0.500	0.250	110	50
	Silt deposits	1.7	0.400	0.200	110	50
	Megabreccias	2.0	1.620	0.900	220	100
	Upper Lacustrine	1.9	0.950	0.500	220	100
	Lower Lacustrine	1.9	1.240	0.650	220	100
	Sandstone	2.0	1.620	0.900	220	100
	Marls	2.2	2.610	1.450	220	100
	Calcarenite	2.2	2.610	1.450	220	100

Figure 8. Computing cross section at L'Aquila through the AQK station (location in Fig. 7) with the physical parameters attributed to lithotypes.

Along the cross section, the outcropping units are represented by megabreccias except in the Aterno river, where recent fluvial sediments are present. At engineering scale, after the 2009 seismic sequence, geological and geophysical studies, beside several drillings with down-hole tests generally reaching depths around 25-30 m, have been performed at L'Aquila to reconstruct the shallow 200-300 m of subsoil [10]. The new interpretation of available gravity data indicated a meso-cenozoic carbonate unit (average density=2.6 g/cm³) lying at maximum depth of 200-300 m in the Aterno valley, below flysch or breccia unit (average density=2.4 g/cm³) and Quaternary products including alluvial and lacustrine deposits and fan alluvial material (average density=1.9 g/cm³). Strong lateral and vertical geological heterogeneities have been evidenced which, in the center of L'Aquila, are mainly due to a discontinuous stiff top layer of breccias, called megabreccias, overlying soft lacustrine sediments. The geometry of the vertical and lateral passage of the lacustrine to megabreccia deposits is still poorly known due to the shallow drillings. Taking into account the available geological cross sections [10], a computing cross section has been prepared (Fig. 8). We have attributed V_S of 0.2 km/s to the thin silt deposits according to surface measurements at Coppito (Fig. 6). Taking into account the V_S profile obtained for the path 6 (Fig. 3), we have assigned velocities of 0.9 km/s both to megabreccias and sandstones, and of 1.45 km/s both to marls and calcarenites. We are in agreement with [17] as regards velocities of megabreccias, instead a strong discrepancy results for velocities of calcarenites (our 1.45 km/s against 2.5 km/s). Literature V_S [17] have been attributed to Aterno river recent deposits (colluvium and alluvial deposits) and to upper and lower lacustrine soils.

The 1-D reference model has been chosen according to the regional model [11]. A parametric study has been done for the best dip angle between 43° and 60°. A dip of 56° turned out to be able to fit the observed response spectra and is in agreement with the geometry of the seismogenic fault [29]. A small valley of colluvium had to be hypothesized beneath the AQK station, realistically assumed on the basis of geological considerations, in order to fit the observed response spectra and the frequency of the main peak of the H/V spectral ratio obtained from the main shock recording at the AQK station (Fig. 9). A good fit results between observed and synthetic response spectra despite the fictitious greater distance (6 km) of the cross section in the computation of the ground motion for a point source.

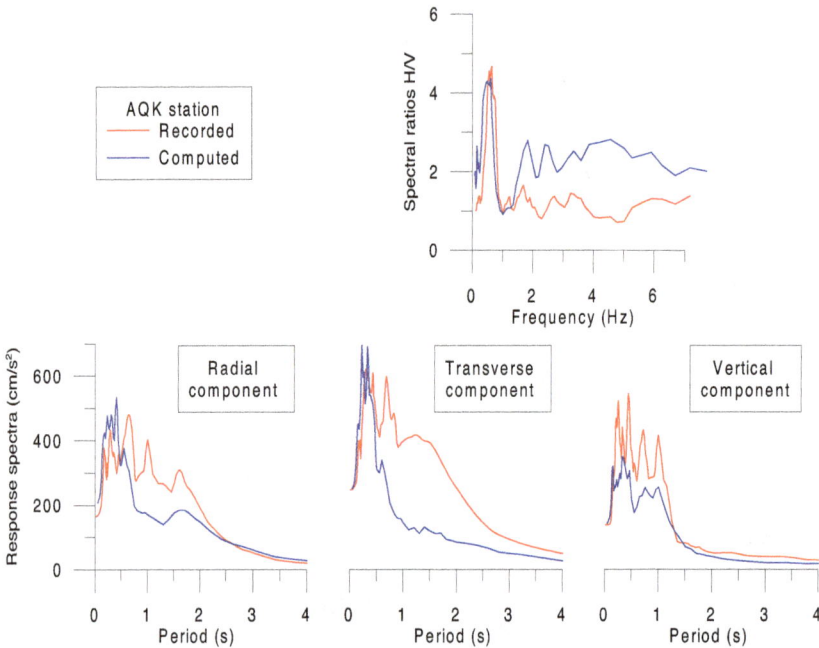

Figure 9. Comparison at the AQK station of the computed and recorded main shock: (top) H/V spectral ratio; (bottom) response spectra computed for 5% damping.

Acceleration time series for P-SV and SH-waves, including surface waves, have been computed along the cross section (Fig. 10), using a grid spacing of 3 m since the lowest seismic velocity is 200 m/s and 10 points are requested for a good sampling of the minimum wavelength corresponding to 7 Hz frequency. They are shown at an array of sites along the cross section, with 50 m spacing. The alluvial and colluvial soil cover is responsible for the amplification of the peak values and duration of accelerations along the radial and transverse components. This amplification is higher along the vertical component.

Spectral amplifications computed for the vertical component of ground motion show maximum values of 10 at frequencies lower than 1 Hz, in correspondence of the layer of megabreccias and of about 5 at 4 Hz in correspondence of the Aterno river alluvial sediments (Fig. 11). As regards the radial and transverse components, spectral amplifications of 2-3 are computed for a wide frequency range (1-7 Hz), along the cross section. Taking into account that the majority of the buildings, generally 2-5 floor, at the historical center of L'Aquila suffered serious damage, we can argue that structures lying on soils suffered amplifications of 2-3 along the horizontal components and up to 5 along the vertical component of the ground motion.

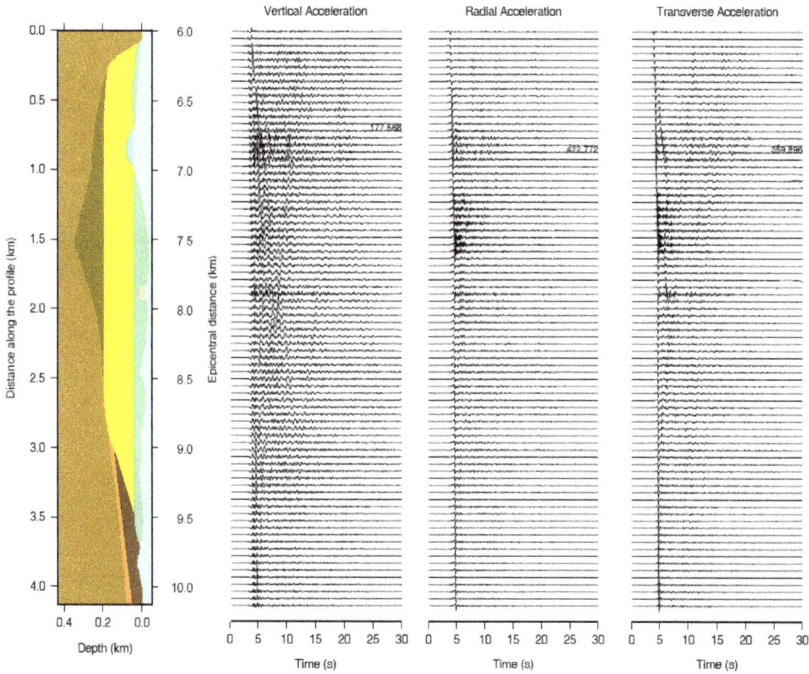

Figure 10. Acceleration time series for SH and P-SV waves computed for the 2-D structural model.

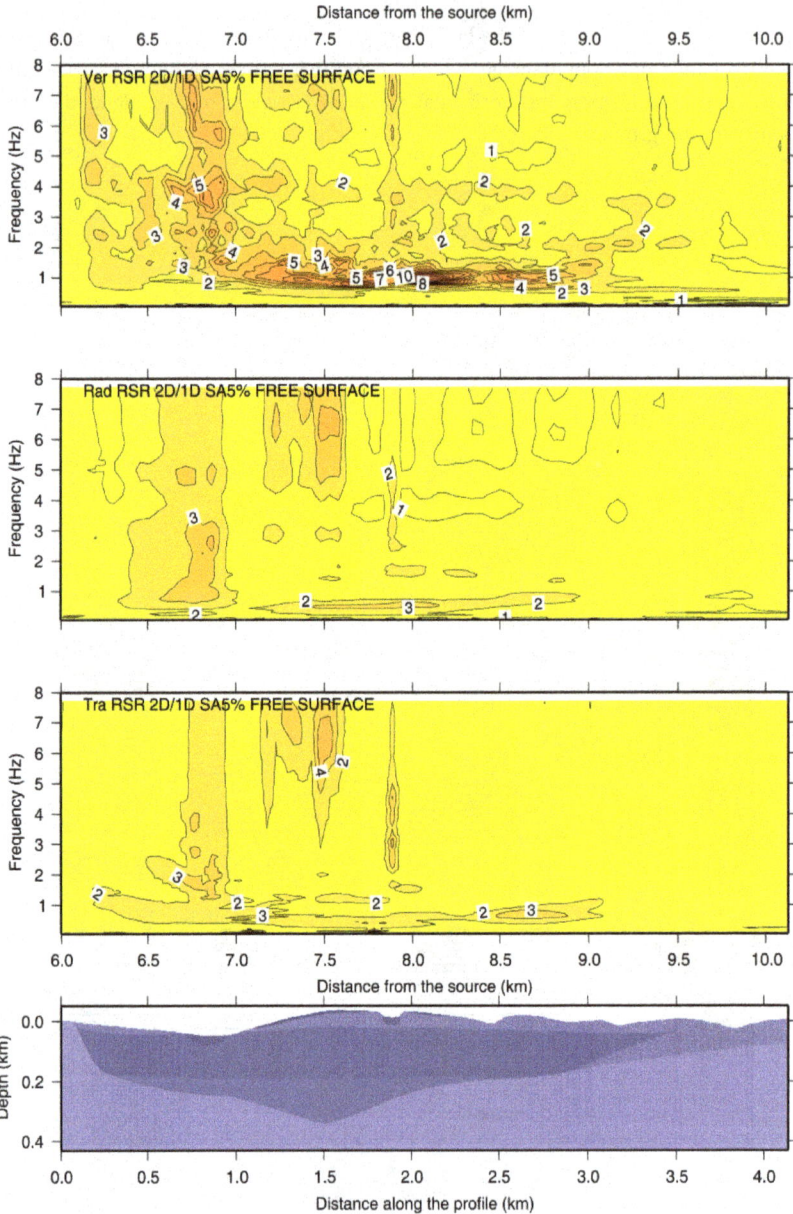

Figure 11. Spectral amplifications (RSR 2D/1D) along the cross section at L'Aquila. Response spectra are computed for 5% damping. From the top vertical, radial and transverse components of the computed ground motion.

5. Conclusions

A realistic estimation of the ground motion at L'Aquila for the M_W 6.3 earthquake is obtained by the NDSHA approach, an innovative modeling technique that takes into account source, propagation and local site effects [8,9]. A key point is the definition of V_S models representative of the seismic path, like those obtained from the non-linear inversion of Rayleigh group velocities of the fundamental mode extracted with the FTAN method from earthquake recordings and active seismic surveys. Very fractured carbonatic rocks with V_S of ~1.4 km/s, covered by alluvial soils with a maximum thickness of ~0.4 km in the center of the Aterno valley are retrieved. The top of the carbonates with the average velocities of 5 and 2.8 km/s, compression and shear respectively, lays at 2-2.5 km of depth and rises to 1 km in the NW part of the valley. This result contradicts available seismic [17] and gravity [10] modeling of the carbonate horizon with V_S=2.5 km/s and density ϱ =2.6 g/cm^3 at some hundred meters of depth. The carbonate horizon with V_S of ~3 km/s is found at 4 km of depth. Moreover, the shallowest 30 m of alluvial soils have average V_S of ~0.2 km/s against ~0.5 km/s as obtained from cross-hole measurements at the AQV station, about 500 m distant. Such velocity difference evidences that strong lateral and vertical geological heterogeneities are present and that the cross-hole (and down-hole) point-like measurements, even though quite precise, may not be representative of the average seismic path.

The soundness of the synthetics is in the good fitting of the recorded H/V spectral ratio and response spectra, despite the point-source approximation. The lateral and vertical geological variability mainly due to the covering of megabreccias on soft soils are responsible of spectral amplifications, mostly for the vertical component, for a wide frequency range (0.5-7 Hz). Taking into account that the majority of the buildings, generally 2-5 floor, at the historical center of L'Aquila suffered serious damage, we can argue that spectral amplifications might have been responsible for damage, beside the near-field conditions. This study shows that realistic ground motion can be computed in advance for the several active faults of the L'Aquila district and that a sounded building code can be formulated for the restoration of the existing damaged buidings and for new building design.

Acknowledgements

We are grateful to Prof. G.F. Panza for the use of computer programs. Many thanks to Dr. C. Donadio for help in drawing geological cross section.

Author details

M.R. Costanzo, C. Nunziata and V. Gambale

Dipartimento di Scienze della Terra, dell' Ambiente e delle Risorse, Univ. Napoli Federico II, Italy

References

[1] Chiarabba C, Amato A, Anselmi M, Baccheschi P, Bianchi I, Cattaneo M, Cecere G, Chiaraluce L, Ciaccio MG, De Gori P, De Luca G, Di Bona M, Di Stefano R, Faenza L, Govoni A, Improta L, Lucente FP, Marchetti A, Margheriti L, Mele F, Michelini A, Monachesi G, Moretti M, Pastori M, Piana Agostinetti N, Piccinini D, Roselli P, Seccia D and Valoroso L. The 2009 L'Aquila (central Italy) Mw6.3 earthquake: main shock and aftershocks. Geophysical Research Letters 2009;36(18) L18308.

[2] Galli P, Camassi R. Rapporto sugli effetti del terremoto aquilano del 6 aprile 2009. Dipartimento della Protezione Civile Istituto Nazionale di Geofisica e Vulcanologia. QUEST Team 2009. http://www.emidius.mi.ingv.it/DBMI08/aquilano/query_eq/ (accessed 17 July 2012).

[3] Gruppo di Lavoro MPS. Redazione della mappa di pericolosità sismica prevista dall'Ordinanza PCM 3274 del 20 marzo 2003. Final report, INGV Milano−Roma 2004; 65 pp. http://zonesismiche.mi.ingv.it (accessed 17 July 2012).

[4] CPTI WORKING Group. Catalogo Parametrico dei Terremoti Italiani, versione 2004 (CPTI04). INGV, Bologna 2004. http://emidius.mi.ingv.it/CPTI04/ (accessed 17 July 2012).

[5] Galli P, Galadini F, Pantosti D. Twenty years of paleoseismology in Italy. Earth Science Reviews 2008;88: 89-117.

[6] Cultrera G, Mucciarelli M, Parolai S. The L'Aquila Earthquake - A View of Site Effects and Structural Behavior from Temporary Networks. Special Issue Bulletin Earthquake Engeneering 2011;9(3): 691–892.

[7] Nunziata C., Costanzo M.R., Vaccari F., Panza G.F. Evaluation of linear and nonlinear site effects for the M_W 6.3, 2009 L'Aquila earthquake. In: D'Amico S. (ed.) Earthquake Research and Analysis - New frontiers in Seismology. Intech; 2012. p155-176. Avaible from http://www.intechopen.com/books/earthquake-research-and-analysis-new-frontiers-in-seismology (accessed 17 July 2012).

[8] Panza GF, Romanelli F, Vaccari F. Seismic wave propagation in laterally heterogeneous anelastic media: theory and applications to seismic zonation. Advances in Geophysics 2001;43: 1-95.

[9] Panza GF, Irikura K, Kouteva M, Peresan A, Wang Z, Saragoni R (eds). Advanced seismic hazard assessment. Pure and Applied Geophysics 2011; Topical Volume 168(1-2): 366pp.

[10] Gruppo di Lavoro MS-AQ. Microzonazione sismica per la ricostruzione dell'area aquilana. Regione Abruzzo, Dipartimento della Protezione Civile 2010 http://www.protezionecivile.it/jcms/it/microzonazione_aquilano.wp (accessed 17 July 2012).

[11] Brandmayr E, Raykova RB, Zuri M, Romanelli F, Doglioni C, Panza GF. The litho-sphere in Italy: structure and seismicity. Journal of the Virtual Explorer 2010. 36: paper 1. DOI: 10.3809/jvirtex.2009.00224.

[12] Levshin AL, Yanovskaya TB, Lander AV, Bukchin BG, Barmin MP, Ratnikova LI, Its EN. Seismic Surface Waves in a Laterally Inhomogeneous Earth. Keilis-Borok V.I., editor. Norwell:Kluwer;1989.

[13] Nunziata C. FTAN method for detailed shallow V_S profiles. Geologia Tecnica e ambientale 2005;3: 25-43.

[14] Nunziata C. Low shear-velocity zone in the Neapolitan-area crust between the Campi Flegrei and Vesuvio volcanic areas. Terra Nova 2010;22: 208–217.

[15] Boncio P, Lavecchia G. A structural model for active extension in central Italy. Journal of Geodynamics 2000;29 (3–5): 233–244.

[16] Chiarabba C, Bagh S, Bianchi I, De Gori P, Barchi M. Deep structural heterogeneities and the tectonic evolution of the Abruzzi region (Central Apennines, Italy) revealed by microseismicity, seismic tomography, and teleseismic receiver functions. Earth Planetary Science Letters 2010;295 (3-4): 462-476.

[17] De Luca G, Marcucci S, Milana G, Sanò T. Evidence of Low-Frequency Amplification in the City of L'Aquila, Central Italy, through a Multidisciplinary Approach Including Strong- and Weak-Motion Data, Ambient Noise, and Numerical Modeling. Bulletin of the Seismological Society of America 2005;95: 1469–1481.

[18] Vezzani L, Ghisetti F. Carta Geologica dell' Abruzzo, Scala 1:100,000. S.EL.CA. 1998, Firenze.

[19] Natale M, Nunziata C, Panza GF. Average shear wave velocity models of the crustal structure at Mt. Vesuvius. Physics of the Earth and Planetary Interiors 2005;152: 7-21.

[20] Panza G.F. The resolving power of seismic surface wave with respect to crust and upper mantle structural models. In: Cassinis R. (ed.). The solution of the inverse problem in Geophysical Interpretation. Plenum press; 1981. p39-77.

[21] Boyadzhiev G, Brandmayr E, Pinat T, Panza GF. Optimization for non-linear inverse problems. Rendiconti Lincei 2009;19: 17–43. DOI: 10.1007/s12210-008-0002-z.

[22] Puglia R, Ditommaso R, Pacor F, Mucciarelli M, Luzi L, Bianca M. Frequency variation in site response over long and short time scales, as observed from strong motion data of the L'Aquila (2009) seismic sequence. Bulletin of Earthquake Engeneering 2011;9: 869-892. DOI: 10.1007/s10518-011-9266-2.

[23] Nuove Norme Tecniche per le Costruzioni, D.M. 14.01.2008, G.U. n.29 del 04.02.2008, Suppl. Ordinario n.30.

[24] Schnabel P.B., Lysmer J., Seed H.B. SHAKE: a computer program for earthquake response analysis of horizontally layered sites. Report No. UCB/EERC 72/12, Earthquake Engineering Research Center 1972, University of California, Berkeley.

[25] Nunziata C, Natale M, Panza GF. Seismic characterization of neapolitan soils. Pure and Applied Geophysics 2004;161(5-6) 1285-1300.

[26] Nunziata C. A physically sound way of using noise measurements in seismic microzonation, applied to the urban area of Napoli. Engineering Geology 2007;93(1-2): 17-30.

[27] Gusev AA. Descriptive statistical model of earthquake source radiation and its application to an estimation of short period strong motion. Geophysical Journal of the Royal Astronomical Society 1983;74: 787-800.

[28] Aki K. Strong motion seismology: Strong Ground Motion Seismology. NATO ASI Series C., Erdik M.O., Toksoz M.N. (eds.). D. Reidel Publishing Company; 1987. 204: 3-39.

[29] Galli P, Camassi R, Azzaro R, Bernardini F, Castenetto S, Molin D, Peronace E, Rossi A, Vecchi M, Tertulliani A. Il terremoto aquilano del 6 Aprile 2009: rilievo macrosismico, effetti di superficie ed implicazioni sismotettoniche. Il Quaternario, Italian Journal of Quaternary Sciences 2009;22(2): 235-246.

Chapter 4

Simulation of Near-Field Strong Ground Motions Using Hybrid Method

Babak Ebrahimian

Additional information is available at the end of the chapter

1. Introduction

Earthquake disaster investigations have shown that numerous strong earthquakes are caused by remobilization of active faults. Many casualties and severe damages to structures as well as huge economic losses have resulted from ground motions of strong earthquakes caused by active faults buried under urban areas. Recently, both potential hazard and defenses of active faults concealed under urban area has become a grand research subject paid highly attention to by the seismologists. Near-field strong ground motions, especially their high frequency content, are intensively affected by both slip heterogeneity on fault plane and rupture process of an earthquake fault. In the simulations of near-field strong ground motions, modeling effective finite fault source is very important. The gradually increasing number of recorded near source time histories has recently enabled strong motion seismologists to analyze more precisely the character of the near-fault ground motions and therefore contribute to the physical understanding of those features that control them (Malagnini et al. 2002, 2011; Akinci et al. 2010; D'Amico et al. 2010). Mavroeidis and Papageorgiou (2002) presented a comprehensive review and study of the factors that influence the near-source ground motions.

The stochastic method of synthesizing ground motion based on seismology interests engineers specifically in simulating higher-frequency ground motions (Akinci et al. 2001). The method is widely used to predict ground motions for regions, in which ground motion recordings from past earthquake are not available (Boore 2003). For far field, the point source model of Boore (1983) is very effective; however, for near-field, the method can not incorporate the factors which have significant effect on the near-field strong ground motions, and yields an overestimation of such ground motions. The stochastic ground motion modelling technique, also known as the band limited white-noise method, has been first described by Boore (1983). Ever since, many researchers have applied the method to simulate ground motions from point sources (e.g., Boore and Atkinson 1987; Atkinson and Boore 1995; Zafarani et al. 2005; D'Amico et al. 2012).

On the other hand, realistic acceleration time-histories should be employed in structural analysis to reduce the uncertainties in estimating the standard engineering parameters (Hutchings 1994), particularly for non-linear seismic behavior of structures. Thus, designers need to know the dynamic characteristics of predicted ground motion consistent with source rupture for a particular site to be able to adequately design an earthquake-resistant structure. Hall et al. (1995), Makris (1997), Chopra and Chintanapakdee (2001), Zhang and Iwan (2002) have experimentally as well as analytically studied the elastic and inelastic response of engineering structures subjected to actual near-fault records or simplified waveforms intending to represent the typical ground motion pulses observed in near-field regions.

During the past decades, much effort has been given in reliable simulation of strong ground motion from finite faults through methodologies that include theoretical or semi-empirical modeling of the parameters affecting shape, duration and frequency content of the strong motion records. Due to unavailability of strong recorded ground motion, simulation of ground motion has been carried out using the stochastic method proposed by Boore (2003). The ground motion spectrum has been generated by Atkinson and Boore model (1995). Even though the success of the point-source model has been pointed out repeatedly, it is also well known that it often breaks down, especially near the sources of large earthquakes. Recently, Beresnev and Atkinson (1997) have proposed a technique that overcomes the limitation posed by the hypothesis of a point source. Their technique is based on the original idea of Hartzell (1979) to model large events by the summation of smaller ones. In Beresnev and Atkinson (1997), the high-frequency seismic field near the epicentre of a large earthquake is modeled by subdividing the fault plane into a certain number of sub-elements and summing their contributions, with appropriate time delays, at the observation point. Each element is treated as a point source. A stochastic model is used to calculate the ground motion contribution from each sub-element, while the propagation effects are empirically modeled. Combining the stochastic method with the finite fault source model, Silva (1997), Beresnev and Atkinson (1998), Motazedian and Atkinson (2005) have proposed different methods, which could be effective for simulating or predicting near-field ground motions.

Two Californian earthquake events may be characterized as historical milestones related to near-source ground motions: the 1966 Parkfield and the 1971 San Fernando earthquakes. The 1966 Parkfield, California, event provided the now famous Station 2 (C02) record at a distance of only 80 m from the fault break (Housner and Trifunac 1967). Modern quantitative analysis of strong ground motion observations was started with this record. Aki (1968) and Haskell (1969) demonstrated that the observed transverse (i.e., fault-normal) displacement component of this ground motion record, which exhibited a simple impulsive form, was precisely what is expected for a right-lateral strike-slip rupture propagating from northwest to southeast. The 1971 San Fernando, California, earthquake provided the equally well-known Pacoima Dam (PCD) record. The strike-normal velocity component of this record also exhibited an impulsive character that several investigators attempted to model (e.g., Boore and Zoback 1974; Niazy 1975; Bouchon 1978). In addition, this record was the one that made earthquake engineers recognize the severe implications of the impulsive characteristics of near-source ground motions on flexible structures.

At high frequencies ($f > 1$ Hz), ground motions become increasingly stochastic in nature. The stochastic methods are generally capable of matching the spectral amplitudes of high frequency ground motions, but are generally not capable of matching the recorded waveforms (Somerville 1998). Firstly, ground motions are estimated by identifying the major regional faults and propagating seismic waves generated at these potential sources to the site of interest. The two commonly used techniques, finite-fault and point source methods of Boore and Atkinson (1987) and Beresnev and Atkinson (1997, 1998) are used for simulation of earthquakes. Both techniques have an omega-squire spectrum.

The main objective of the chapter is to simulate the near-fault strong motion records. Simulation of ground motions is carried out using the hybrid method proposed by Mavroeidis and Papageorgiou (2003) and the stochastic model of Boore (2003). Due to unavailability of strong recorded ground motion, the stochastic method proposed by Boore (2003) is applied to simulate the acceleration time histories. The ground motion spectrum is generated by Atkinson and Boore model (1995). Firstly, macro-source parameters characterizing the whole source area, i.e. global source parameters such as fault length, fault width, rupture area, and average slip on the fault plane are estimated; secondly, slip distributions characterizing heterogeneity or roughness on the fault plane, i.e. local source parameters are reproduced by the hybrid slip model; finally, the finite fault source model, which is developed based on the global and local source parameters is combined with the stochastic method. A simple, yet effective, analytical model proposed by Mavroeidis and Papageorgiou (2003) is also used to adequately describe the impulsive character of near-fault ground motions both qualitatively and quantitatively. The calculated response spectra are compared with those, mentioned in International Building Code (IBC 2000) and Iranian Code of Practice for Seismic Resistance Design Building (Standard No. 2800) to validate the availability and practicability of the proposed method for near-field Tombak site at south-eastern part of Iran. This site includes massive LNG storage plants near to fault. Then, the response of mentioned site under simulated ground motion has been studied by conducting one dimensional ground response analysis. According to the above study, the bed rock and ground surface accelerations of the site are provided. The results can be used in hazard analysis of specific sites in the considered region, particularly for the performance analysis of structures.

2. Simulation method

A simple and powerful method for simulating ground motions is to combine parametric or functional descriptions of the ground motion's amplitude spectrum with a random phase spectrum modified such that the motion is distributed over a duration related to the earthquake magnitude and to the distance from the source. This method of simulating ground motions often goes by the name "Stochastic method". It is particularly useful for simulating the higher-frequency ground motions of most interest to engineers (generally, $f > 1$ Hz), and it is widely used to predict ground motions for regions of the world in which recordings of motion from potentially damaging earthquakes are not available. One of the essential characteristics of the method is that it distills what is known about the various factors affecting ground motions (source, path, and site) into simple functional forms.

2.1. Stochastic finite-fault simulation method

In this study, the Stochastic Method is used for simulating the strong ground motion. The method assumes that the far-field accelerations on an elastic half space are band-limited, finite-duration, white Gaussian noise, and that the source spectra are described by a single corner-frequency model whose corner frequency depends on earthquake size (Mayeda and Malagnini 2009). The ground spectrum Y (M_0, R, f) is conveniently broken into several simple functions – the Earthquake Source (E); the Path (P); the Site (G) and the instrument or type of motion (I):

$$Y(M_0, R, f) = E(M_0, f) P(R, f) G(f) I(f) \tag{1}$$

where, M_0 is the seismic moment, R is the shortest distance from the fault to the site and f is the frequency. Atkinson and Boore model (1995) is used to obtain the ground motion spectrum. The process of strong ground motion simulation is depicted as a flowchart in Figure 1.

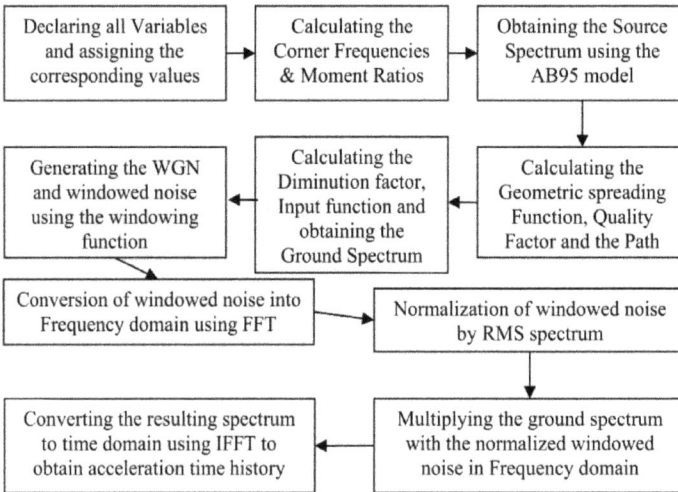

Figure 1. Flow chart showing the structure of the FOTRAN program for Atkinson and Boore model (1995)

The source spectrum, E, is obtained by the following equations specifying both the shape and the amplitude as a function of the earthquake size:

$$E(M_0, f) = C M_0 S(M_0, f) \tag{2}$$

$$S(M_0, f) = S_a(M_0, f) \times S_b(M_0, f)$$ (3)

By adopting the source spectrum model AB95 (Atkinson and Boore model 1995), the above equation for source spectrum is rewritten considering the seismic moment dependence of the above factors S_a in terms of corner frequencies f_a and f_b:

$$E(M_0, f) = CM_0 \left\{ \frac{1 - \varepsilon}{1 + \left[f/f_a \right]^2} + \frac{\varepsilon}{1 + \left[f/f_b \right]^2} \right\}$$ (4)

where, C is a constant given by

$$C = \frac{< R_{\Theta\Phi} > VF}{4\pi \rho_s \beta_s R_o}$$ (5)

Here, $< R_{\Theta\Phi} >$ accounts for the radiation pattern (≈ 0.55); V represents the partition of total shear wave energy into horizontal components ($= 0.707$); F accounts the effect of free surface (≈ 2); ρ_s and β_s are the density and shear wave velocity of the bedrock; R_o is a reference distance and usually taken as 1 km. The corner frequencies f_a and f_b are obtained from the seismic moment using the following relations

$$\log f_a = 2.41 - 0.533 M_0$$ (6)

$$\log f_b = 1.431 - 0.188 M_0$$ (7)

The Source duration is evaluated as $0.5/f_a$.

The path effects are represented by simple functions that describe the geometric spreading function, attenuation (intrinsic and scattering attenuation), and the general increase of duration with distance due to wave propagation and scattering. The simplified path effect, P, is given by the multiplication of the geometrical spreading and Q functions:

$$P = Z(R) \exp\left\{ \frac{-\pi R f}{Q(f) C_Q} \right\}$$ (8)

The relation between distance and geometrical spreading function, $Z(R)$, is given by the following function

$$Z(R) = \frac{1}{70}\sqrt{\frac{130}{R}} \tag{9}$$

and $Q(f)$ is the frequency dependent quality factor which is given by the following equation

$$Q(f) = 100 f^{0.8} \tag{10}$$

The path duration function of 0.05R is calculated from Atkinson and Boore (1995).

The attenuation or diminution operator $D(f)$ accounts for the path-independent loss of high frequency in the ground motions. A simple multiplicative filter can account for the diminution of the high frequency motions. Here, f_{max} is 10 Hz. The diminution factor is calculated based on the following equation

$$D(f) = \left\{1 + \left(\frac{f}{10}\right)^8\right\}^{-0.5} \tag{11}$$

The particular type of ground motion resulting from the simulation is controlled by the filter $I(f)$. If ground motion is desired, then

$$I = -\left(2\pi f\right)^n \tag{12}$$

where, $I = (-1)^{0.5}$. n = 0, 1, 2 for ground displacement, velocity and acceleration, respectively.

A time domain simulation is carried out to get the actual Fourier amplitude spectrum. A White Gaussian Noise (WGN) is produced and windowed off using a windowing function given below

$$W\left(t; \varepsilon, \eta, t_\eta\right) = a\left(t/t_\eta\right)^b \exp\left\{-c\left(t/t_\eta\right)\right\} \tag{13}$$

where,

$$\begin{aligned}
a &= \left\{\exp(i)/e\right\}^b \\
b &= -\left(\varepsilon \ln \eta\right)\big/\left[1 + \varepsilon\left(\ln \varepsilon - 1\right)\right] \\
c &= b/\varepsilon \\
t_\eta &= f_{Tgm} \times T_{gm}
\end{aligned} \tag{14}$$

Boore suggested the values of ε and η to be 0.2 and 0.05, respectively and $f_{T_{gm}}$ = 2 based on Saragoni and Hart (1974). The windowed WGN is converted to frequency domain and normalized by its root mean square amplitude. The entire process of obtaining WGN is shown in Figure 2. Then, the ground motion spectrum, shown in Figure 2, is multiplied with the normalized windowed noise to get the Fourier amplitude spectrum as shown in Figure 2.

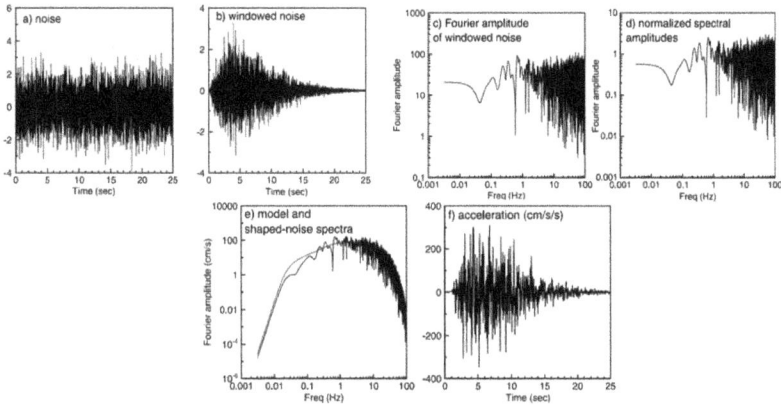

Figure 2. Basis of procedure for simulating ground motions using the stochastic method

2.2. Analytical model proposed by Mavroeidis and Papageorgiou (2003)

For near-field strong ground motions, most of the elastic energy arrives coherently in a single, intense, relatively long period pulse at the beginning of record, representing the cumulative effect of almost all the seismic radiation from the fault. The phenomenon is even more pronounced when the direction of slip on the fault plane points toward the site as well. The Mavroeidis and Papageorgiou model (2003) adequately describes the impulsive character of near-faults ground motions both qualitatively and quantitatively. In addition, it can be used to analytically reproduce empirical observations that are based on available near-source records. The input parameters of the model have an unambiguous physical meaning. The proposed analytical model has been calibrated using a large number of actual near-field ground motion records. It successfully simulates the entire set of available near-fault displacement, velocity, and (in many cases) acceleration time histories, as well as the corresponding deformation, velocity, and acceleration response spectra. An "objective" definition of the pulse duration is given based on model input parameters. In addition, Mavroeidis and Papageorgiou (2003) investigate the scaling characteristics of the model parameters with earthquake magnitude. Also, Mavroeidis et al. (2004) derive the Fourier transform of the analytical model and identify the parameters that have the most significant effect on the spectral characteristics of the model. Finally, a simplified (adequate for engineering purposes) method is proposed for the synthesis of near-fault ground motions. The pulse duration (or period), the pulse amplitude, as well as the number and phase of half cycles are the key parameters that define the waveform characteristics of near-fault velocity pulses. Therefore, an analytical model with four parameters in principle should suffice to describe the entire set of velocity pulses generated due to forward directivity or permanent translation effects. Seismologists have used "wavelets" (also referred to as "signals," "signatures," or "pulses"), particularly in fields such as seismic filtering, wavelet processing, wave-propagation modelling, and trace inversion (Hubral and Tygel 1989). Although, various wavelets have been proposed in the literature, only a limited number of them are popular and frequently used in practice. In this study, the analytical wavelet signal proposed by Mavroeidis and Papageorgiou (2003) is chosen and expressed by

$$f(t) = A \frac{1}{2} \left[1 + \cos\left(\frac{2\pi f_p}{\gamma} t \right) \right] \cos\left(2\pi f_p t + v \right) \tag{15}$$

This problem is easily resolved by limiting the time interval of the signal as follows

$$-\frac{\gamma}{2f_p} \leq t \leq \frac{\gamma}{2f_p} \tag{16}$$

The period of the harmonic oscillation should be smaller than the period of the envelope represented by the elevated cosine function in order to produce physically acceptable signals; that is,

$$\frac{1}{f_p} < \frac{\gamma}{f_p} \Rightarrow \gamma > 1 \tag{17}$$

The combination of equations (15) to (17) yields the formulation of the proposed analytical model for the near-fault ground velocity pulses:

$$v(t) = \begin{cases} A \frac{1}{2} \left[1 + \cos\left(\frac{2\pi f_p}{\gamma} (t - t_0) \right) \right] \cos\left(2\pi f_p (t - t_0) + v \right), \\ 0 \qquad\qquad\qquad , \ otherwise \end{cases} \tag{18}$$

where, parameter A controls the amplitude of the signal; f_p is the frequency of the amplitude-modulated harmonic (or the prevailing frequency of the signal); v is the phase of the amplitude-modulated harmonic (i.e., $v = 0$ and $v = \pm \pi / 2$ define symmetric and antisymmetric signals, respectively); γ is a parameter that defines the oscillatory character (i.e., zero crossings) of the signal (i.e., for small γ the signal approaches a deltalike pulse; as γ increases, the number of zero crossings increases); and t_0 specifies the epoch of the envelope's peak. The analytical expressions for the ground acceleration and displacement time histories, compatible with the ground velocity given by equation (18), are

$$
a(t) = \begin{cases} -\dfrac{A\pi f_p}{\gamma}\begin{bmatrix} \sin\left(\dfrac{2\pi f_p}{\gamma}(t-t_0)\right)\cos\left[2\pi f_p(t-t_0)+v\right]+ \\ \gamma\sin\left[2\pi f_p(t-t_0)+v\right]\left[1+\cos\left(\dfrac{2\pi f_p}{\gamma}(t-t_0)\right)\right] \end{bmatrix}, t_0-\dfrac{\gamma}{2f_p}\le t\le t_0+\dfrac{\gamma}{2f_p}\ \ with\ \gamma>1 \\[6pt] 0, \hspace{6cm} otherwisw \end{cases} \tag{19}
$$

$$
d(t) = \begin{cases} \dfrac{A}{4\pi f_p}\begin{bmatrix} \sin\left(2\pi f_p(t-t_0)+v\right)+\dfrac{1}{2}\dfrac{\gamma}{1-\gamma}\sin\left[\dfrac{2\pi f_p(1-\gamma)}{\gamma}(t-t_0)+v\right]+ \\ \dfrac{1}{2}\dfrac{\gamma}{1+\gamma}\gamma\sin\left[\dfrac{2\pi f_p(1+\gamma)}{\gamma}(t-t_0)+v\right] \end{bmatrix}+C,\ t_0-\dfrac{\gamma}{2f_p}\le t\le t_0+\dfrac{\gamma}{2f_p}\ \ with\ \gamma>1 \\[6pt] \dfrac{A}{4\pi f_p}\dfrac{1}{\left(1-\gamma^2\right)}\sin(v-\pi\gamma)+C,\ t<t_0-\dfrac{\gamma}{2f_p} \\[6pt] \dfrac{A}{4\pi f_p}\dfrac{1}{\left(1-\gamma^2\right)}\sin(v+\pi\gamma)+C,\ t>t_0+\dfrac{\gamma}{2f_p} \end{cases} \tag{20}
$$

A parametric study in terms of v and γ of the normalized (with respect to f_p and A) acceleration, velocity, and displacement pulses can be performed based on the equations presented previously. We define the normalized time variable as

$$
\bar{t} = 2\pi f_p(t-t_0) \tag{21}
$$

Then, the normalized acceleration and displacement time histories can be expressed by rewriting equations (19) and (20), as

$$
\bar{a}(\bar{t}) = \dfrac{a(t)}{Af_p} = \begin{cases} -\dfrac{\pi}{\gamma}\left[\sin\left(\dfrac{\bar{t}}{\gamma}\right)\cos(\bar{t}+v)+\gamma\sin(\bar{t}+v)\left(1+\cos\left(\dfrac{\bar{t}}{\gamma}\right)\right)\right], -\pi\gamma\le\bar{t}\le\pi\gamma\ \ with\ \gamma>1 \\[6pt] 0, \hspace{6cm} otherwisw \end{cases} \tag{22}
$$

$$
\bar{d}(\bar{t}) = \dfrac{d(t)}{\left(A/f_p\right)} = \begin{cases} \dfrac{1}{4\pi}\left[\sin(\bar{t}+v)+\dfrac{1}{2}\dfrac{\gamma}{\gamma-1}\sin\left(\dfrac{\gamma-1}{\gamma}\bar{t}+v\right)+\dfrac{1}{2}\dfrac{\gamma}{\gamma+1}\sin\left(\dfrac{\gamma+1}{\gamma}\bar{t}+v\right)\right],\ -\pi\gamma\le\bar{t}\le\pi\gamma\ \ with\ \gamma>1 \\[6pt] \dfrac{1}{4\pi}\dfrac{1}{\left(1-\gamma^2\right)}\sin(v-\pi\gamma),\ \bar{t}<-\pi\gamma \\[6pt] \dfrac{1}{4\pi}\dfrac{1}{\left(1-\gamma^2\right)}\sin(v+\pi\gamma),\ \bar{t}>\pi\gamma \end{cases} \tag{23}
$$

Assuming that the duration of the pulse is independent of the source–station distance for stations located within ~10 km from the causative fault, the pulse period and the moment magnitude are related through the following empirical relationship obtained by least-squares fit analysis:

$$\log T_p = -2.2 + 0.4 M_W \tag{24}$$

In this section, we propose a very simplified methodology for generating realistic synthetic ground motions that are adequate for engineering analysis and design. We exploit the simple analytical model introduced in the present work to describe the coherent (long-period) component of motion and the stochastic (or engineering) approach to synthesize the incoherent (high-frequency) seismic radiation (for a review of the stochastic approach of ground motion synthesis; see Boore (1983) and Shinozuka (1988)). For the latter component of motion, due to the proximity of the point of observation to the source, it is necessary to use a source model that provides guidance as how to distribute the available seismic moment of the simulated event on the fault plane. Such a source model is the specific barrier model of Papageorgiou and Aki (1983). According to this model, an earthquake is visualized as a sequence of equal-size sub-events uniformly distributed on a rectangular fault plane. At the present time, the proposed mathematical model along with its scaling laws can take into account (with confidence) only for the forward directivity effect. Even though the analytical expression can replicate near-fault ground motion records that manifest the permanent-translation effect as well, the limited number of recordings with permanent translation does not permit the derivation of appropriate scaling laws. Therefore, the proposed analytical model should be utilized with caution for the generation of synthetic long-period ground motions that intend to incorporate the permanent translation effect. In these cases, the permanent offsets of the synthetic displacement time histories should be compatible with the tectonic environment and earthquake magnitude of the simulated event. The proposed methodology is written in MATLAB with the following steps:

1. Select the moment magnitude, M_W, of the potential earthquake and calculate the prevailing frequency, f_P, by $f_P=1/T_P$. For selected values of the parameters A, γ and ν (or for a suite of values of these three parameters), generate the coherent component of acceleration time history (or a suite of time histories) using equation (19).

2. For the selected fault–station geometry, generate the synthetic acceleration time histories for the moment magnitude, M_W, specified previously, using the specific barrier model.

3. Calculate the Fourier transform of the synthetic acceleration time histories generated in steps 1 and 2.

4. Subtract the Fourier amplitude spectrum of the synthetic time history generated in step 1 from the Fourier amplitude spectrum of the synthetic time history produced in step 2.

5. Construct a synthetic acceleration time history so that (a) its Fourier amplitude spectrum is the difference of the Fourier amplitude spectra calculated in step 4; and (b) its phase

coincides with the phase of the Fourier transform of the synthetic time history generated in step 2.

6.   Superimpose the time histories generated in steps 1 and 5. The near-source pulse is shifted in time so that the peak of its envelope coincides with the time that the rupture front passes in front of the station.

3. The seismotectonic and seismicity of Tombak region

The Zagros region is one of the most seismically active regions in Iran. The Tombak LNG terminal is located along the Persian Gulf northern coast, south of the Zagros Mountains, which mark the deforming zone separating Arabia (Arabian plate) and Central Iran (Eurasian plate) (Figure 3(a)). Location of Tombak area is presented in Figure 3(b). The massive LNG storage tanks exist in this terminal. These tanks have high importance from engineering and economical point of view so seismic loads should be considered in their analysis and design. The relevant codes of LNG storage containers emphasize that a comprehensive seismic hazard investigation should be conducted for regional seismicity and earthquake events of known near-fault.

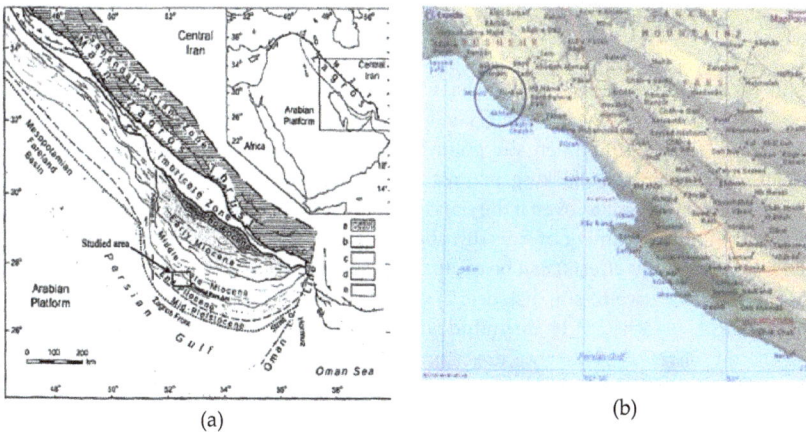

(a)   (b)

Figure 3. Location map: (a) Zagros folded zone, and (b) Tombak area

West of the Makran coast, where oceanic crust is subducting beneath Eurasia, the collision of the Arabian shield with Iran has uplifted the Zagros Mountains. The Zagros Mountains belt represents the early stage of a continental collision between the Arabian plate and the central Iran continental blocks. The Zagros Mountains are a seismically active region. Seismicity is restricted to the region between the Main Zagros Thrust and the Persian Gulf. Strong earthquakes are thought to occur on blind active thrust faults, which do not reach the surface. Fault plane solutions of these earthquakes indicate displacement mainly on low to high-angle reverse faults at depth of 6-12 km in the uppermost part of the basement. Most of the earthquakes for the region have generally M = 5.0 to 6.5, and have originated on sources beneath the decollement (Berberian 1995). Subduction on the main Zagros thrust has now ceased and it is seismically inactive (Ni and Barazangi 1986) except for the northern Zagros, where the surface trace of the thrust has been reactivated as right-slip main recent fault. The Zagros active fold-thrust belt lies on the north-eastern margin of the Arabian plate, on Precambrian (Pan-African) basement. It is composed of Cambrian to Neogene's folded series and is the result of five major tectonic events (Berberian and King 1981; Berberian 1983). The Zagros fold-thrust belt is composed of five units. The folds are parallel to the thrust faults. The axial part of the folds, striking NW SE, appears as broad asymmetrical folds with axial planes dipping to the NE and North. Their north-eastern limbs gently dip (20°) to the NW whereas their south-western limbs are steeper (40°) to the SE reaching 60 to 80°down slope and in some cases are nearly vertical, overturned or thrusted. The Main Zagros Thrust Fault (MZTF) indicates a fundamental change in sedimentary and structural evolution and seismicity. It marks the geosuture between the two colliding plates of the Eurasia and the Arabia. The global zone taken into account lies between 32°N and 26°N in latitude and 50°E and 58°E in longitude.

4. Estimation of the model parameters

The source and earthquake parameters have been obtained from the pervious seismic hazard study which has been conducted for Tombak area. The most of the models based on the stochastic method are fundamentally point-source models. Although it is true that near and intermediate-field terms are lacking, in most applications the frequencies are high enough that the far-field terms dominate, even if the site is near the fault. Furthermore, the effects of a finite-fault averaged over a number of sites distributed around the fault (to average over radiation pattern and directivity effects) can be captured in several ways: 1) using the closest distance to faulting as the source-to-site distance; 2) using a two-corner source spectrum; 3) allowing the geometrical spreading to be magnitude dependent. The material properties described by density ρ, and shear wave velocity β, are estimated to be 2.8 gr/cm^3 and 3.5 km/sec, respectively. All parameters used for simulation are summarized in Table (1).

In the methodology of Beresnev and Atkinson (1997,1998), modelling of finite source requires information of the orientation and dimensions of fault plane, as well as information of the dimensions of sub-faults and the location of hypocenter. The trends of epicenteral and hypocenteral distribution are in accordance with the strike and dip angle of the focal mecha-

Parameters	Values
$\rho_s, \beta_s, V, <R_{\Theta\Phi}>, F, R_0$	2.8, 3.5, 0.707, 0.55, 2.0, 1.0
Geometrical spreading (including factors to insure continuity of function)	$r<40$ km: $1/r$ $r\geq 40$ km: $(1/40)(40/r)^{0.5}$
Q, c_Q	$180f^{0.45}, 3.5$ km/s
Source duration	$0.5/f_a$
Path duration	$0.05\,R$
Site amplification	Boore and Joyner (1997) generic rock
Site diminution parameters (f_{max}, κ)	100.0, 0.03

Table 1. Model parameters

nism (strike, dip, slip) = (175, 85, 153) of the mainshock (Yamanaka 2003). The source dimension is therefore roughly estimated to be 20 km x 16 km (Yamanaka 2003).

Source parameters can be classified into two types (Irikura 2000): global source parameters and local source parameters. They represent different features of the fault source and are determined by different methods. The global source parameters characterize the macro feature of the entire source area and include spatial orientation of fault (location, attitude, buried depth), fault size (length, width, area), and both average slip and average rupture velocity on the fault plane. In the global source parameters both the slip type and spatial orientation are determined by seismogeology investigation and geophysical exploration; while the moment magnitude of the scenario earthquake caused by an active fault is estimated from its seismic hazard assessment. Fault size and average slip on the fault plane are also estimated by seismic scaling laws. In this study, the information for generating near-field strong ground motion such as magnitude related to each return period and the epicentral and hypocentral distances for stochastic method have been extracted from the seismogeology investigation that have been presented in Table (2). Table (3) lists the basic parameters used in the strong ground motion predictions.

Zone	Seismic source	Tombak			
		Magnitude	Epicentral distance (km)	Depth (km)	Hypocentral distance (km)
ZFF-C	OBE (475 years)	6.5	5	11	12
	SSE (5000 years)	7.0	5	14	15

Table 2. Parameter values of finite fault source model

Parameters	Values
Fault orientation	Strike 122°, Dip 40°
Fault dimensions along strike and dip (km)	28 × 16
Burial depth of upper limit of the fault (km)	5.0
Moment magnitude (M_w)	6.7
Subfault dimensions along strike and dip (km)	1 × 1
Stress drop (bar)	50
Q(f)	$150f^{0.5}$
Geometrical spreading	$1/R$
Windowing function	Saragoni-Hart
Kappa	0.05
Crustal shear wave velocity (km/s)	3.7
Rupture velocity (km/s)	0.8 × shear wave velocity
Crustal density (g/cm^3)	2.8

Table 3. Basic parameters used in the strong ground motion predictions

Using the magnitude for each return period, the duration of pulse is defined for both levels of earthquakes. The other parameters for the pulse are extracted from Mavroeidis and Papageorgiou (2003). The data has been obtained by the calibration procedure of actual near-fault strong ground motion records. The chosen parameter values have been summarized in Table (4).

ω	Return period (year)	Amplitude of pulse A (cm/s)	Oscillatory character of the signal γ	Phase of Pulse ν (degree)	Time shift t_0 (second)
6.5	475	50	1.5	134°	5
7.0	5000	90	2	100°	6.4

Table 4. Input parameters for defining long period pulse

5. Results and discussion

A simple and powerful method for simulating ground motions is based on the assumption that the amplitude of ground motion at a site can be specified in a deterministic way, with a random phase spectrum modified such that the motion is distributed over a duration related to the earthquake magnitude and to distance from the source. This stochastic method is particularly useful for simulating the higher-frequency ground motions, and it is used to predict ground motions for regions of the world in which recordings of motion from damaging earthquakes are not available. This simple method has been successful in matching a variety of ground motion measures for earthquakes with seismic moments spanning more than 12 orders of magnitude. SMSIM (StochasticModel SIMulation or Strong Motion SIMulation) is a set of programs for simulating ground motions based on the stochastic method. Programs are included both for time-domain and for random vibration simulations. In addition, programs are included to produce Fourier amplitude spectra for the models used in the simulations and to convert shear velocity versus depth into frequency-dependent amplification. The necessary parameters in these models are distinguished for defining the theoretical relationships. In this study, the near-field strong motion time histories, obtained from the stochastic method, are presented for both levels of earthquakes (475y and 5000y) as shown in Figure 4. The long period pulse has been calculated and presented in Figure 5 for both return periods.

In this stage, the pulse acceleration superimposes to the synthetic acceleration time history. The near-source pulse is shifted in time so that the peak of its envelope coincides with the time of rupture front of station. The final acceleration time histories for both levels of earthquakes are shown in Figure 6. The response spectra, obtained from simulated strong ground motion analysis, for 5% damping ratio are shown in Figure 7 for return periods 475 and 5000 years. As it is seen in Figures 7(a) and 7(b), the response spectrum has a sudden increasing for period ranges of 1 to 4 and 2 to 6 for 475y and 5000y, respectively. Therefore, structures which their periods settle in these ranges are influenced from near-field due to the long period pulse ground motion. In Figure 8, the smoothed response spectrum, obtained for return period of 475 years, is compared with IBC 2000 and Standard No. 2800. This figure shows that the response spectra are close to each other in the period range of 0 to 1 second. When the period exceeds than 1 second, the effect of long period pulse becomes apparent in the response spectra.

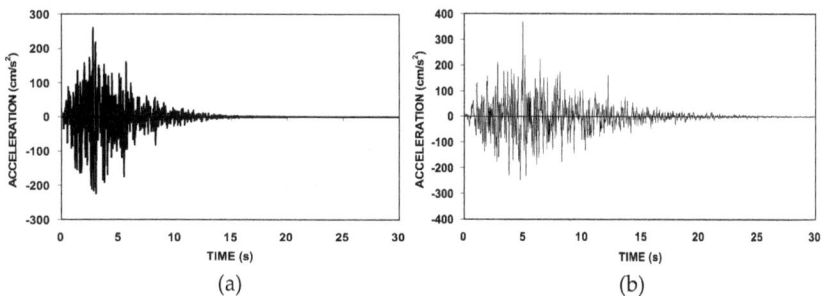

(a)   (b)

Figure 4. Synthetic acceleration time histories for return periods: (a) 475 years, and (b) 5000 years

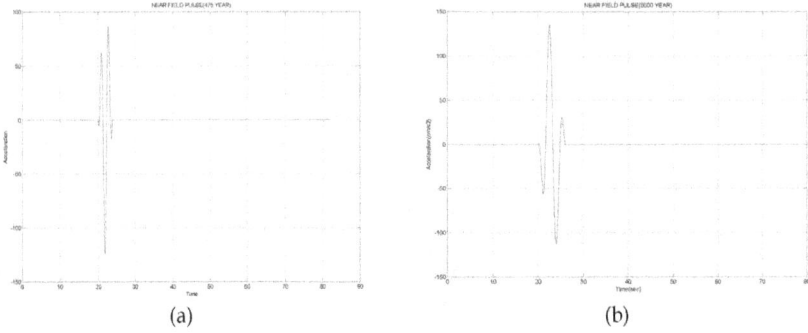

(a)   (b)

Figure 5. Long period pulses for return periods: (a) 475 years, and (b) 5000 years

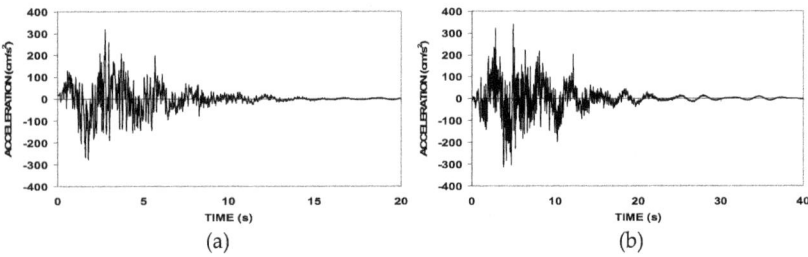

(a)   (b)

Figure 6. Final acceleration time histories for return periods: (a) 475 years, and (b) 5000 years

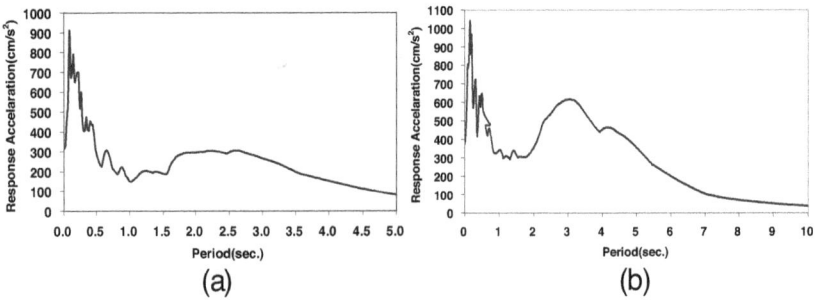

(a)   (b)

Figure 7. Response spectra of final acceleration time histories for return periods: (a) 475 years, and (b) 5000 years

DESIGN RESPONSE SPECTRA (475year)

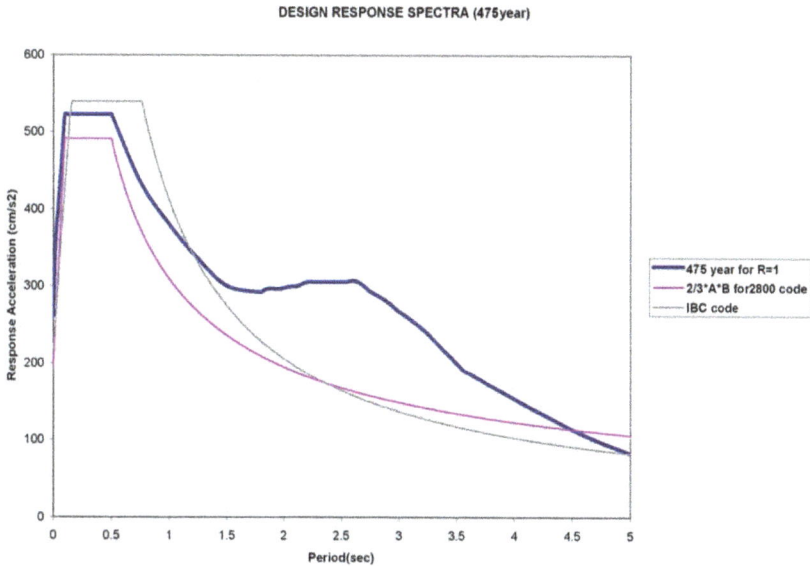

Figure 8. Comparison between response spectra (Blue Curve: 475y, Red Curve: Standard No. 2800, Gray Curve: IBC 2000)

6. Site response analysis

Soil investigation and borings are carried out for the detail design of storage tank foundations and spill basin structures of NIOC LNG Project. The site is located at the Persian Gulf coast in Tombak region, approximately 60 *km* away from Assaluyeh city, in Bushehr province. Standard penetration test has been running since initial stages of drilling operation. SPT tests have been performed approximately in each 1.5 *m* advance of drilling. Thus, there is a full set of data covering whole area with SPT results. Seismic tests of down-hole are done for full depth (80 *m*). Shear and compression waves velocities (*Vs* & *Vp*) are determined, accordingly. The results are presented in Table (5). The longitudinal wave velocity increases due to water table. The water table in borehole is approximately 9.0 *m* below the ground. Generally, the shear wave velocity increases versus depth due to an increase of soil density. According to Standard No. 2800, the shear wave velocity more than 760 *m/s* is assigned as rock; therefore the seismic bed rock is located at 10 *m* below the ground. The mentioned standard is used to classify the soil type which is type II and class C in this project. One-dimensional ground response analysis of the site is carried out by the equivalent linear approach using SHAKE 91 program (Idriss et al. 1992)

Depth	Density	Vp	Vs	E	G	K	υ
m	gr/cm³	m/s	m/s	MPa	MPa	MPa	
0-1.3	2.1	570	330	571	229	377	0.25
1.3-4	2.1	850	500	1297	525	817	0.24
4-5.5	2.1	1040	600	1891	756	1263	0.25
5.5-10	2.1	1500	700	2801	1029	3353	0.36
10-20	2.25	1520	760	3466	1300	3466	0.33
20-30	2.3	1600	800	3925	1472	3925	0.33
30-42	2.15	1600	835	3936	1499	3505	0.31
42-52	1.96	1650	870	3879	1484	3358	0.31
52-62	2.2	1600	820	3911	1479	3660	0.32
62-78	2.17	1650	900	4529	1758	3564	0.29

Table 5. Soil properties

According to the geotechnical site investigation, the soil type and the thickness of each layer are defined. From depth 0.0 m to 8.0 m, there are very diverse layers including boulder, gravel with some silty sand or sandy silt. SPT tests for these layers are refused because of coarse size of grains. From 8.0 m to 41.6 m, a very dense light brown sandy gravel with some cobbles is observed. SPT values are very high for this layer and some trace silt is seen from depth 32.0 m to 32.5 m. After this layer, a very dense light brown silty sand with some gravel is exist with 2.0 m thick. From 43.5 m to 50.0 m, we can see a very dense light brown sandy gravel with some cobbles and trace silt and clay. This layer changes to very dense gray silty sand with trace boulder (at depth: 50.0 m to 52.0 m). From 52.0 m to 52.6 m, borehole drilling machine interfaces to a piece of rock presenting by "boulder" phrase in borehole log. From 52.6 m to 54.0 m, borehole log shows a very dense to medium dense of gray silty sand. This layer is become very dense and its color changes to light brown from depth 54.0 m to 61.0 m. From depth 61.0 m to 64.0 m, some gravel and trace silt is added to last mentioned layer, so we can see a silty sand with some gravel and trace silt. From 64.0 m to 73.6 m, borehole log shows a very dense light brown sandy gravel with trace cobbles and clay. At the end of borehole, to depth 80.0 m, borehole log shows a very dense light brown silty sand with trace gravel and clay. Considering the results, particularly the seismic down-hole and SPT data, it is found that the average shear wave velocity equals approximately to 650 m/s and the SPT number is more than 50. For site specific response analysis, it is recommended to use shear wave velocity measured in the field. The average shear wave velocity of soil within 30 m was found to be around 650 m/s. The simulated ground motion is used as the input motion for ground response analysis. The results of ground response analysis are presented in Figure 9. The variation of maximum acceleration with depth as shown in Figure 9 indicates that the increase of PGA at the surface is about 1.06 and 1.11 times higher than those on the bed rock for return periods of 475 years and 5000 years, respectively. The response spectra obtained from ground response analysis for 5% damping ratio are shown in Figure 10.

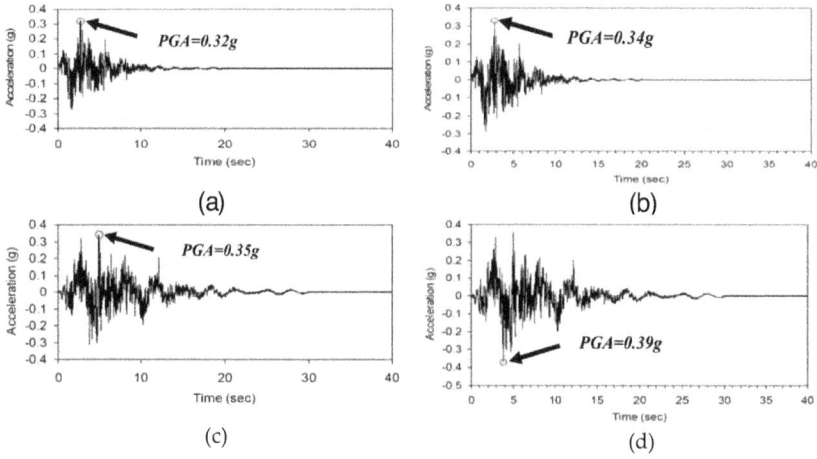

Figure 9. Acceleration time histories: (a) simulated on bedrock for return period = 475 years, (b) obtained on ground surface for return period = 475 years, and (c) simulated on bedrock for return period = 5000 years, (d) obtained on ground surface for return period = 5000 years

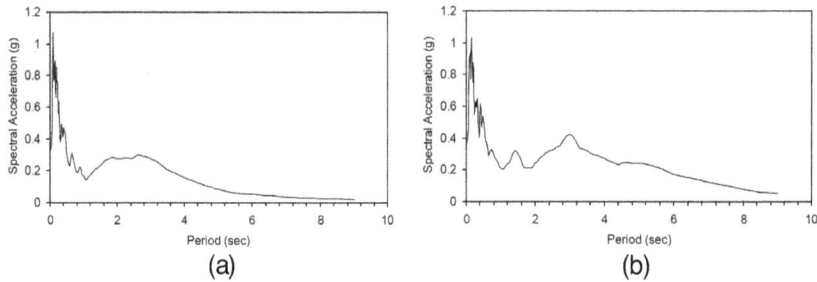

Figure 10. Spectral accelerations obtained by response analyses for return periods: (a) 475 years, and (b) 5000 years

7. Conclusions

The present study has been focused on simulating near-field strong ground motions using hybrid method for Tombak area in south-eastern part of Iran. The simulation of ground motion has been carried out using the stochastic method proposed by Boore (2003). Afterwards, the analytical model proposed by Mavroeidis and Papageorgiou (2003) is applied to consider the impulsive character of near-fault ground motion. Then, the response of mentioned site under simulated ground motion has been studied by conducting one dimensional ground response analysis. The results can be used for estimating the probable ground motion acceleration time-histories to be used in the hazard analysis of specific sites in the region under study, particularly for performance analysis of exiting structures. The ability of this hybrid method in simulating strong motions is also shown in this study. The simulation parameters, obtained in this study, can be used to asses the strong-motion level at a much larger number of sites, where no record is available, to investigate how different characteristics of motion affect the damage distribution in the Tombak region. Based on the above study the following conclusions are derived:

- The maximum PGAs on bedrock at the Tombak site are established as 0.32 g and 0.35 g for return periods of 475 and 5000 years, respectively.

- Ground response analysis indicates small increases of PGAs at the surface which are about 1.06 and 1.11 times higher than those on the bedrock for return periods of 475 and 5000 years, respectively.

- The response spectra obtained from the analyses indicate that the effect of long period pulse appears in the period ranges of 1 to 4 and 2 to 6 for 475y and 5000y, respectively, which are the typical ranges of period for LNG storage tank structures constructed in the Tomabk region.

- The obtained results demonstrate that the employed models adequately describe the nature of the impulsive near-fault ground motions both qualitatively and quantitatively.

- The proposed procedure mentioned in the present work will facilitate the study of the elastic and inelastic response of structures subjected to near-source seismic excitations within the considered region.

Author details

Babak Ebrahimian

School of Civil Engineering, Faculty of Engineering, University of Tehran, Tehran, Iran

References

[1] Aki, K. (1968). Seismic displacement near a fault. Journal of Geophysical Research, , 73, 5359-5376.

[2] Akinci, A, Malagnini, L, Pino, N. A, Scognamiglio, L, Herrmann, R. B, & Eyidogan, H. (2001). High-frequency ground motion in the Erzincan region. Turkey: inferences from small earthquakes. *Bulletin Seismological Society of America*, , 91, 1446-1455.

[3] Akinci, A, Malagnini, L, & Sabetta, F. (2010). Characteristics of the strong ground motions from the 6 April 2009 L'Aquila earthquake, Italy. *Soil Dynamics and Earthquake Engineering*, , 30, 320-335.

[4] Atkinson, G. M, & Boore, D. M. (1995). Ground motion relations for Eastern North America. *Bulletin Seismological Society of America*, , 85, 17-30.

[5] Berberian, M. (1983). Generalized tectonic map of Iran. in M. Berberian (Editor), Continental Deformation in the Iranian Plateau, Contribution to Seismotectonics of Iran, Part IV. *Geological Survey Iran*, 52, 625 pp.

[6] Berberian, M, & King, G. C. P. (1981). Toward a paleogeography and tectonic evalution of Iran. *Canadian Journal of Earth Science*, , 18, 210-265.

[7] Berberian, M. (1995). Master blind thrust faults hidden under the Zagros folds: active basement tectonics and surface morphotectonics. *Tectonophysics*, , 241, 193-224.

[8] Beresnev, I. A, & Atkinson, G. M. (1997). Modeling finite-fault radiation from the ω^n spectrum. *Bulletin Seismological Society of America*, , 87, 67-84.

[9] Beresnev, I. A, & Atkinson, G. M. (1998). Stochastic finite-fault modeling of ground motions from the 1994 Northridge, California Earthquake. I. validation on rock sites. *Bulletin Seismological Society of America*, , 88, 1392-1401.

[10] Boore, D. M, & Zoback, M. D. (1974). Two-dimensional kinematic fault modeling of the Pacoima Dam strong-motion recordings of February 9, 1971, San Fernando earthquake. *Bulletin Seismological Society of America*, , 64, 555-570.

[11] Boore, D. M. (1983). Stochastic simulation of high-frequency ground motions based on seismological models of the radiated spectra. *Bulletin Seismological Society of America*, , 73, 1865-1894.

[12] Boore, D. M, & Atkinson, G. M. (1987). Stochastic prediction of ground motion and spectral response parameters at hard-rock sites in Eastern North America. *Bulletin Seismological Society of America*, , 77, 440-467.

[13] Boore, D. M. (2003). Simulation of ground motion using stochastic method. *Journal of pure and applied Geophysics*, , 160, 635-676.

[14] Bouchon, M. (1978). A dynamic crack model for the San Fernando earthquake. *Bulletin Seismological Society of America*, , 68, 1555-1576.

[15] Chopra, A. K, & Chintanapakdee, C. (2001). Comparing response of SDOF systems to near-fault and far-fault earthquake motions in the context of spectral regions. *Journal of Earthquake Engineering and structural Dynamics*, , 30, 1769-1789.

[16] Amico, D, Caccamo, S, Parrillo, D, Lagana, F, Barbieri, C, & The, F. th September 1999 Chi-Chi earthquake (Taiwan): a case of study for its aftershock seismic sequence. *Izvestiya-Physics of the Solid Earth*, 46 (4), 317-326.

[17] Amico, D, Akinci, S, & Malagnini, A. L. (2012). Predictions of high-frequency ground-motion in Taiwan based on weak motion data. *Geophysical Journal International*, , 189, 611-628.

[18] Hall, J. F, Heaton, T. H, Halling, M. W, & Wald, D. J. (1995). Near-source ground motion and its effects on flexible buildings. *Earthquake Spectra*, , 11, 569-606.

[19] Hartzell, S. (1979). Analysis of the Bucharest strong ground motion record for the March 4, 1977 Romanian earthquake. *Bulletin Seismological Society of America*, , 69, 513-530.

[20] Haskell, N. A. (1969). Elastic displacements in the near-field of a propagating fault. *Bulletin Seismological Society of America*, , 59, 865-908.

[21] Housner, G. W, & Trifunac, M. D. (1967). Analysis of accelerograms: Parkfield earthquake. *Bulletin Seismological Society of America*, , 57, 1193-1220.

[22] Hubral, P, & Tygel, M. (1989). Analysis of the Rayleigh pulse. *Geophysics*, , 54, 654-658.

[23] Hutchings, L. (1994). Kinematic earthquake models and synthesized ground motion using empirical Green's functions. *Bulletin Seismological Society of America*, , 84, 1028-1050.

[24] Idriss, I. M, & Joseph, I. S. (1992). *User manual for SHAKE 91*.

[25] IBCInternationalk Building Code. (2000). *International Code Council, Inc.*, Country Club Hills, IL.

[26] Irikura, K. (2000). Prediction of strong ground motions from future earthquakes caused by active faults-Case of the Osaka Basin. *Proceedings of the 12th World Conference on Earthquake Engineering*, paper 2687.

[27] Makris, N. (1997). Rigidity-plasticity-viscosity: can electrorheological dampers protect base-isolated structures from near-source ground motions? *Journal of Earthquake Engineering and structural Dynamics*, , 26, 571-591.

[28] Malagnini, L, Akinci, A, Herrmann, R. B, Pino, N. A, & Scognamiglio, L. (2002). Characteristics of the ground motion in Northeastern Italy. *Bulletin Seismological Society of America*, , 92, 2186-2204.

[29] Malagnini, L, Akinci, A, Mayeda, K, Munafo, I, Herrmann, R. B, & Mercuri, A. (2011). Characterization of earthquake-induced ground motion from the L'Aquila seismic sequence of 2009, Italy. *Geophysical Journal International*, , 184, 325-337.

[30] Mavroeidis, G. P, & Papageorgiou, A. S. (2002). Near-source strong ground motion: characteristics and design issues. *Proceedings of the Seventh U.S. National Conference on Earthquake Engineering (7NCEE)*, Boston, Massachusetts, July 2002., 21-25.

[31] Mavroeidis, G. P, & Papageorgiou, A. S. (2003). A mathematical representation of near-fault ground motions. *Bulletin Seismological Society of America*, , 93, 1099-1131.

[32] Mavroeidis, G. P, Dong, G, & Papageorgiou, A. S. (2004). Near-fault ground motions, and the response of elastic and inelastic single-degree-of-freedom (SDOF) systems. *Journal of Earthquake Engineering and structural Dynamics*, 33(9), 1023-1049.

[33] Mayeda, K, & Malagnini, L. (2009). Apparent stress and corner frequency variations in the 1999 Taiwan (Chi-Chi) sequence: evidence for a step-wise increase at Mw ~5.5. *Geophysical Research Letters*, 36, L10308.

[34] Motazedian, D, & Atkinson, G. M. (2005). Stochastic finite-fault modeling based on dynamic corner frequency. *Bulletin Seismological Society of America*, , 95, 995-1010.

[35] Niazy, A. (1975). An exact solution for a finite, two-dimensional moving dislocation in an elastic half-space with application to the San Fernando earthquake of 1971. *Bulletin Seismological Society of America*, , 65, 1797-1826.

[36] Ni, J, & Barazangi, M. (1986). Seismotectonics of the Zagros continental collision zone and a comparison with the Himalayas. *Journal of Geophysical Research*, 91(88), 8205-8218.

[37] Papageorgiou, A. S, & Aki, K. c barrier model for the quantitative description of inhomogeneous faulting and the prediction of strong ground motion. I. Description of the model. *Bulletin Seismological Society of America*, , 73, 693-722.

[38] Saragoni, G. R, & Hart, G. C. (1974). Simulation of artificial earthquakes. *Journal of Earthquake Engineering and structural Dynamics*, , 2, 249-267.

[39] Shinozuka, M. (1988). State-of-the-art report: engineering modeling of ground motion. *Proceedings of the Ninth World Conference on Earthquake Engineering (9WCEE)*, Tokyo, Japan, August 1988., 2-9.

[40] Silva, W. J. (1997). Characteristics of vertical strong ground motions for applications to engineering design. *Proceedings of the FHWA/NCEER Workshop on the National Representation of Seismic Ground Motion for New and Existing Highway Facilities (I.M. Friedland, M.S. Power, and R.L. Mayes, eds.)*, Technical Report NCEER-, 97-0010.

[41] Somerville, P. G. (1998). Emerging art: earthquake ground motion. *Geotechnical Earthquake Engineering and Soil Dynamics III, Proceedings of a Specialty Conference held in Seattle, Washington*, August 3-6, 1998. Geotechnical special publication, , 75(75), 1-38.

[42] Standard No(2008). Iranian Code of Practice for Seismic Resistance Design Buildings. 3rd Edition, *Building and Housing Research Center*, PN S 253.

[43] Yamanaka, Y. (2003). Seismological Note. *Earthquake Information Center*, Earthquake Research Institute, University of Tokyo.(145)

[44] Zafarani, H. Noorzad As. & Ansari, A. (2005). Generation of near-fault response spectrum for a large dam in Iran. *Hydropower and Dams*, 12 (4), 51-55.

[45] Zhang, Y, & Iwan, W. D. (2002). Active interaction control of tall buildings subjected to near-field ground motions. *Journal of Structural Engineering ASCE,* , 128, 69-79.

Chapter 5

Speedy Techniques to Evaluate Seismic Site Effects in Particular Geomorphologic Conditions: Faults, Cavities, Landslides and Topographic Irregularities

F. Panzera, G. Lombardo, S. D'Amico and P. Galea

Additional information is available at the end of the chapter

1. Introduction

The ground motion that can be recorded at the free surface of a terrain is the final result of a series of phenomena that can be grouped into three fundamental typologies: the source mechanism, the seismic wave propagation till the bedrock interface below the investigated site and the site effects (Fig. 1). The first two features define the kind of seismic input whereas the third represents all modifications that can occur as a consequence of the interaction between seismic waves and local characteristics of the investigated site. The physical and mechanical properties of terrains as well as their morphologic and stratigraphic features appreciably affect the characteristics of the ground motion observed at the surface. The whole process of modifications undergone by a given seismic input in terms of amplitude, frequency content and duration, as a consequence of local characteristics, is generally termed the "local seismic response". It is indeed well known that the spectral composition of a seismic event is modified first during the source-bedrock path (attenuation function), and second, when the seismic input interacts with the soft terrains layered between the bedrock and the free surface (Fig. 1a). This latter effect, significantly changes the spectral content so that it is extremely important for estimating the final input to which all structures built in the study area will be subjected.

The influence of local geologic features on the ground motion peculiarities and damage due to earthquakes is well known since years. Studies of Wood (1908) and Baratta (1910) concerning the San Francisco 1906 and the Messina 1908 earthquakes, respectively, pointed out, since the beginning of the last century, that the damage distribution is a function of different site conditions existing in various areas affected by the same shock. Similar effects have been observed by several authors during all destructive earthquakes occurred up till the present.

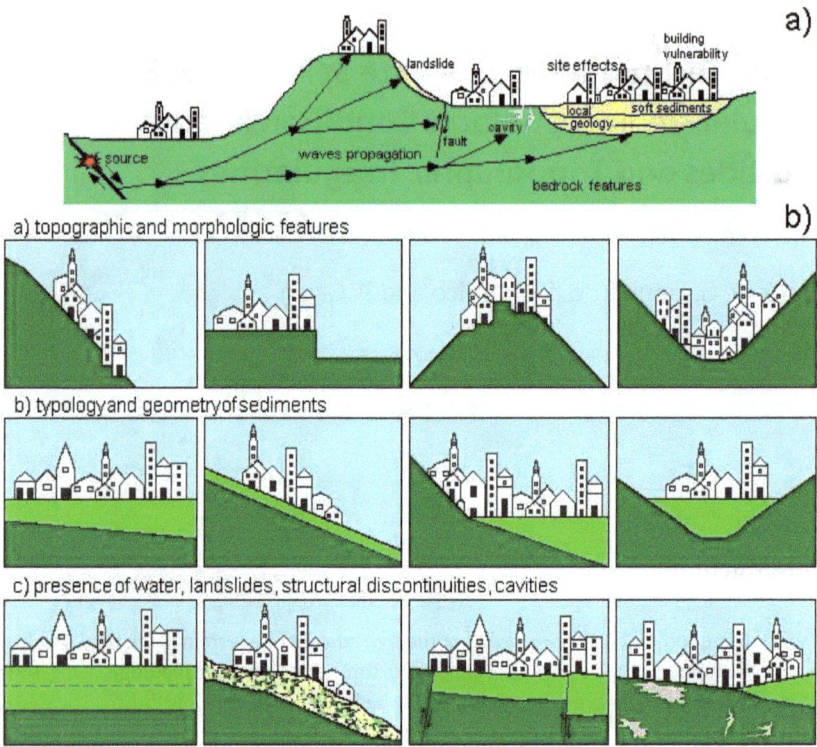

Figure 1. a) Sketch showing the influences that can affect a seismic signal propagating from the source to the free surface of a terrain. b) Scenaria for local seismic response.

Site effects occur as a result of several physical phenomena such as multiple reflections, diffraction, focusing, resonance etc., to which the incoming wavefront is subjected. This is a consequence of the various mechanical properties of terrains, the presence of heterogeneities and discontinuities, as well as the geometry of shallower layers and the existence of topographic irregularities both in the basement and the surface. In Figure 1b are shown the principal morphologic and/or structural features that contribute to characterize the local hazard scenario. They are grouped in effects linked to the layers' geometry (a), effects linked to the possible presence of water-bearing strata, landslides, structural discontinuities and cavities (b) and effects linked to the topography (c).

Generally stratigraphic effects are schematized as the modifications affecting a seismic motion that propagates almost vertically inside a deposit having a flat free surface, horizontal layers and negligible lateral heterogeneities. The theoretical analysis of such problem was tackled by Kramer (1996), and however, considering the described simplified assumptions, it can be postulated that the incident waves at the base of the deposit that are reflected at its free surface, are partially reflected again at the deposit-basement interface. The amount of reflected energy that is "trapped" into the deposit increases with the seismic impedance contrast between the terrains forming the deposit and the basement. Besides, the trapped waves interfere between themselves and the incident waves as a consequence of the geometric features of the deposit, the physical properties of the terrains and the frequency content of the seismic input. Stratigraphic site effects are therefore mainly connected to seismic wave trapping phenomena inside the deposit due to reflections as well as interference and resonance effects between incident and reflected waves. The local seismic response, becomes, of course, more complex when the basement-deposit interface has a more irregular geometry, or in the presence of faults, cavities and particular topographic conditions.

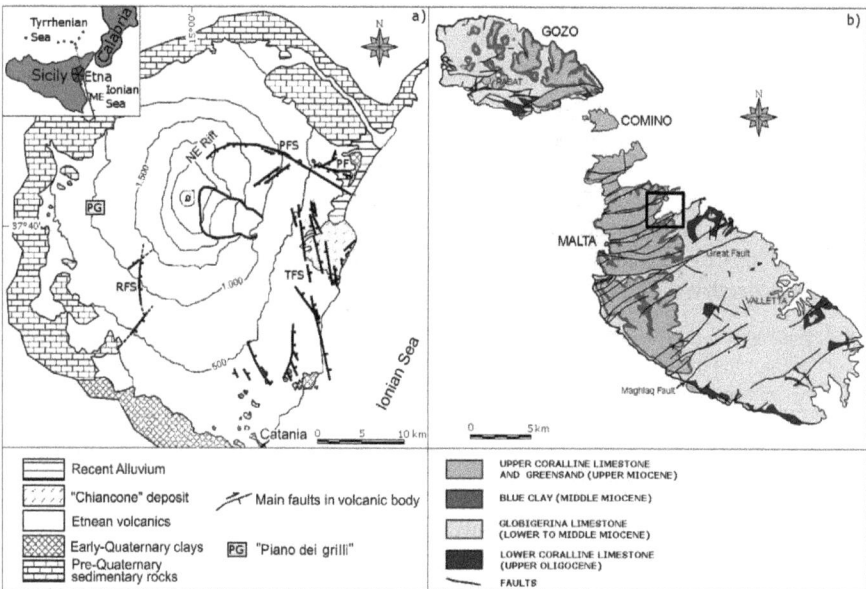

Figure 2. a) Simplified geological map of Mt Etna showing the main structural features (modified from Neri et al., 2007), RFS = Ragalna fault system, PFS = Pernicana fault system, PF = Piedimonte fault, TFS = Timpe fault system. In the inset map, the Malta Escarpment (ME) is shown. (b) Sketch geological map of the Maltese Islands (modified from Various Authors, 1993). The black square indicates the investigated area.

In this study the characteristics of the local seismic response, linked in particular to the presence of discontinuities such as faults and cavities, as well as topographic irregularities and landslide phenomena, are investigated. Case-studies of sites located both in South-eastern Sicily and in Malta are described, illustrating, besides local amplification phenomena, the possible presence of directional effects.

2. Methodologies

Several different methodological approaches are commonly adopted to quantitatively assess the local seismic response. In practice, it is evaluated with respect to a reference site represented by the outcrop of a rocky basement (either real or supposed) existing in the investigated area. In other words, the local ground motion is compared with the one relative to a reference bedrock outcrop.

The site response can be evaluated through various approaches, also collateral between themselves, each of them having specific advantages and/or drawbacks well known in literature (see Pitilakis, 2004). Main methodologies can be grouped into two categories: numerical methods and experimental methods.

Numerical methods are founded on the use of computer codes that simulate wave propagation through soft deposits, from the bedrock to the free surface. Such codes allow the modelling of the dynamic behaviour of a terrain by adopting linear, equivalent-linear or non- linear models (e.g. SHAKE, Schnabel et al, 1972; EERA, Bardet et al, 2000; DESDRA, Lee e Finn, 1978) that can be either mono or multi-dimensional (Hudson et al, 1994). These methods provide in output the time history of involved seismic parameters and require as input data a detailed knowledge of the site geometry, the geotechnical properties of terrains and the stress-strain relationships.

The experimental methods allow us to evaluate the local seismic response using the records of seismic signals that be generated by earthquakes, artificial seismic sources or ambient noise. They are only moderately expensive and take implicitly into account all site effects, although their drawback is linked to the use of low or very low energy events, so that the seismic response evaluation is performed at low deformation levels and entirely in the linear field.

The results described in the present study, draw from the use of spectral ratios evaluated through comparison between the investigated site and the reference one (SSR *Standard Spectral Ratio* technique) and/or by calculating the spectral ratios between the horizontal and the vertical components of motion at the investigated site (HVSR *Horizontal to Vertical Spectral Ratio* and HVNR *Horizontal to Vertical Noise Ratio* or Nakamura method).

The SSR technique (Borcherdt 1970) consists in computing the Fourier spectral ratio of the same seismic waves (generally S waves) simultaneously recorded by the horizontal components of two seismic stations, one of which is located on a bedrock outcrop. The main difficulty associated with this technique is a proper choice of the reference site that has to be a flat outcrop of the bedrock. Moreover, the correct use of the SSR technique requires that the distance between test and reference sites has to be significantly smaller than the epicentral distance.

The earthquake HVSR, or receiver function technique, does not need a reference station and consists in the computation of the horizontal-to-vertical spectral ratio of the components of motion recorded at one seismic station only (Lermo and Chavez-Garcia 1993). This technique is founded on the assumption that the vertical component of motion is not affected by the local geological conditions. It is applied both to the time window of shear waves and to the entire seismic record and has shown to be a good approach for the evaluation of the site fundamental frequency whereas it appears less reliable for the estimate of the amplitude values.

The Nakamura technique (HVNR) (Nakamura, 1989) uses as a seismic input the ambient noise and computes the spectral ratio between the horizontal and the vertical components of motion. Ambient noise has, in recent years, become widely used for site amplification studies. Its use appears opportune for significant reductions in field data acquisition time and costs. The evaluation of site response using the HVNR technique is largely adopted since it requires only one mobile seismic station with no additional measurements at rock sites for comparison. Besides, it does not require the long and simultaneous deployment of several instruments which is necessary to collect a useful earthquake data set. The basic hypothesis of using ambient noise is that the resonance of a soft layer corresponds to the fundamental mode of Rayleigh waves, which is associated with an inversion of the direction of Rayleigh waves rotation (Nogoshi and Igarashi, 1970; Lachet and Bard, 1994). Thus, the ratio between the horizontal and vertical spectral components of motion can reveal the fundamental resonance frequency of the site. Reliability of such approach has been asserted by many authors (e.g. Lermo and Chavez-Garcia, 1993; Bard, 1999) who have stressed its significant stability in local seismic response estimates. It is commonly accepted that, although the single components of ambient noise can show large spectral variations as a function of natural and cultural disturbances, the H/V spectral ratio tends to remain invariant, therefore preserving the fundamental frequency peak (Cara et al., 2003).

In the present study, ambient noise records were performed using a Tromino instrument (www.tromino.it), a compact 3-component velocimeter with a reliable instrumental response in the frequency range 0.5-10 Hz. The signals were processed by evaluating the horizontal-to-vertical noise spectral ratios (HVNR). Following the guidelines suggested by the European project Site EffectS assessment using AMbient Excitations (SESAME, 2004), time windows of 30 s were considered, selecting the most stationary part and excluding transients associated to very close sources. Fourier spectra were calculated and smoothed using a triangular average on frequency intervals of ± 5% of the central frequency.

The potential presence of directional effects in the ground motion recorded at the surface was also investigated. Such investigations can be done by computing the spectral ratios (SSR, HVSR, HVNR) after rotating the horizontal components by steps of 10° starting from 0° (north) to 180° (south) and plotting the contours of the spectral ratio amplitudes as a function of frequency and direction of motion. This approach (Spudich et al., 1996) is powerful in enhancing, if any, the occurrence of site specific directional effects. A direct estimate of the polarization angle, for noise data, can be achieved through two different methods. The time domain method (TD) by Jurkevics (1988) and the time-frequency (TF) polarization analysis by Burjánek et al. (2010 and 2012). The results obtained through polarization techniques are quite robust since these approaches are very efficient in overcoming the bias linked to the denominator behavior that could occur in the HVNR's technique and at the same time, longer time-series are processed therefore reducing the problems that may be linked to signal-to-noise ratio. In the TD approach, a direct estimate of the polarization angle is achieved by computing the polarization ellipsoid through the eigenvalues and eigenvectors of the covariance matrix obtained by three-component data (Jurkevics, 1988). The polarization ellipsoid of the analyzed signal is estimated by band-pass filtering it in the interval 1.0 - 10.0 Hz, using the whole recordings and considering a moving window of 1 s with 20% overlap, therefore obtaining the strike of maximum polarization for each moving time window. The results are finally plotted in circular histograms (rose diagrams) showing the polarization azimuths in intervals of ten degrees. In the TF method, a continuous wavelet transform for signal time-frequency decomposition, is firstly used. Subsequently, the polarization analysis on the complex wavelet amplitude for each time-frequency pair, is applied. In particular, histograms of the polarization parameters are created over time for each frequency. Polar plots are then adopted for depicting the final results, which illustrate the combined angular and frequency dependence.

3. Geologic and tectonic features of the studied areas

South-Eastern Sicily is located in a complex tectonic region being at the boundary between African and European plates (see inset map in Fig. 2a). Along this border, Mt. Etna, a basaltic volcano more than 3300 m high and with a diameter of about 40 km, resulting from the interaction of regional tectonics and local scale volcano-related processes (McGuire and Pullen, 1989), is sited. The island of Malta is placed in the Hyblean foreland, belonging to the African plate. Most of the formations here outcropping were deposited during the Oligocene and Miocene when the whole area was part of the Malta - Ragusa platform and, as such, attached to the African margin (Pedley et al., 1978).

The whole study area is delineated by the crossing of lithosphere structures that give rise to the origin of Mt. Etna and by the presence of the Malta Hyblean fault system that runs down the Sicilian coast towards the Ionian sea (ME in the inset map of Fig. 2a). A series of horst and graben structures, NW-SE and NNW-SSE oriented, that are linked to the Malta-Hyblean escarpment, characterize indeed the tectonic setting of this area.

On Malta, the geo-structural pattern is dominated by two intersecting fault systems which alternate in tectonic activity. An older ENE-WSW trending fault, the Victoria Lines Fault (or Great Fault), traverses the islands and is crossed by a younger NW-SE trending fault, the Maghlaq Fault (Fig. 2b), parallel to the Malta trough which is the easternmost graben of the Pantelleria Rift system. The faults belonging to the older set, all vertical or sub vertical, are part of a horst and graben system of relatively small vertical displacement (Illies, 1981; Reuther et al. 1985).

As concerns Mt. Etna, its eastern flank is the more tectonically active part. Here, several NNW and NNE-trending fault segments (Timpe fault system, TFS), arranged in a 30 km long system (Fig. 2a), control the present topography and show steep escarpments with very sharp morphology (Monaco et al., 1997). This system represents the northernmost prolongation of the Malta Escarpment and forms a system of parallel step-faults having vertical offsets up to 200 m that down-throw towards the sea. Most of these faults are highly seismogenic and generate shallow earthquakes as well as co-seismic cracks in the soil and creep phenomena (Azzaro, 1999). In the north eastern part of the area, the active Pernicana fault system (PFS) represent the most significant tectonic structure. It is a strike-slip fault roughly E-W oriented with a length of about 18 km from the NE rift to the coastline (Neri *et al.*, 2004; Azzaro *et al.*, 2001; Acocella and Neri, 2005). At the end of this structure, close to the coast line, the Calatabiano and the Piedimonte faults (PF) can be considered, following Lentini *et al.* (2006), as the neotectonic structures of the basement outcropping in north-eastern Sicily (Fig. 2a).

The western flank of the volcano is affected by a moderate tectonic activity, the Ragalna fault system (RFS) being the main structure (Fig. 2a). This system is formed by three distinct fault segments the Calcerana and the Ragalna faults trending NE-SW and the N–S striking Masseria Cavaliere fault (Azzaro, 1999; Rust and Neri, 1996). This latter structure is a fresh east-facing escarpment up to 20 m high and 5 km long. Less evident compared to the previous one, the NE-SW striking Calcerana fault and the NE-SW trending structure, reported by some authors in the area between Ragalna and Biancavilla, do not show strong field manifestations.

4. Effects connected to the presence of faults

Fault zones are generally characterized by a highly fractured low-velocity belt (damage zone), hundreds of meter wide, bounded by higher-velocity area (host rock) that can broaden for some kilometres (Ben-Zion *et al.* 2003; Ben-Zion and Sammis 2003, 2009 and references therein). Such geometrical setting and impedance contrast is theoretically similar to the well known situations, widely studied in engineering geology and seismology, when soft sediments overly stiff rock. In the presently depicted case, the discontinuity is almost vertically oriented and, as described by Irikura and Kawanaka (1980), it is in principle proficient to produce local amplification of ground motion (Peng and Ben-Zion, 2006; Calderoni *et al.*, 2010; Cultrera *et al.*, 2003; Seeber *et al.*, 2000), as well as to support the development of fault zone trapped waves (e.g. Li *et al.*, 1994; Mizuno and Nishigami, 2006).

There is a large number of papers that describe propagation properties of fault-guided waves in terms of ground motion amplification having a propensity to be maximum along the fault-parallel direction. These observations, both in theoretical and experimental approaches deal with almost pure strike slip faults such as the S. Andreas and the Anatolian faults (see Li *et al.*, 2000; Ben Zion *et al.*, 2003). In the Anatolian fault, Ben Zion *et al.* (2003) observed fault guided waves and an almost constant time delay in the shear waves arrival for different epicentral distances. The authors interpreted such observation as a consequence of a trapping mechanism in the shallower part of the fault between the first 3 and 5 kilometers. Lewis *et al.* (2005) come to a comparable conclusion through the observation of hundreds of seismograms of small magnitude events recorded close to the San Jacinto fault in California. In Italy, Rovelli *et al.* (2002), investigating the Nocera Umbra fault, observed a shallow trapping zone (1-1.5 km) with evidence of measurable amplification effects, as well as pronounced polarization of the ground motion in a direction parallel to the fault strike. On the other hand, Boore *et al.* (2004) in the Calaveras fault, California, noticed that directional effects observed during earthquakes, were occurring parallel to the fault strike and not perpendicularly as it would be expected for a strike-slip mechanism of faulting. Studies about local seismic response nearby fault zones have been performed in Italy and in California by Cultrera *et al.* (2003), Calderoni *et al.* (2010), Pischiutta *et al.* (2012) who observed evidence of ground motion amplification in the fault zone environments and strong directional effects with high angle to the fault strike. Similar studies, performed by Rigano *et al.* (2008) and Di Giulio *et al.* (2009) documented the presence of a systematic polarization of horizontal ground motion, near faults located on the eastern part of the Etnean area, that was never coincident with the strike of the tectonic structures. These directional effects were observed both during local and regional earthquakes, as well as using ambient noise measurements, therefore suggesting the use of microtremors for investigating ground motion polarization properties along and across the main tectonic structures.

In the present study, the results of studies performed in the Etnean area and in South eastern Sicily by Rigano and Lombardo (2005), Lombardo and Rigano (2006), Rigano *et al.* (2008), Di Giulio *et al.* (2009) will be briefly summarized and the outcomes from new investigations carried out in fault areas located both in the Etnean area and in Malta will be shown. Moreover, several measurements were performed in areas significantly distant from the studied tectonic structures (Piano dei Grilli, Etna and the Malta area), in order to observe how directional effects can change at increasing distance from the fault lines.

4.1. Results and discussion

In Figure 3, some examples of polarization plots obtained by Rigano *et al.* (2008), from microtremor measurements performed along and across the Pernicana and Tremestieri faults, are reported. The authors analyzed both earthquake and ambient noise records and found that in all investigated sites the ground motion polarization azimuths were principally NW-SE and NE-SW oriented, for the Pernicana (Fig. 3a) and the Tremestieri (Fig. 3b) faults respectively. It is interesting to observe that such directions were never found coincident with the strike of the investigated faults. Tests were also performed by the authors in order to verify that the observed polarizations were not affected by features of local sources, such as the volcanic tremor, and were stable in time.

The new ambient noise measurements described in the present study, were performed along four short profiles, having recording points spaced about 50m from each other, crossing the Masseria Cavaliere and the Ragalna fault (see respectively Tr#1, Tr#2 and Tr#3, Tr#4 in Fig. 4a). Results of directional effects investigations (Fig. 5) show that at these sampling sites, located in close proximity to the tectonic structures, the H/V spectral ratios have a tendency to increase in amplitude, in the frequency range 1.0-6.0 Hz, at angles of about 80°-90° for Masseria Cavaliere and 60°-70° for Ragalna faults (specify whether angles are from N, or with fault trace). In general, H/V spectral ratios show a broad band frequency effect with multiple adjacent peaks pointing out a preferential direction which is the typical behavior of directional resonances.

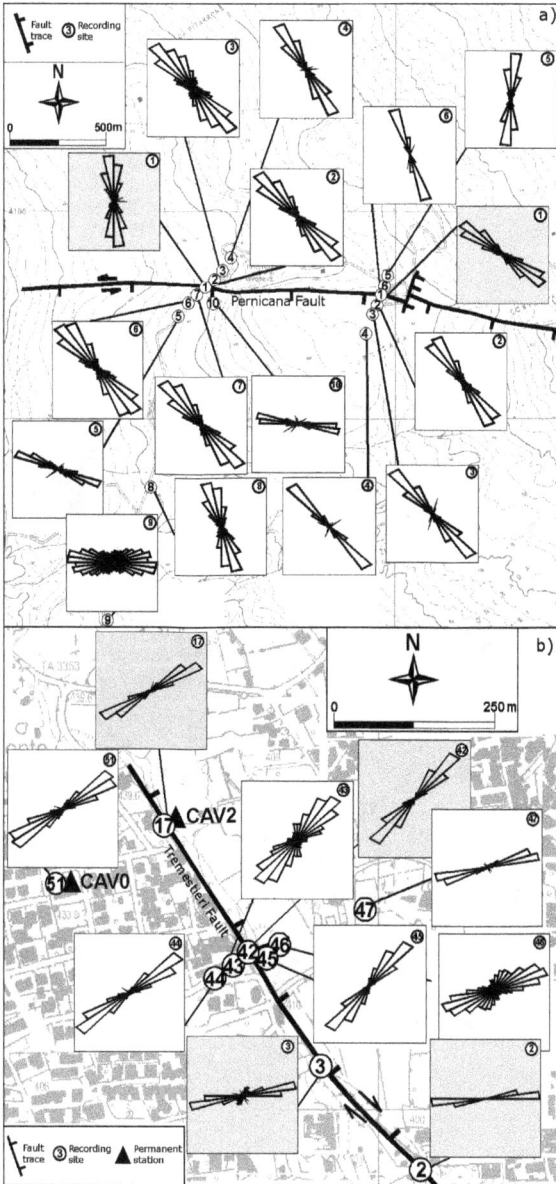

Figure 3. Examples of horizontal polarization angles obtained from ambient noise recorded along and nearby the Pernicana (a) and the Tremestieri (b) faults (modified from Rigano et al., 2008).

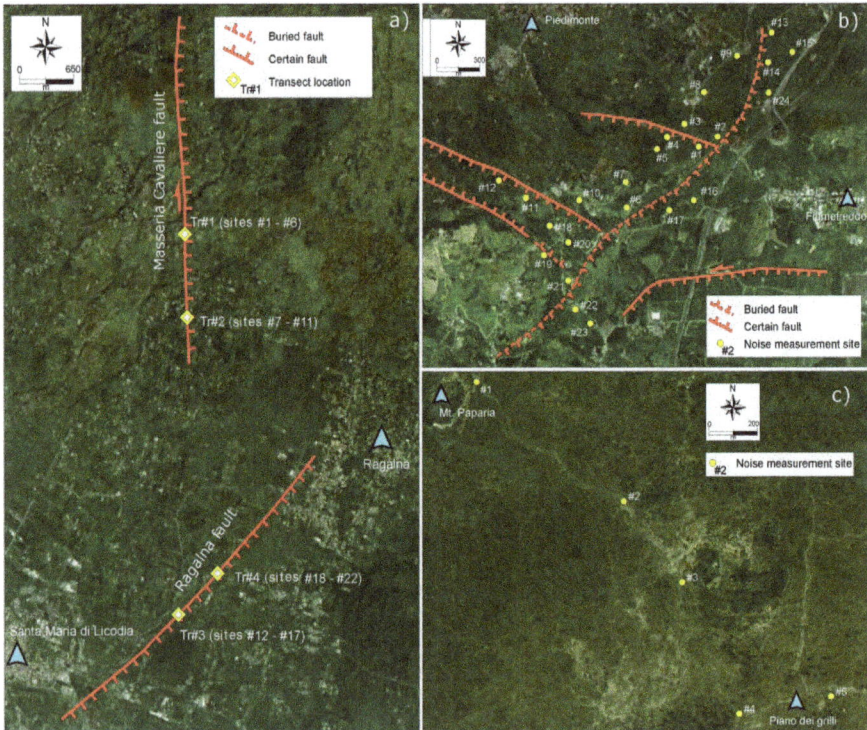

Figure 4. Location of the noise measurement sites. a) transects performed on the Ragalna fault system; b) measurement points on the Piedimonte fault; c) ambient noise recording sites in an area far from the investigated faults.

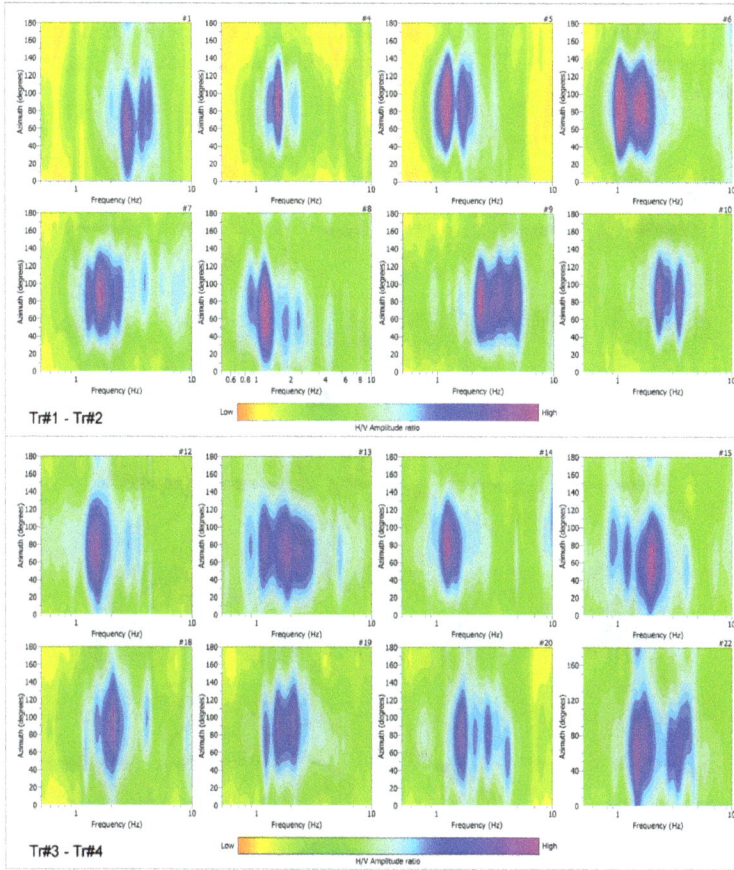

Figure 5. Examples of the contours of the geometric mean of the spectral ratios as a function of frequency (x axis) and direction of motion (y axis) obtained at selected ambient noise recording sites located on the Tr#1, Tr#2, Tr#3 and Tr#4 transects performed on the masseria Cavaliere and the Ragalna fault, respectively.

In order to better quantify the horizontal polarization of the ground motion, the covariance matrix method in the time and in the time-frequency domain were applied (Fig. 6). The results give a clear indication that ambient noise is affected by a significant horizontal polarization at the measurement sites along and across the investigated faults. It is interesting to observe that results obtained through the TF method clearly show that the recorded ambient noise is polarized in a narrow frequency band (1.0-6.0 Hz), following a roughly east-west and north-east-southwest trend, for the Masseria Cavaliere and the Ragalna faults respectively. The TD results confirm the same polarization trend although in some cases the rose diagrams show polarizations that are not sharply oriented (see e.g. #1, #2, #16 and #17, in Fig. 6). It is in our opinion clear that possible contributions of high-frequency noise (6.0-10.0 Hz), related to transients or to small scale geologic heterogeneities of a site, may imply incorrect results as shown for instance in Figure 7 where the TD rose diagrams dramatically change their polarization directions whereas the TF polar diagrams indicate that changes occur at higher frequencies only.

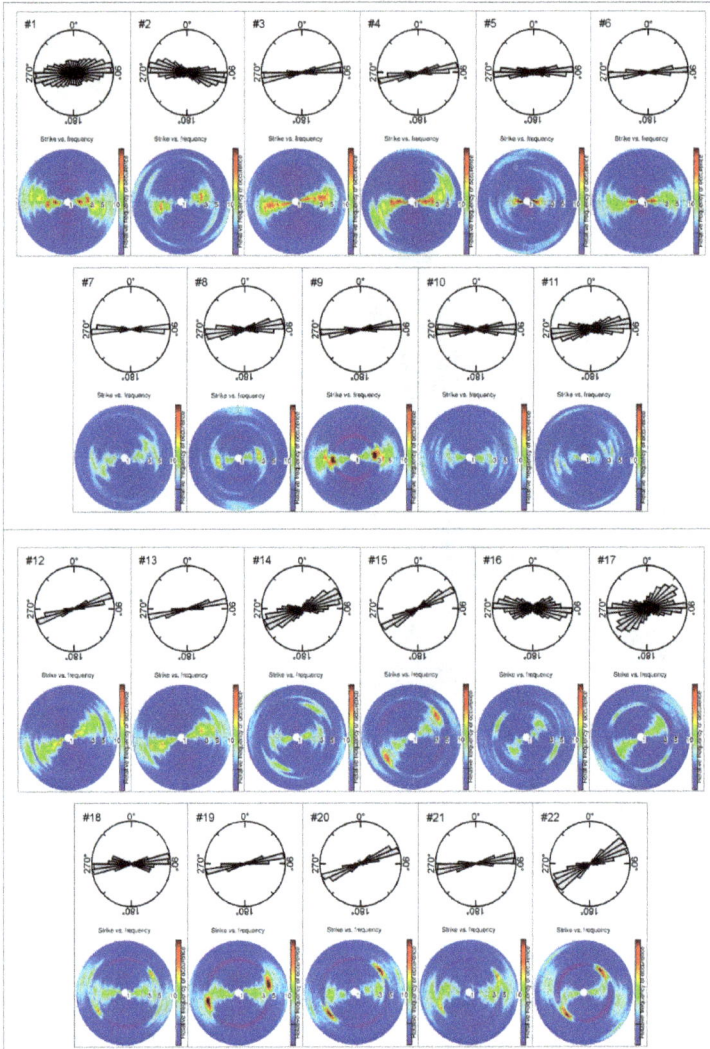

Figure 6. Results of horizontal polarization angles computed through the TD (rose diagrams) and TF analysis considering the frequency range of 1.0-10.0 Hz (polar plots), for Ragalna fault system.

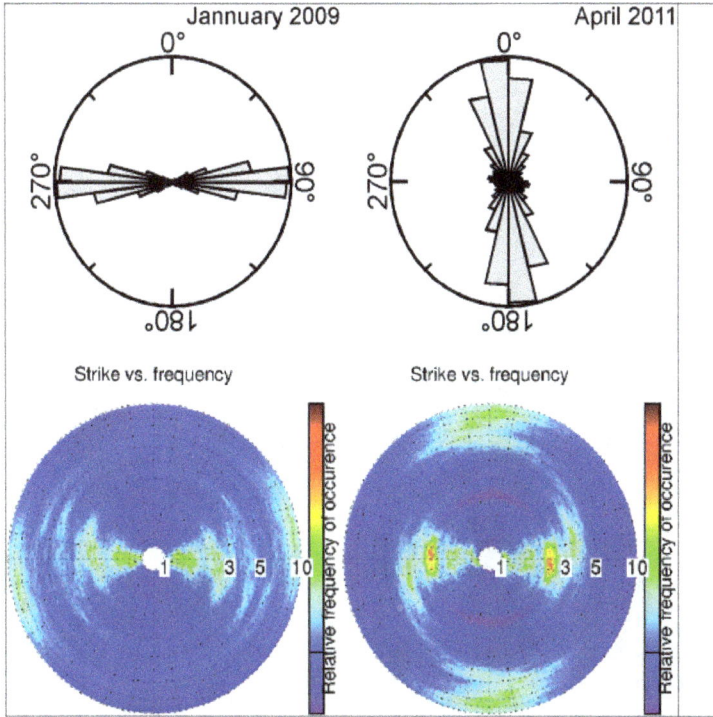

Figure 7. Polarization test on the measurements performed in two different periods (January 2009 and April 2011) in the site #10 of the transect Tr#2 of the Masseria Cavaliere fault.

The other important structure that we investigated is the Piedimonte fault system (PF). The major fault, striking ENE-WSW, is located at the end of the Pernicana fault. It seems to represent an old structural element with a marked morphologic scarp, but no evidence of activity in historical times (Azzaro *et al.*, 2012). The minor faults which spread out from the main structure, striking mostly WNW-ESE, are related with the movement affecting the Pernicana fault. The directional resonance plots (Fig. 8), obtained by rotating the NS and EW components of motion seems to highlight the presence of two different structural behaviors. The results of measurements performed on the footwall of the fault (Fig. 8a) highlight strong directional effects in the frequency range 1.0-6.0 Hz with a NE- SW strike. In the hanging wall (Fig. 8b) the azimuths, the frequency bands and the amplitudes of the HVNRs vary at each measurement point in relation with the small scale geologic framework of each site. Polarization results (Fig. 9) confirm the frequency range and azimuths observed through the rotated spectral ratios. Indeed, on the footwall of the major Piedimonte fault the recorded ambient noise is polarized in a narrow frequency band (1.0-6.0 Hz), similarly to the faults in the western area of the volcano, but with an angle of about 45°. In the hanging wall a rather scattered distribution of polarizations is observed (e.g. #13, #15, #16, #23).

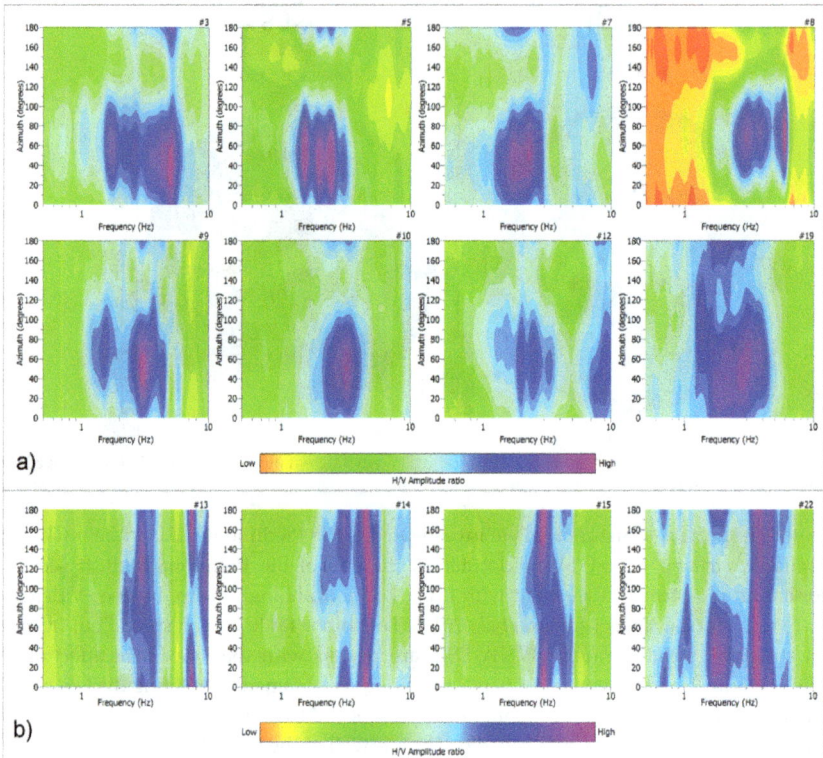

Figure 8. Examples of the contours of the geometric mean of the spectral ratios as a function of frequency (x axis) and direction of motion (y axis) obtained at selected ambient noise recording sites located on the footwall a) and hanging wall b) of the Piedimonte fault.

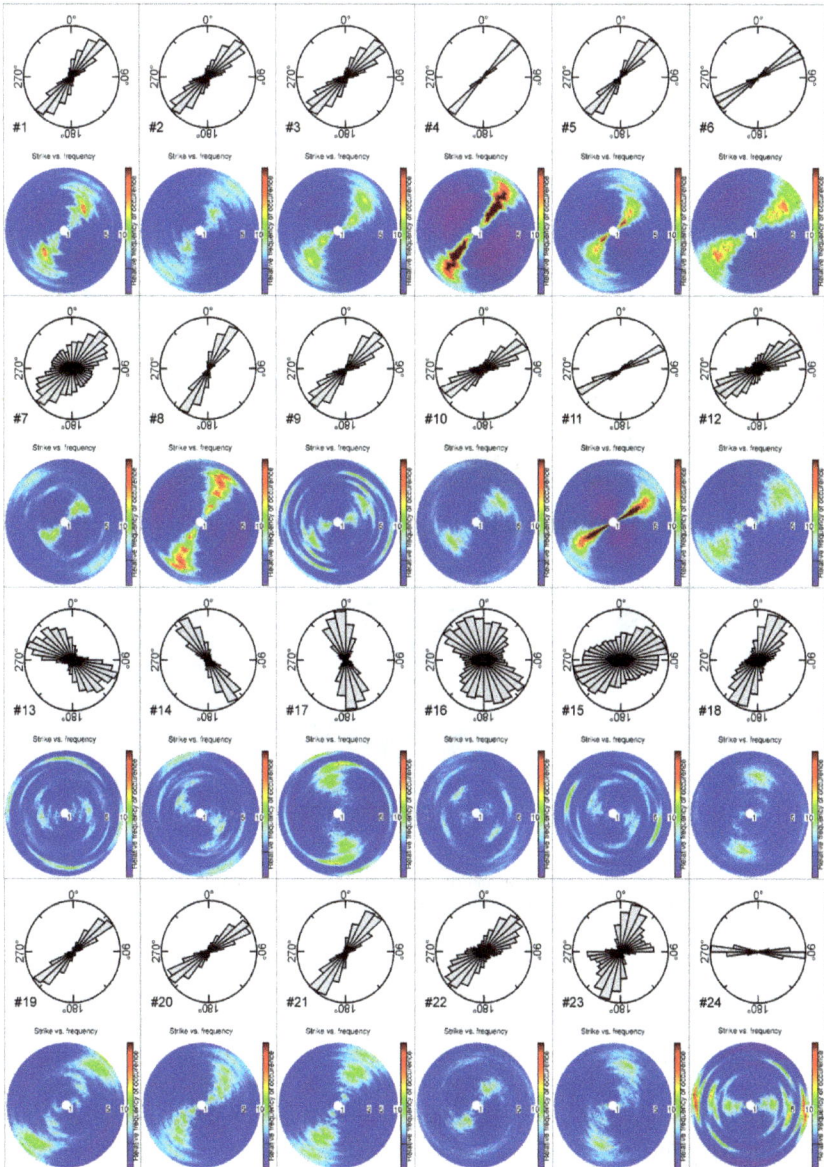

Figure 9. Horizontal polarization angles computed through the TD (rose diagrams) and TF analysis considering the frequency range of 1.0-10.0 Hz (polar plots), for the Piedimonte fault.

Ambient noise recordings were also performed in some fault segments belonging to the Great Fault system of the Malta area, in order to investigate site effect features of tectonic structures located in a non volcanic area. Results obtained (Fig. 10) indicate the existence of directional effects similar to those found in the investigated Etnean structures, although a slightly more complicated pattern, probably linked to the influence of the complex lithology existing at both sides of the fault, is observed. However, the polarization plots obtained from records near the fault show a prevailing direction that is not parallel to the structure strike.

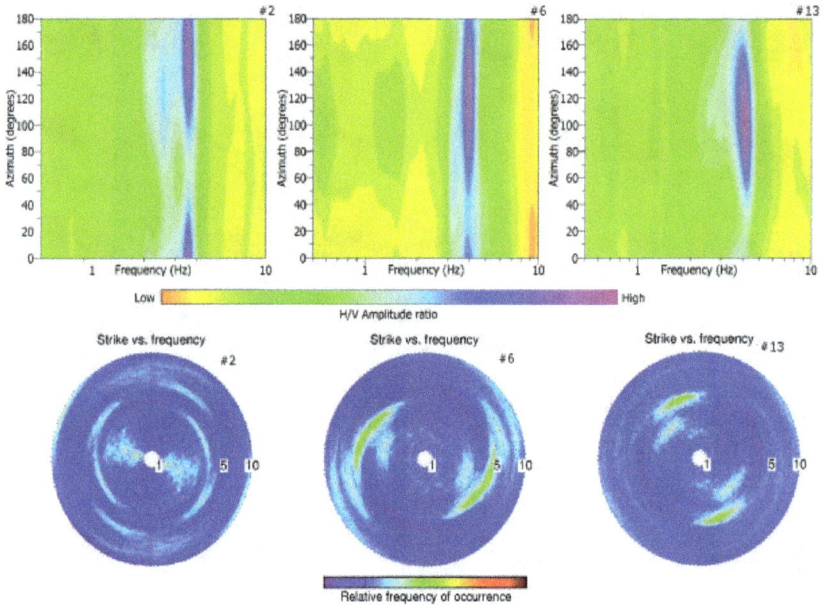

Figure 10. Examples of directional resonances and horizontal polarization azimuths obtained at selected ambient noise recording sites located on the Malta Great Fault system.

It is not an easy task to interpret all experimental observations when these exhibit prevailing polarization directions that are sharply changing and are always non-coincident with the fault strike. In a recent paper, Pischiutta *et al.* (2012), find that the mean polarization azimuth turns out to be perpendicular to the expected synthetic cleavages. We have to remember that four types of fractures can develop in fault zone (Riedel, 1929; Tchalenko, 1968): a) extensional fracture (T); b) synthetic cleavage (R); c) antithetic cleavage (R'); d) pressure solution surfaces (P). The orientation of these fractures depends on the direction of the resulting stress localized around the fault. Therefore, an interpretation of the present results could, in our opinion, be proposed by focusing our findings in the frame of the brittle rheology of the hosting rock (Pischiutta *et al.*, 2012).

Finally, tests were performed to check if the observed directional effects keep the same orientation in areas significantly distant from fault lines. For this reason, ambient noise was recorded in the PG (Piano dei Grill) area, located in the western flank of Mt. Etna, as well as in an area in the northern Malta island. The results (Fig. 11) show that ground motion directions coming from both rotated spectral ratios and polarization diagrams tend to become randomly distributed and/or uniformly scattered. Such findings further support the observations of Rigano and Lombardo (2005) that performed ambient noise measurements in proximity and at few kilometers distance from the Mt. Tauro fault, located in south-eastern Sicily (Fig. 12).

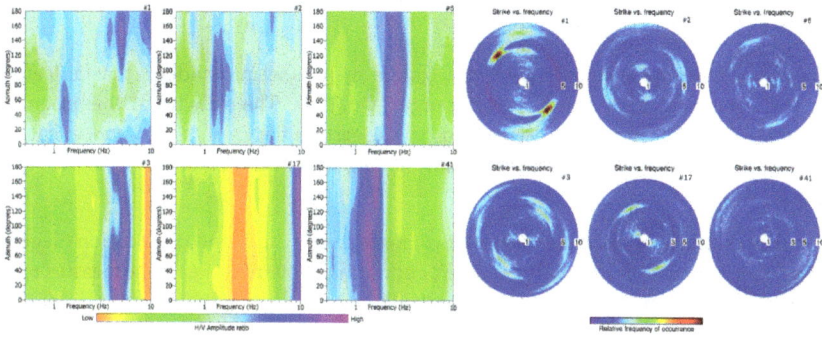

Figure 11. Examples of directional resonances and horizontal polarization azimuths obtained at selected ambient noise recording sites located on the PG area (#1, #2, #5) and in the northern Malta area (#3, #17, #41).

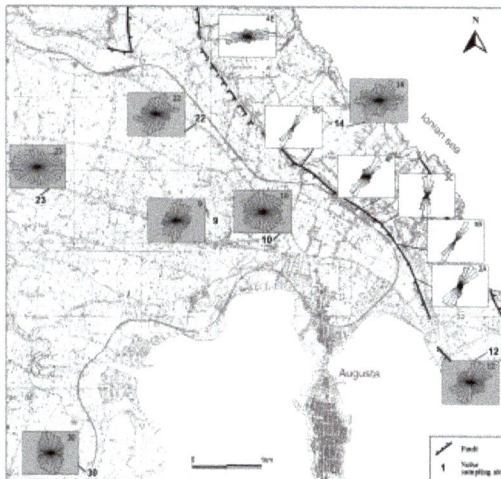

Figure 12. Ambient noise polarization plots in the Augusta area (modified from Rigano and Lombardo, 2005).

The outcomes of the present research allowed us to draw the following considerations:

- In the neighborhood of fault areas, the presence of a damage zone implies the existence of ground motion amplifications and persistent directional effects of the horizontal component of motion, set into evidence by both earthquake and ambient noise records, that are observed till several hundred meters distance from the fault line;

- The directional site effects and the polarization angles observed for all the investigated structures are always non-parallel to the fault strike making a simple explanation in terms of fault-trapped waves not convincing. To attempt a possible explanation for this recurrent ground motion property we postulate the existence of a tight relationship with the expected synthetic cleavages.

- The directional resonance and the TF polarization analysis set into evidence that fault effects appear concentrated in the frequency range 1.0-6.0 Hz. The stability of this frequency interval, observed both in the western and in the eastern flanks of the volcano, encourage us to affirm that it is a possible marker for observing site effects in fault zones, at least in the Etnean area.

- Polarization directions coming from both rotated spectral ratios and polarization diagrams tend to become randomly distributed and/or uniformly scattered when noise measurements are performed in areas where no fault s are evident.

Finally, it seems important to point out that present results give further support to findings from previous studies (e.g. Rigano *et al.*, 2008; Di Giulio *et al.*, 2009; Pischiutta *et al.*, 2012) concerning site effects in fault zones and promote the use of ambient noise recordings as a fast technique for preliminary investigations about angular relations between fractures field and directions of amplified ground motion and for preliminary quick surveys in area where the urbanization or the presence of shallow sedimentary deposits hide the evidence of tectonic structures.

5. Site effects linked to the presence of cavities

The presence of either natural or artificial cavities in the shallower part of various lithotypes is an important aspect whose effects, in terms of the local seismic response evaluation, are still not fully investigated. Grottos can originate from different processes and affect rocky lithotypes that at the surface appear very stiff and characterized by good elastic properties. It is for instance possible to observe the development of cavities, several meters wide and hundreds of meters long, inside basaltic lava flows, that are related to the cooling of the shallower part of lavas while the still fluid portion flows underneath. Also typical is the presence of extensive cavities in calcareous formations, due to the development of karstic phenomena. Local seismic effects related to such conditions therefore need to be investigatedsince, although the lithology often belongs to the bedrock type, they cannot be considered free from significant modifications of both amplitude and frequency content of the seismic input.

The scientific literature concerning these phenomena is rather poor. Studies were performed by Nunziata *et al.* (1999) in some cavities existing inside the pyroclastic terrains of the Napoli downtown area where the authors observed, through numerical modeling, an amplitude decrease of the ground motion at the top of the investigated cavities. Experimental studies, using both earthquake and ambient noise records were recently performed in south-eastern Sicily by Lombardo and Rigano (2010) and Sgarlato *et al.* (2011) and their results will now be briefly summarized.

The influence of cavities in the evaluation of the local seismic response was studied in some selected sites located in the Hyblean region (in the cities of Lentini, Melilli, Siracusa and Modica) and the urban area of Catania, taking into account both natural and artificial cavities such as railway tunnels. In total, the measurements were carried out in about fifteen cavities. They were selected according to criteria of relatively easy access, different geometric features and possibility of having detailed underground surveys. The majority of the investigated grottoes develops in heavily urbanised areas and in some cases, houses and small edifices are built over, or neighbouring them. About 400 time histories of microtremors were recorded in 90 measurements sites that were located inside and over the vault of each grotto, as well as in its neighbourhood, along short profiles having a few tens of meters length, evaluating the horizontal-to-vertical noise spectral ratio as well as the polarization angle of the horizontal component of motion. Besides, in the Catania area, a cavity (Petralia grotto) was selected to install four seismic stations for recording earthquakes. The grotto is located in the northern part of Catania and roughly trends in E–W direction. It develops at a depth of about 3 m from the topographic surface where its easternmost part is open (Fig. 13). Its cross section has a variable size ranging between 10 and 15 m in width and is about 2.5 m high on average. This cavity is formed by several chambers connected by tight passages and it shows evidence of several collapses, the first of which took place at a few tens of meters from the opening of the cavity. Its origin is connected to the flow, cooling and drainage of a pre-historical Etnean lava that, similarly to other lava flows have covered, till historical times, the Catania urban area terrains. The stations were deployed inside, over the vault, in the vicinity of the grotto (named *incave, upcave* e *outcave*, respectively) and in a reference site (*uni*), about 500 m away, located on the bedrock. Data were processed using the earthquake's horizontal to vertical spectral ratio (HVSR) or receiver function technique, and the standard spectral ratio (SSR) to a reference site. A set of 34 seismic events, showing a good signal-to-noise ratio were recorded for five months.

5.1. Results and discussion

Examples of the HVNR results obtained in the various investigated cavities are reported in Fig. 14a. As the plots summarize, it is not possible to observe a unique behavior. The results show the lack of H/V spectral ratio peaks inside some cavities, as shown in the examples #6, and #1* where spectral ratio peaks do not reach the amplitude of three units. On the other hand, H/V obtained from measurements performed in some other cavities (#1, #3, #2* and #3*) tend to reach a significant amplitude (>2 units) in some frequency bands. Such behavior appears related to the size of each cavity. It is in fact observed that a tendency towards H/V significant peaks is evident in cavities whose height is not less than 3-4 metres. It is also

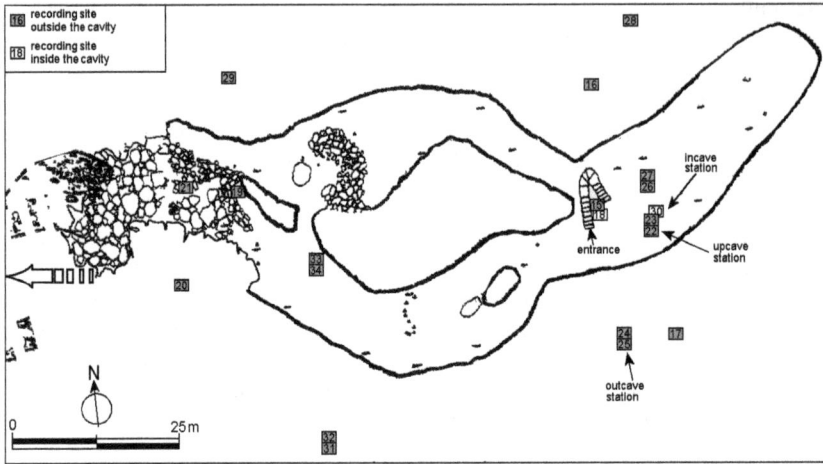

Figure 13. Sketch map of the eastern end of the Petralia grotto and location of permanent and mobile stations; white and grey squares refer to ambient noise recording sites located respectively inside or over and outside the grotto.

interesting to note that in some cases considerable effects are observed in H/V spectral ratios from measurements performed over the vault of the cavity (i.e. #3, #3*), while in other cases (#1 and #2*) pronounced peaks are observed in measurements performed both inside and over the grotto.

Some interesting considerations can be inferred from investigating possible directional effects. The plots in Fig. 14b show that the peaks centered at 2.5 Hz and 1.5 Hz, for cavities #1 and #2* respectively, as well as the peak in the range 4.0 – 6.0 Hz, observed for the cavities #6 and #3*, are markedly directional. The peak values increase up to 3 - 4 units, at directions of 90° and 180° that are nearly coincident with the strikes of the investigated cavities (see the strikes reported in the panels of Fig. 14a).

No.	Name	Locality	H (m)	No.	Name	Locality	H (m)
1	Della Chiesa	Catania	5	1*	De Cristoforis	Lentini	2
2	Micio Conti	Catania	3	2*	C.le Palma	Lentini	5
3	Di Bella	Catania	8	3*	Speri	Lentini	6
4	Caflish	Catania	2	1	Ipogeo	Siracusa	8
5	Ciancio	Catania	4	1	Mastro Pietro	Melilli	8
6	Petralia	Catania	2	1	Barriera	Melilli	10
7	Novalucello	Catania	1	1	Cava	Modica	6
8	Magna	Catania	2.5				

Table 1. List of investigated cavities.

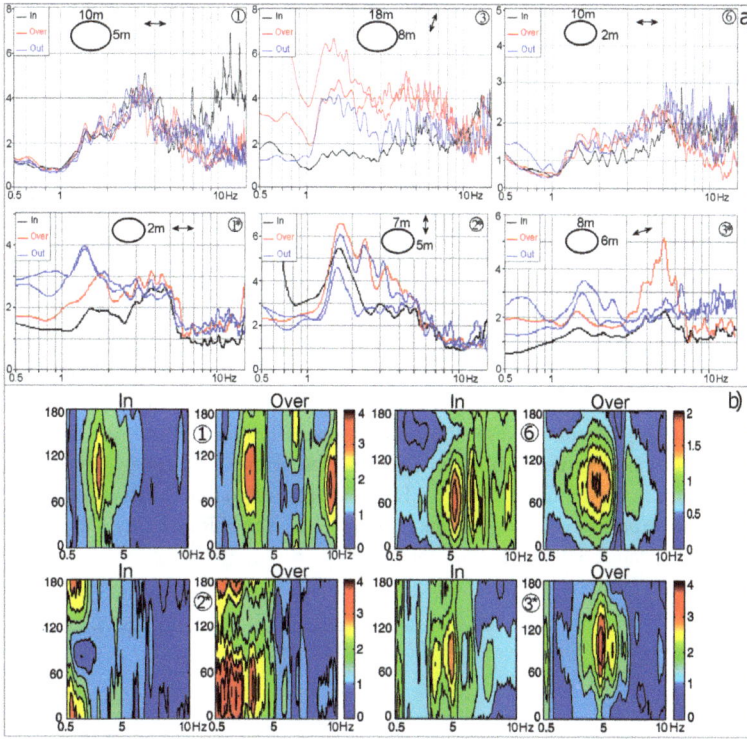

Figure 14. a) H/V spectral ratios obtained from ambient noise measurements performed in different grottoes (numbers refer to the grottoes listed in Tab. 1); the ellipse represents the dimensions of the vertical section of each cavity while the arrows indicate in which direction the grotto develops underground. (b) Contours of the geometric mean of ambient noise spectral ratios as a function of frequency (x-axis) and direction of motion (y-axis) for recordings performed at *Della Chiesa* (# 1), *Petralia* (# 6), *C.le Palma* (# 2*) and *Speri* (# 3*) grottoes.

To validate the reliability of these ambient noise measurements, a comparison was made with findings from the HVSR and SSR of earthquake data recorded in a test site (Petralia grotto). All spectral ratios (Fig. 15a, b) obtained from records at incave, upcave and outcave stations, show moderate peaks that reach at most a value of 3 units. The HVSR show peaks in two frequency ranges, namely 1.2–1.8 Hz and 3.0–7.0 Hz. It has to be noted that in complex situations the identification of main resonance frequencies through HVSR analysis can be biased by the presence of deamplification phenomena in the vertical component of the ground motion. For this reason the ratios between the vertical component spectra of records at the local permanent stations and at the reference one were calculated (Fig. 15c) in order to highlight the frequency band at which the results can be considered reliable. In Fig. 15c no evident deamplification phenomena are observed, but the vertical component at outcave station, especially in the frequency range 3.0 – 7.0 Hz, shows a slight tendency to deamplification which could explain the spectral peaks observed, in the same frequency range, in the HVSR. The comparison of HVSR obtained at the three sites shows a tendency toward slightly more pronounced peaks at outcave station with respect to stations located both over and inside the cavity, the last one, in particular, showing always smaller spectral peaks for the EW component in the frequency band 3.0–7.0 Hz (Fig. 15a). A similar behavior is also observed in the SSR (Fig. 15b) where the spectral ratios obtained for the incave station show, especially in the same frequency range, smaller amplifications in both EW and NS components. Such a tendency is also shown in the results of ambient noise measurements (Fig. 15d and e). Moreover, it is interesting to point out that both HVSR and HVNR show, at upcave station, a striking amplitude decrease, in the frequency band 7.0–10.0 Hz, of EW component of motion (see Fig. 15a, d, e). Such behavior appears related to the amplitude increase of the vertical component of motion at the upcave station, as confirmed by the V/Vref shown in Fig. 15c. This implies that, at highest frequencies, both the HVSR and the HVNR show higher values at incave station rather than at upcave station. On the other hand, being evident that in the above mentioned frequency band, at upcave station the EW component of motion has a low amplitude, the SSR shows also a striking decrement (Fig. 15b) and the spectral ratios at incave are lower than the same spectral ratios obtained at upcave. This last observation is a consequence of the low amplitude values at the denominator of the SSR, with respect to that of HVSR, as expected for the horizontal component of motion at the reference site. The SSR however shows amplification mostly in the range 3.0–7.0 Hz (Fig. 15b).

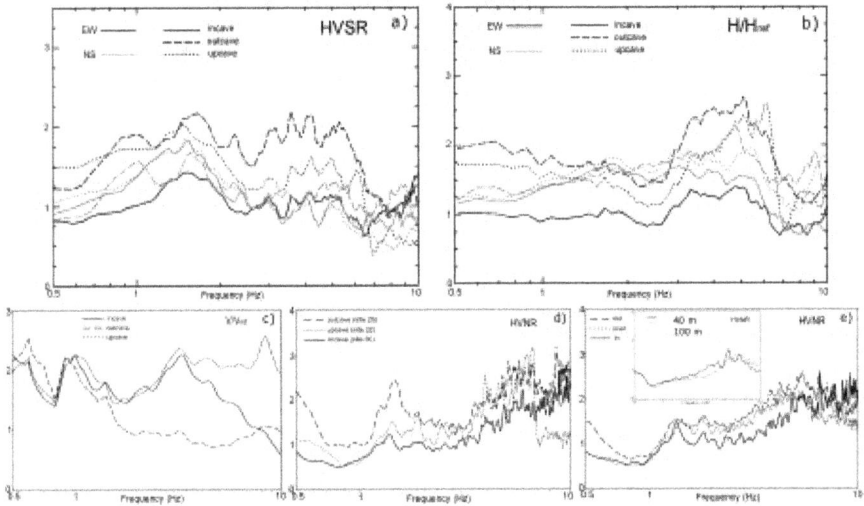

Figure 15. Spectral ratios HVSR (a) and standard spectral ratios SSR (b) of all events recorded at Petralia grotto; V/ Vref) of the vertical component of seismic events recorded at the local permanent stations and at the reference one (c); HVNR recorded at the sites of permanent stations (d) and average of all measurements performed in different sites located inside, over and outside the investigated grotto (e), the location of all recording sites is shown in Fig. 14. The inset shows the HVNR from ambient noise recorded in two sites located at distance of about 40m (grey curve) and 100 m (black curve) from the grotto.

Other noise measurements were performed to the south of the cavity at distances of about 40 m and 100 m from the cavity entrance in order to test how the spectral features previously described appear at increasing distance from the grotto. The obtained HVNR (see inset in Fig. 15e) shows that the 1.2-1.6 Hz spectral ratio peaks disappear already at a distance of about 40 m from the cavity, therefore indicating that they seem to belong mostly to specific features of the grotto area. On the other hand, the peaks at 3.0 – 7.0 Hz, although less pronounced, are still observed, implying that they are linked, at least in part, also to structures extending in a wider area around the grotto.

A polarization analysis was performed to investigate directivity effects related to the cavity. The azimuthal directions of the horizontal component of motion were obtained after filtering the signal in three frequency bands (1.0 – 3.0, 3.0 – 7.0 and 7.0 – 10.0 Hz) aiming to investigate the frequency ranges observed in the H/V spectral ratios. It can be clearly observed (Fig. 16) that almost all rose diagrams show a sharp polarization in the EW direction. Only polarizations obtained at the incave station by filtering the signal in the range 1–3 Hz appear highly scattered. Polarization effects were also investigated, using ambient noise records, in the sites located at 40 and 100 m away from Petralia grotto (see bottom panels in Fig. 16). The results obtained show that such polarization effects become less evident as the distance from the cavity increases.

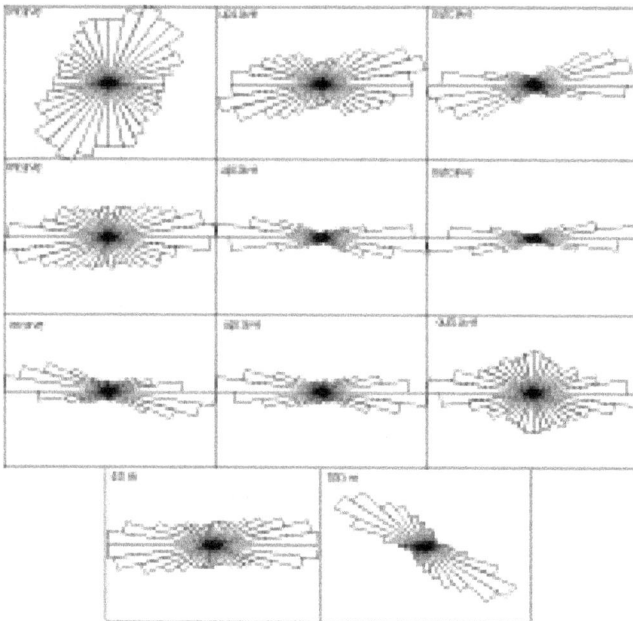

Figure 16. Polarization of the horizontal component of motion obtained by filtering the recorded seismic events at 1.0–3.0 Hz (upper panels), at 3.0–7.0 Hz (middle panels) and 7.0–10.0 Hz (lower panels); the bottom panels refer to polarization from ambient noise measurements performed at sites located 40 and 100 m away from the grotto.

In order to compare the results from recordings in a natural cavity with those from records performed in a cavity having a simpler geometry, ambient noise was also recorded inside, over, and in the neighborhood (≈30 m) of two artificial tunnels. Both tunnels are located close to the Catania urban area, dug at about 4 m from the topographic surface and having a length of about 100 m, but different height. One of them (height of about 4 m) is dug in massive lavas and the other (height of about 7 m) is excavated in altered lavas. HVNRs show that in the smaller tunnel (Fig. 17a), dug in massive lavas, spectral peaks are significantly less pronounced than those observed in the tunnel, dug in altered lavas, having a greater height (Fig. 17b). In this tunnel, H/V spectral peaks, in the frequency range 4.0–7.0 Hz, obtained from measurements performed over and inside, attain values of about 8 and 4 units, respectively, therefore confirming the observation, aforementioned for the Petralia grotto, that inside the cavity spectral peaks are less pronounced than those observed when measurement is performed over the cavity. It is noteworthy that the spectral ratios from measurements in both tunnels show peaks that however are more pronounced than those observed from H/V performed in and over the study grotto. This is possibly related to the height that in both tunnels is significantly greater than in the studied cavity, considering also that the lava characteristics (altered lavas) of the Petralia grotto area are comparable to those of the lava where the greater artificial tunnel is dug.

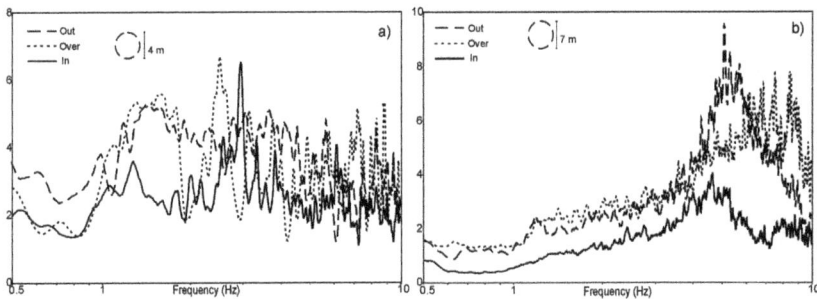

Figure 17. Spectral ratios HVNR from measurements performed in two underground tunnels having different height.

The results so far described are quite complex but nevertheless some interesting considerations can be inferred:

- The size of the vertical section and geometry of the cavities seem to play an important role. According to experimental data, it appears evident that only cavities having height greater than about 4 meters show significant H/V spectral peaks.

- Results of spectral ratios at Petralia grotto (having height of about 2.5 m) show spectral ratio peaks less pronounced than those observed in grottoes and tunnels having height of about 4 and 7 m. Such behavior, in our opinion, need to be tested and interpreted with the support of numerical modeling, to investigate also about possible amplifications/deamplification of the vertical component of motion as highlighted by processing earthquake data.

- Findings from HVSR and SSR at the test site #6 point out the lack, or the modest presence, of amplification effects at the recording site located inside the cavity with respect to stations placed over and outside the grotto. This effect could be explained in the frame of the constructive interferences between direct and reflected waves that can take place at the free surface, so that, as it would be generally expected, a decrement of amplitude oscillations can be observed at depth.

- In all investigated cavities, significant directivity effects were observed, pointing to the existence of a marked polarization of the horizontal component of the ground motion in a direction parallel to the main axis of the grotto. Such evidence is not negligible in the planning of buildings to be erected in the neighboring areas. It is however remarkable that experimental data show that at distances greater than about 100 m from the cavity, the polarization analysis shows azimuths no more coincident with the strike of the grotto.

It is however important to denote that further investigations need to be performed in other cavities having various height, performing also 2D/3D modeling in order to simulate the variations that a wavefront undergoes when propagating through a terrain having a strong impedance contrast due to the existence of a hollow space.

6. Site effects in landslide zones

Landslide phenomena, besides exposing the affected areas to a considerable natural risk, imply the occurrence of significant variations in the local seismic response. To investigate such features, Fekruna Bay, in the area of Xemxija (Fig. 18), was selected. Xemxija is a seaside village and marina on the northeastern part of Malta and it is a very important site for touristic attractions, as well as cultural and historical heritage. The study area spans a couple of square kilometres. More than half of it is intensely built, while the remaining area consists of meadows and agricultural land. The area is characterized by a geology and topography that varies over small spatial scales. Its geomorphologic features are the result of the combined effect of the lithology, tectonics and coastal nature that shaped the region, and such features contribute towards the degree of geological instability of the whole area and particularly to the cliff sections.

The outcropping local geology in the Fekruna bay (Pedley *et al.*, 2002) is characterised by the Upper Coralline Limestone (UCL) and the Blue Clay (BC) formations. Underneath the BC, a carbonatic formation, the Globigerina Limestone formation (GL) consisting mainly of loosely aggregated planktonic foraminifers, and a Lower Coralline Limestone formation (LCL), which consists of massive biogenic limestone beds, are present. This geolithologic sequence gives rise, in the north coast of Malta, to lateral spreading phenomena which take place within the brittle and heavily jointed and faulted UCL formation overlying the BC which consists of softer and unconsolidated material (Mantovani *et al.*, 2012). The UCL formation is characterized by a prominent plateau scarp face, whereas BC produces slopes extending from the base of the UCL scarp face to sea level. It is well known that lateral spreading usually takes place in the

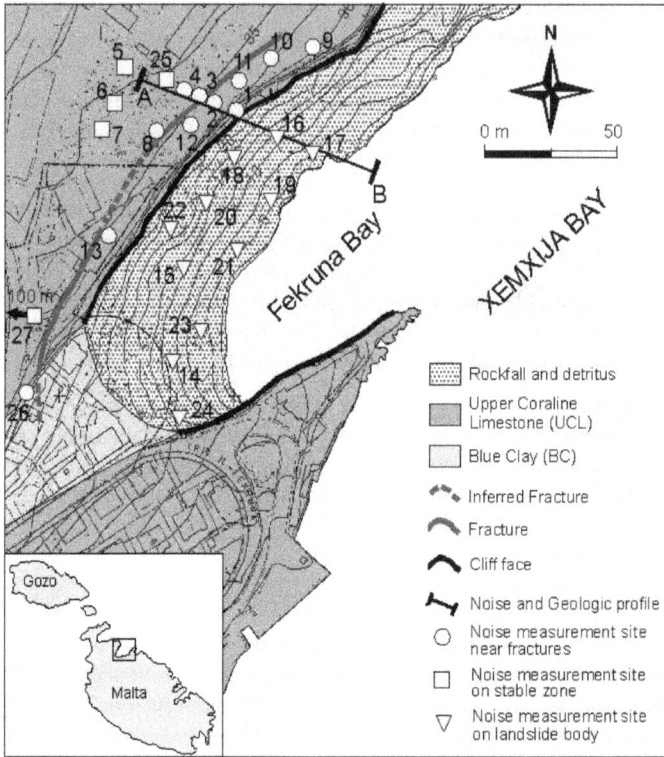

Figure 18. Geo-lithologic map of the north-eastern part of the Xemxija bay (modified from MEPA Various Authors, 2004). The insets show the location of the study area and a sketch of lateral spreading effects along the Xemxija coast.

lateral extension of cohesive rock masses lying over a deforming mass of softer material where the controlling basal shear surface is often not well defined.

A preliminary study of the area, focusing our attention on the risk of landsliding and rockfalls, was carried out using the ambient noise HVNR technique in order to characterize the rock masses behavior in the presence of fractures linked to the landslide body. This type of measurements can be done quickly and with a high spatial density, providing a fast tool for setting the dynamic behavior of the rock outcropping. Generally, this spectral ratio exhibits a peak, corresponding to the fundamental frequency of the site whereas, it is less reliable as regards its amplitude. Nevertheless, the HVNR curve contains valuable information about the underlying structure, especially as concerns the relationship between V_S of the sediments and their thickness (Ibs-Vonseth and Wholenberg 1999; Scherbaum *et al.* 2003). Recordings of ambient noise and the use of the HVNR technique has recently had widespread use in studying landslides (*e.g.* Del Gaudio *et al.*, 2008; Burjánek *et al.*, 2010; Del Gaudio and Wasowski, 2011; Burjánek *et al.*, 2012).

We recorded ambient noise at 27 sites using a 3-component seismometer (Tromino, www.tromino.eu) and directional effects were investigated by rotating the NS and EW components of motion by steps of 10 degrees starting from 0° (north) to 180° (south). Moreover, a direct estimate of the polarization angle by using the covariance matrix method (Jurkevics, 1988) and the time-frequency (TF) polarization method (Burjánek et al., 2010; 2012) was achieved.

6.1. Results and discussion

A dense microtremor measurement survey was carried out focusing attention on the NE part of the bay, in which there is major evidence of slope instability and in which a high level of cliff fracturing is evident. Recording sites were located in order to sample the area as uniformly as possible. Moreover, several recording sites where chosen on a linear deployment for investigating the role of the fractures in the HVNR behaviour. The measurements were performed in three different zones: indeed, we carried out one set of recordings in a stable platform and far away from the cliff edge, a set of measurements in the fractured area along the cliff, and another set on the landslide body (see Fig. 18 for measurement points location). Examples of the spectral ratios obtained in the three different zones are shown in Figure 19a. The results allowed us to identify a region, away from the cliff edge, where the HVNR peaks are around a stable frequency of about 1.5 Hz. These fundamental peaks may be generally associated with the interface separating the BC layer from the underlying carbonates of the Globigerina Limestone formation (GL). Moreover, the presence of the BC layer gives rise to a velocity inversion since it has a lower shear wave velocity with respect to the overlying UCL formation. This causes the HVNR values to drop below 1 unit over a wide frequency range (Di Giacomo et al., 2005). The origin of the resonance peak as linked to the BC/GL interface was confirmed by the results obtained through a 1D modelling, performed by computing the synthetic HVNR curves (Fig. 19b). To compute the synthetic spectral ratios we considered that ambient vibrations wavefield can be represented by the superimposition of random multi modal plane waves moving in all the directions at the surface of a flat 1-D layered visco-elastic solid, as in Herrmann (2002) formulation, extending the modal summation up to the fifth mode. We also applied initial constraints on the thicknesses and elastic parameters of the layers using borehole logs data and we took shear waves velocity values from a separate preliminary study carried out in the same area using the ReMi, MASW and Refraction methods (Panzera et al., 2011a). Similar behaviour to the above is observed in the spectral ratio obtained at sites close to the cliff edge and all around the identified fractures, but with slightly different features at the high-frequency interval. We observe indeed a clear and predominant peak at around 1.5 Hz, which is associated with the interface between BC and GL, and several smaller peaks at higher frequency (> 9.0 Hz). The fact that these peaks are not visible in the unfractured region leads us to postulate that they may be associated with the presence of fractures and of blocks almost detached from the cliff and therefore free to oscillate. Finally, the sites on the rock-fall area show a different HVNR behaviour with respect to the measurements taken on the plateau. In this area it is possible to identify HVNRs showing bimodal dominant peaks at low frequency, in the range 1.0-3.0 Hz, as well as pronounced peaks at about 3.0 Hz and at frequency higher than 9.0 Hz. The bimodal peaks at low frequency (1.0-3.0 Hz) can, in our opinion, be associated with the contact between the rockfall and detritus unit and the BC formation, as well as to the interface between BC and the underlying GL formation.

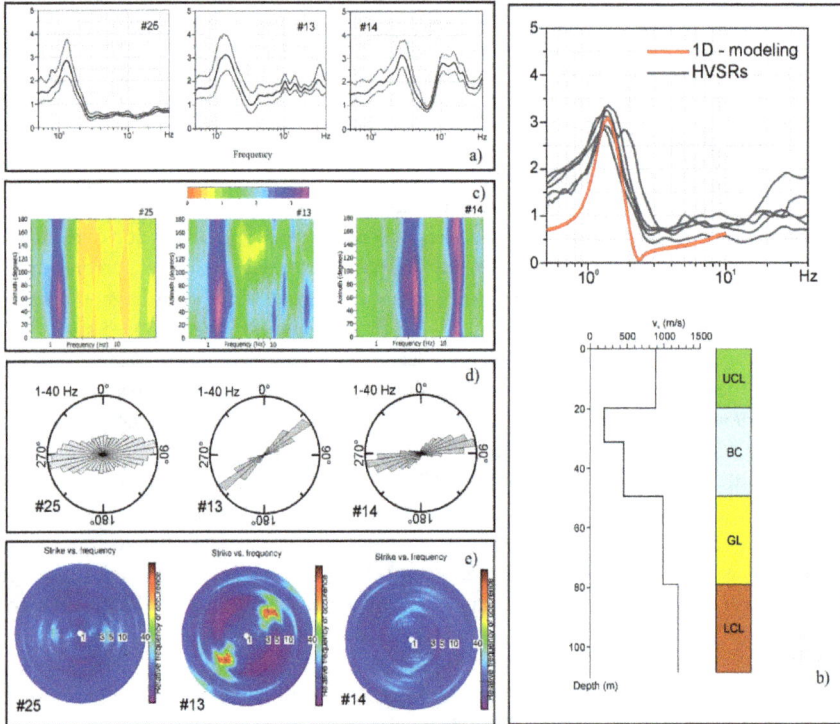

Figure 19. a) Example of HVNR results obtained at some recording sites located in the not fractured zone (#25), in the cliff area (#13), and in the landslide zone (#14); (b) results obtained through a 1D modelling, performed in sites located on the non-fractured zone, by computing the synthetic HVNR curve; (c) rotated HVNR, (d) rose diagrams and (e) polar plots obtained at recording sites #25, #13 and #14.

Inspection of directional effects (see examples in Fig. 19c) show that they are clearly evident in the cliff fracture zones,at an angle of about 40°-60° N in all the considered frequency range, although some variability in azimuth is observed at high frequency (>9 Hz) at site #13. On moving away from the cliff edge, the rotated HVNR show a slight change of the directional resonance angle and an amplitude decrease of the rotated spectral ratios at high frequency. Such a behavior could be linked to the increase of rock stiffness and a reduction of the blocks' freedom to oscillate. Finally, it is evident that the directionality pattern observed in the rotated HVSRs performed on the landslide body is quite complex. The general trend has a prevailing direction of about 40°-60° N at low frequency (1.0-9.0 Hz), similarly to what is observed in the fractured zone, whereas different resonant frequencies and directions that could be ascribed to the vibration of smaller blocks can be observed at higher frequencies (9.0-40.0 Hz). Furthermore, we obtained a direct estimate of the polarization angle through the full use of the three-component vector of the noise wave-field. General behavior of the noise wave-field, in this frequency range is shown through rose diagrams, whereas, in order to distinguish between properties of low and high frequency components of the signal, strike versus frequency polar plots were obtained. The examples shown in Figure 19d and e show that the maxima of the horizontal polarization occur in the north-east to east-north- east direction, although in some cases the high frequency directionality is more complex. As observed by Burjánek et al. (2010), high-frequency ground motion can indeed reflect the vibration of smaller blocks that imply both different resonant frequencies and directions. The polarization observed for the sites located away from the unstable areas (see e.g. #25 in Figure 19) show a trend with more dispersed and variable directions. The boundaries of the landslide area therefore appear well defined by the polarization pattern and as postulated by Kolesnikov et al. (2003) the landslide activity is characterized by strong horizontal polarization in a broad frequency band. In our study, a tendency seems evident for the entire landslide body to generally vibrate with a north-east azimuth and it may be assumed that, during a strong earthquake the ground motion would be amplified in this direction. Studies of Burjánek et al. (2010; 2012) point out that the ambient noise polarization take place at about 90 degree angle to the observed fractures which are perpendicular to the sliding direction. In the present study the polarization angle is parallel to the opening cracks, which appears in contrast to the above mentioned results. A possible explanation of our findings is that there exists a prevailing north-easterly sliding direction of the landslide body which is strongly affecting the polarization direction especially in the 1-10 Hz frequency range.

In Figure 20 we summarize the above results into a tentative draft profile, located as shown in Fig. 18, which illustrates the main geological features and hypothesizes the shape of the landslide body. The bottom panel shows a 2-D diagram obtained by combining all the ambient noise measurements along the profile. Moving along the profile from measurement point #5 to #17, it is interesting to observe the increasing amplitude of the HVNR at frequencies greater than 6.0-7.0 Hz. It can also be noticed that, especially between 50 and 60 m along the profile (sites #1, #2 and #3) the influence of the fracture zone is evident (see dashed area in Fig. 20). This is associated with the vibrational mode of the almost detached blocks. Along the cross section, at distances ranging from 60 to 100 m, it is possible to note both the presence of the bimodal peak associated with the two interfaces detritus/BC and BC/GL, as well as the high frequency peaks most probably associated with the vibration of large blocks that have been detached from the cliff-face and are now partially or totally included in the BC.

As final considerations, it is noteworthy to observe that this study highlights the importance of evaluating the local seismic response in presence of slope instabilities related to landslide hazard. The instability processes that could be potentially triggered are linked to both slow mass movements, which might normally occur in tens or hundreds of years, and to sudden rockfall in the case of ground shaking due to moderate-to-strong earthquake activity. It is worth noting that:

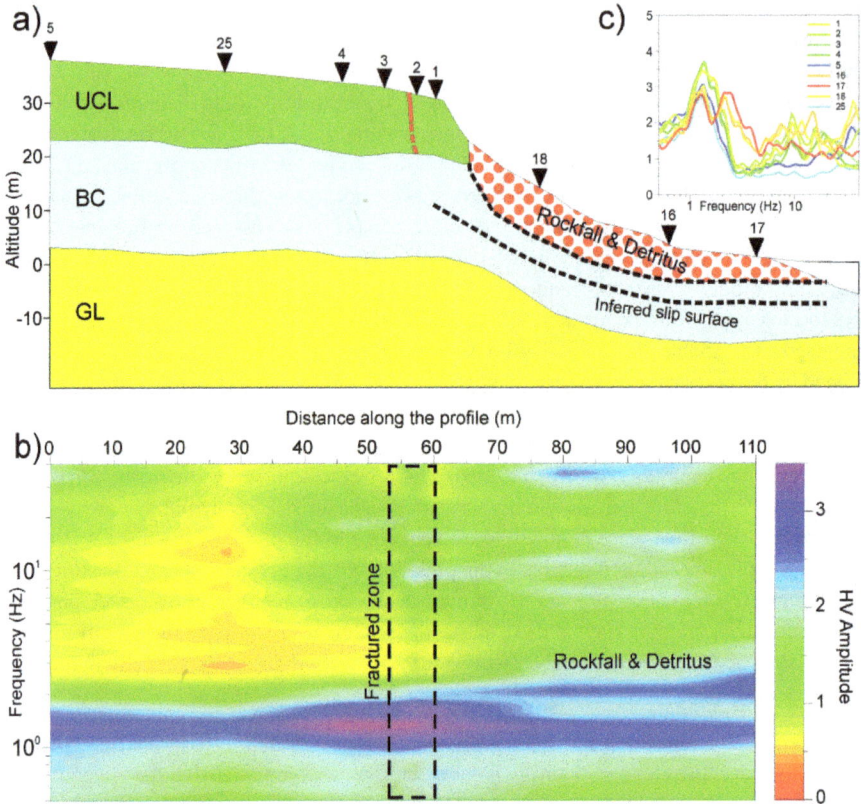

Figure 20. a) Cross section along the A-B profile in Fig. 27; b) 2-D diagram obtained combining all the ambient noise measurements along the profile as a function of distance (x axis) and frequency (y axis); c) HVSR results at the recording sites located across the profile.

- The use of noise measurements indicates that, at least in case of rockfall landslides, the existence of different zones characterized by differences in the dominant spectral ratio peaks that can be related to the presence of shallow lithotypes as well as to the existence of the fractures in the rock and active slip surface that allows the slow sliding of the upper landslide body.

- The instrumental observations indicate that the seismic ground motion can be considerably amplified and such amplification has a directional character that appear related with topographic, lithologic and structural features as well as normal mode rock slope vibration.

- The results of horizontal-to-vertical spectral ratio measurements indicate that this method could be useful for the recognition of site response directional phenomena.

7. Local site effects linked to the topography

The evaluation of the local seismic response when affected by the presence of topographic irregularities is particularly important, from the engineering point of view, mostly since a number of historical villages in Italy are erected on the top of natural reliefs. The influence of the topography on ground motion is linked to the sharpness of the ridge crest (Géli *et al.*, 1988; Bard and Riepl-Thomas, 1999). Amplification effects are generally linked to the focalization of seismic waves at topmost part of a hill, due to the existence of diffraction, reflection, and conversion of the incident waves (Bard, 1982). They appear also frequency-dependent so that resonance phenomena occur when the wavelength of the incident wave is comparable to the horizontal dimension of the hill. In addition, significant directional effects, transverse to the major axis of the ridge, are often observed (Spudich *et al.*, 1996).

Several analytical and numerical methods have been developed to study incoming seismic waves when crossing a hill shaped morphology (*e.g.*, LeBrun *et al.*, 1999; Paolucci, 2002). Although experimental studies using earthquake instrumental records are relatively few, the use of earthquake data has shown to be a successful tool for the evaluation of topographic effects, as well as artificial explosions and ambient noise records, processed with the HVNR technique (*e.g.*, Borcherdt, 1970; LeBrun *et al.*, 1999; Poppeliers and Pavlis, 2002; Pagliaroli *et al.*, 2007).

The present study was performed in two reliefs having different morphologic and geologic features aiming to discriminate the topographic from the stratigraphic effects, using experimental techniques based on earthquake and ambient noise recordings in order to test at the same time the reliability of ambient noise recordings, processed through HVNR techniques, to estimate topographic effects. The first relief investigated is the area of Ortigia (downtown Siracusa, Sicily). It is a hill shaped peninsula, mostly formed by a carbonate sequence, elongated in the N-S direction, reaching a length of about 1,500 m, with a maximum height of 30 m a.s.l., having a transverse section width of about 700 m (Fig. 21). The second test area is the university campus (S. Sofia hill), a ridge located in the northern part of Catania (Fig. 22). The S. Sofia hill has a gentle topography with a flat surface at the top, it is elongated for about 700 m in NW-SE direction with a maximum height of 40 m. Its longitudinal section (B-B' in Fig. 22) is asymmetric, consequently the northwestern side is quite gentle with respect to the southeastern part. On the other hand, the transverse section is more regular and symmetric (A-A' in Fig. 22), with a base having width of about 500 m. This area has a more complex geology. The most frequently cropping out lithotype is basaltic lava that in pre-historical and historical times flowed onto the valleys originally existing in the sedimentary formations, formed by sand and gravel, laying over a marly clays basement.

In both areas several ambient noise measurements were performed, processing data with spectral ratio techniques and evaluating the directional effects as well. Moreover, in the S. Sofia area three permanent stations located respectively on the top of the hill, along the slope and at a reference site, about three kilometers away from the study area (see Fig. 22a, b). The area was monitored for about two years and a set of 44 local and regional seismic events, having a good signal-to-noise ratio was selected and processed using HVSR and SSR techniques.

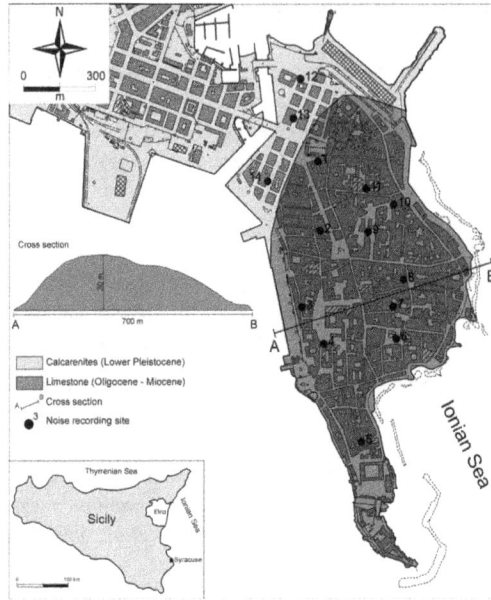

Figure 21. Geolithologic map of Ortigia (downtown Syracuse).

7.1. Results and discussion

Figure 23 shows a direct comparison of the rotated HVNRs, in the frequency band 1.0-10.0 Hz, and the results of noise polarization analysis for the same recording sites, filtering the signal in the range 1.0-3.0 Hz. Both methodologies agree, indicating, particularly in the frequency range 1.0–3.0 Hz, that maxima of HVNR amplitudes take place at 90–100° and maxima of the horizontal polarization strike in the E-W direction. We also compared field data observations with the theoretical resonance frequency (f_0) expected for the topographic effects in Ortigia hill. We adopted the relationship $f_0 = Vs/L$ (Bouchon, 1973; Géli et al., 1988), where L is the width of the hill (about 700 m) and Vs is the shear wave velocity of the limestone outcropping in the peninsula (1,000 m/s). The predicted value, $f_0 = 1.4$ Hz, is consistent with the observed spectral ratio peaks, in the range of 1.0–3.0 Hz. In general, the amplification of ground motion connected to the surface topography is directly related to the sharpness of the topography (Bard, 1994). In such instances topographic effects become clearly detectable with experimental and numerical approaches. In our study, the gentle topography and the homogeneous lithology of the Ortigia peninsula make it an ideal and simple case study for investigating topographic effects using ambient noise records. The Ortigia hill has a natural frequency of about 1.4 Hz and shows an E-W preferential direction of vibration. The specific directional effects in ambient noise, well defined both in space and in a narrow frequency band (1.0–3.0 Hz), are signs of a normal mode of vibration of the hill (Roten et al., 2006).

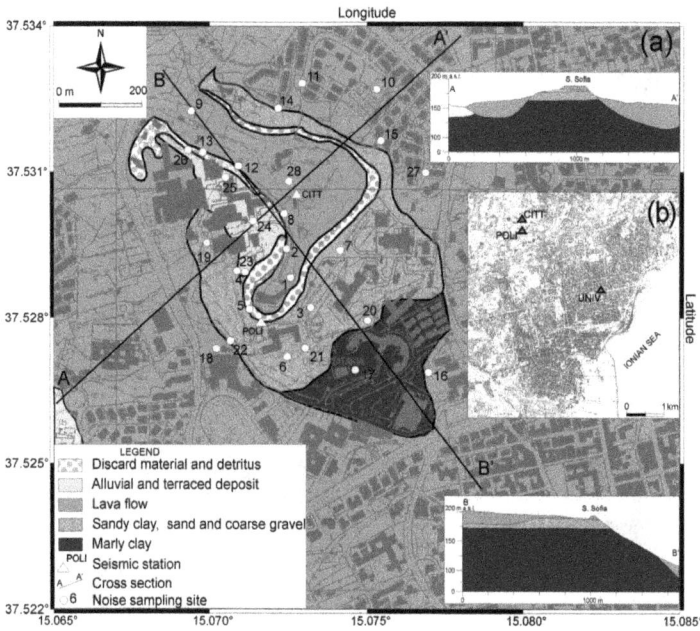

Figure 22. a) Geolithologic map of the S. Sofia hill; the insets show cross sections AA' and BB'. (b) location of the permanent stations with respect to the reference one (UNIV) (modified from Monaco et al., 2000).

The analysis performed in the other study case, (S. Sofia hill) indicate a more composite situation where both the complexity of surficial geology and the morphology significantly affect local amplification and directional effects. The results of SSR and HVSR are reported in Figure 24. Inspection of the Horizontal Standard Spectral ratio (HSSR) (Fig. 24a) point out that at the station CITT less pronounced spectral ratio peaks are observed with respect to the station POLI. The spectral ratio amplitudes obtained through the HVSR method appear underestimated in amplitude with respect to those obtained through the HSSR approach, however, especially at POLI, a good agreement between the two methodologies is observed as regards the frequency range of the dominant peaks (Fig. 24b). Moreover, it is interesting to observe that CITT and POLI stations show flat spectral ratio peaks, with HVSR amplitudes going down to values lower than 1 unit at frequencies higher than about 3.0 and 5.0 Hz, respectively. This appears related to the presence of a significant amplification of the vertical component of motion, as can be observed in the V/Vref plots (Fig. 24c). Such effect could be explained in terms of the complexity of the near-surface morphology and the existence of pronounced stratigraphic heterogeneities that, as postulated by many authors (e.g. Raptakis *et al.*, 2000; Bindi *et al.*, 2009), may affect the vertical component of motion. In any case, greater amplifications are observed in the HSSR at station POLI, in the frequency range 2.0-5.0 Hz and with smaller amplification values at CITT, in the range 1.0-2.0 Hz. This is different from what is usually expected for a topographic ridge, where major spectral ratio amplifications are

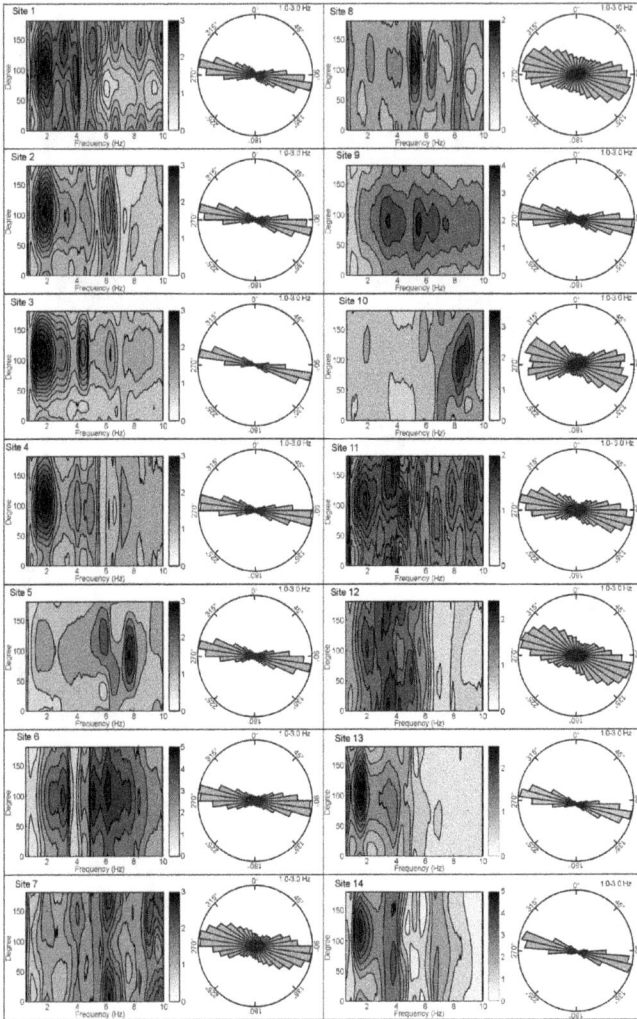

Figure 23. Contours of the geometric mean of the spectral ratios as a function of frequency (x axis) and direction of motion (y axis) and polarization rose diagrams calculated in the ranges 1–3 Hz.

supposed to take place at the top rather than along the slopes of the hill. This behavior, in our opinion, could be related to the gentle slope of the S. Sofia hill. In such conditions the reflection angle between the direction perpendicular to the free surface topography and the upward propagating wavefront is smaller than in the case of a steep slope. Therefore, the focusing effects at the crest are shadowed by laterally propagating waves (Boore, 1973) and it is reasonable to observe only moderate amplifications at the top of the hill. Moreover, it must be remembered that POLI is set on sedimentary terrains whereas CITT is located on a compact lava flow and such lithotype generally does not shows significant spectral ratio peaks as already observed by several authors (Lombardo and Rigano, 2007; Panzera et al, 2011b).

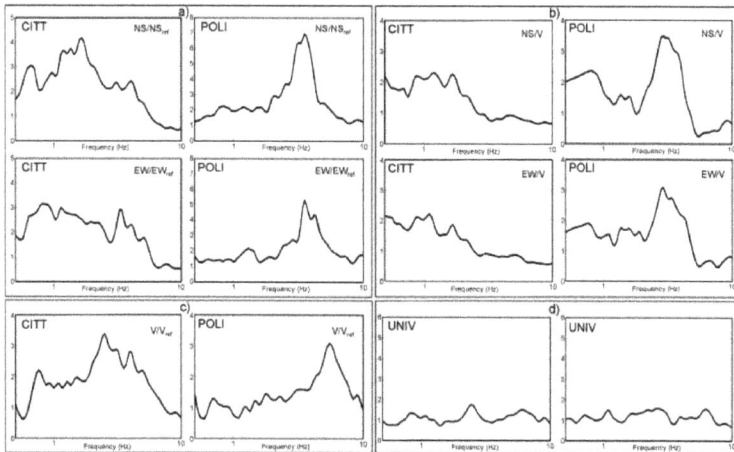

Figure 24. Spectral ratios (HSSR a), HVSR b) and VSSRc)) at permanent stations CITT and POLI; d) HVSR of the NS and EW components of motion at the reference site UNIV.

Ambient noise was recorded, in different lithotypes, along the slopes of the hill at decreasing height from the top (Fig. 22a), to identify possible topographical effects. The HVNRs show flat and, at times, deamplificated spectral ratios in several recording sites (see examples #9, #11, #16, #22 reported in Fig. 25) where the local geology consists in a sequence of thick (10-20 m) massive lava flows that overlie the sand and coarse gravel sediments lying on the marly clay basement. Conversely, when the sedimentary terrains outcrop (see examples #3, #4, #21, #23 reported in Fig. 25), significant spectral ratio peaks, are observed. Results obtained by the HVNR confirm the findings from earthquake data analysis and set into evidence that we are dealing with a geological setting more complex than a simple 1-D layered structure, for which the noise spectral ratio method was originally proposed. The presence of lava flows at the surface imply the existence of possible velocity inversions that give origin to H/V spectral amplitude lower than one unit (Castellaro and Mulargia, 2009; Di Giacomo *et al.*, 2005) and the existence of amplification in the vertical component of the ground motion.

The existence of directional amplification was investigated using both earthquake and ambient noise data. Directional analysis and polarization of the horizontal components of motion show less pronounced directional effect at CITT with respect to POLI station, where clear polarization effects at about 40° appear. The results of polarization analysis are depicted in Figure 26. The hodograms obtained from noise measurements show that the polarization azimuths are similar to those obtained by processing the earthquake data. It appears indeed confirmed that at the top of the studied hill (#9, #12, #14 and #28) the pattern of polarization directions is similar

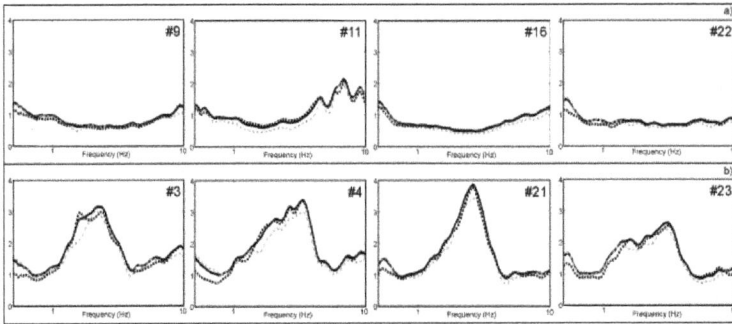

Figure 25. Examples of HVNR results at representative recording sites located on lava flows (a) and on sedimentary terrains (b); solid black lines refer to the average H/V spectra, dotted grey and black lines refers to NS/V and EW/V spectra, respectively.

to what is observed at the station CITT. On the other hand, the noise rose diagrams obtained at the other recording sites point out polarization azimuths that seem to be in agreement with the slope directions of the hill flanks. Only few sites (#17, #21, #22, #25) make an exception to such trend, showing a directional variability that could be linked to the local shallow lithologic features. The investigation on the characteristics of the site response at the S. Sofia hill, therefore set into evidence that the complexity of the near-surface geology, as well as the morphology strongly influence the local amplification of the ground motion and the directional effects. Findings of the present study confirm that major amplification effects do indeed take place on the sedimentary terrains which outcrop along the flanks of the hill. On the contrary, on the lava flows, a significant amplification of the vertical component of motion, is observed as a consequence of velocity inversion effects.

The results coming out from investigations performed in the two test areas allow us to draw the following general considerations about topographic effects:

- Ambient noise is a useful tool that can be widely adopted in site response evaluation and in studies of topographic irregularities as well, especially when a simplified topography and lithology is present. However, in the presence of lateral and vertical heterogeneities, as well as velocity inversions, a combined use of noise, earthquake data and theoretical models is advisable to correctly predict the site response behavior. In such instances topographic effects are more composite and a complex wavefield is observed.

- The wedge angle of the hill appears to play an important role since a wide angle reduces the focusing effects at the crest in favor of laterally propagating waves.

- Both the directional resonance and the polarization analysis confirm the presence of a directional effect transverse to the major axis of the ridge. This behavior is particularly evident in reliefs having a homogeneous simple convex morphology. In such instance the ridge oscillation can be considered similar to what is observed in civil structures, providing there is not significant soil-structure interaction and considering the building as a single-

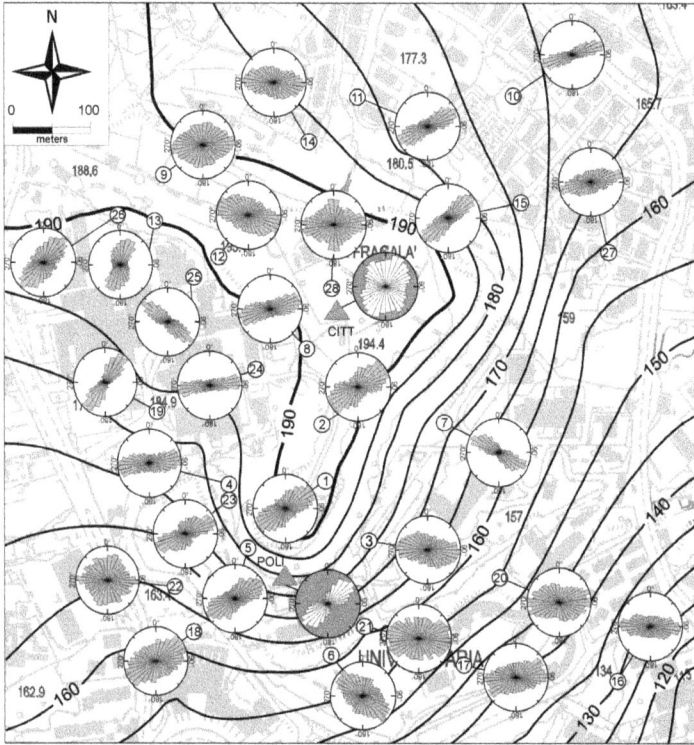

Figure 26. Ground motion polarization from noise measurements in the S. Sofia hill area; hodograms with a grey background refers to results coming from the analysis of earthquakes recorded at CITT and POLI stations.

degree-of-freedom damped oscillator (Gallipoli *et al.*2009). In such instances the vertical component of motion travels through the building without amplification, whereas the horizontal components undergo a significant amplification.

- As a practical implication of the present study it can be observed that the topographic effects cannot be easily evaluated especially when subsurface morphology and lithologic features are predominant.

8. Concluding remarks

The present study has tested the use of ambient noise recordings as a speedy technique for evaluating the local seismic response in several instances where either lithologic and/or morphologic and structural features can significantly affect the response of shallow geologic formations to a seismic input. Our findings further support the reliability of the use of ambient noise recordings for preliminary characterization of dynamic properties of terrains. Its employment has, in recent years, been widely used for site amplification studies since data acquisition time and costs are significantly reduced. Moreover, the HVNR technique can be largely adopted since it requires only one mobile seismic station with no additional measurements at rock sites for comparison. The ratio between the horizontal and vertical spectral components of motion can indeed reveal the fundamental resonance frequency of the site. It also does not require the long and simultaneous deployment of several instruments to collect a useful set of earthquake data.

Our results show also the importance of performing analysis to evaluate directional effects and polarization of the horizontal components of the ground motion. It has indeed to be remembered that directional effects cannot be neglected for a correct planning of edifices and man-made structures in order to reduce the potential risk of building damage as a result of ground motions.

Author details

F. Panzera[1], G. Lombardo[1], S. D'Amico[2*] and P. Galea[2]

*Address all correspondence to: sebdamico@gmail.com

1 Dipartiemnto di Scienze Biologiche, Geologiche ed Ambientali, Universita' di Catania, Italy

2 Department of Physics, University of Malta, Malta

References

[1] Acocella, V, & Neri, M. (2005). Structural features of an active strike-slip fault on the sliding flank of Mt. Etna (Italy). *J. Struct. Geol*, , 27, 343-355.

[2] Azzaro, R. (1999). Earthquake surface faulting at Mount Etna volcano (Sicily) and implications for active tectonics, *J. Geodyn.*, doi:10.1016/S0264-3707(98)00037-4., 28, 193-213.

[3] Azzaro, R, Branca, S, Gwinner, K, & Coltelli, M. (2012). The volcano-tectonic map of Etna volcano, 1:100.000 scale: an integrated approach based on a morphotectonic analysis from high-resolution DEM constrained by geologic, active faulting and seismotectonic data, Ital. J. Geosci., 131(1), 153-170, doi:IJG.2011.29.

[4] Azzaro, R, Mattia, M, & Puglisi, G. (2001). Dynamics of fault creep and kinematics of the eastern segment of the Pernicana fault (Mt. Etna, Sicily) derived from geodetic observations and their tectonic significance. *Tectonophysics*, 333(3-4), 401-415.

[5] Baratta, M. (1910). La catastrofe sismica Calabro-Messinese (28 dicembre 1908), *Società Geografica Italiana, Rom, 458 p.*

[6] Bard, P. Y. (1994). Effects of surface geology on ground motion: recent results and remaining issues. In: *G. Duma (ed), Proc. 10th European Conference on Earthquake Engineering. Wien, 28 Aug.-2 Sept., Balkema, Rotterdam,* , 1, 305-323.

[7] Bard, P. Y. (1999). Microtremor measurements: a tool for site effect estimation? In: *Irikura et al. (ed.), The effects of surface geology on seismic motion. Balkema, Rotterdam,* , 1251-1279.

[8] Bard, P. Y, & Riepl-thomas, J. (1999). Wave propagation in complex geological structures and their effects on strong ground motion. In *Wave motion in Earthquake Engineering*: edited by Kausel and Manolis, WIT Press, , 2, 37-95.

[9] Bardet, J. P, Ichii, K, & Lin, C. H. (2000). EERA, a computer program for Equivalent-linear Earthquake site Response Analyses of layered soil deposits. *University of Southern California, Department of Civil Engineering, user's manual*, 38p.

[10] Ben-zion, Y, Peng, Z, Okaya, D, Seeber, L, Armbruster, J. G, Ozer, N, Michael, A. J, Baris, S, & Aktar, M. (2003). A shallow fault-zone structure illuminated by trapped waves in the Karadere-Duzce branch of the North Anatolian Fault, western Turkey. *Geophys. J. Int.,* , 152, 1-19.

[11] Ben-zion, Y, & Sammis, C. G. (2003). Characterization of fault zones, Pure Appl. Geophys., , 160, 677-715.

[12] Ben-zion, Y, & Sammis, C. G. (2009). Mechanics, Structure and Evolution of Fault Zones, Pure appl. Geophys.,*166*doi:10.1007/s00024-009-0509-y., 1533-1536.

[13] Bindi, D, Parolai, S, & Cara, F. Di Giulio G., Ferretti G., Luzi L., Monachesi G., Pacor F., Rovelli A. ((2009). Site amplifications observed in the Gubbio Basin, Central Italy: Hints for lateral propagation effects. *Bull. Seism. Soc. Am.*, 99(2A), , 741-760.

[14] Boore, D. M. (1973). The effect of simple topography on seismic waves: implications for the accelerations recorded at Pacoima Dam, San Fernando valley, California. *Bull. Seism. Soc. Am.*, 63(5), 1603-1609.

[15] Boore, D. M, Graizer, V. M, Tinsley, J. C, & Shaka, A. F. (2004). A study of possible ground-motion amplification at the Coyote Lake dam, California. *Bull. Seism. Soc. Am.*, doi:l., 94(4), 1327-1342.

[16] Borcherdt, R. D. (1970). Effects of local geology on ground motion near San Francisco Bay. *Bull. Seism. Soc. Am.*, , 60, 29-61.

[17] Bouchon, M. (1973). Effect of topography on surface motion. *Bull. Seism. Soc. Am.*, , 63, 615-632.

[18] Burjánek, J, Gassner-stamm, G, Poggi, V, Moore, J. R, & Fäh, D. (2010). Ambient vi-bration analysis of an unstable mountain slop. *Geophys. J. Int.*, 180(2), 820-828. doi:j.X. 2009.04451.x, 1365-246.

[19] Burjánek, J, Moore, J. R, Molina, F. X. Y, & Fäh, D. (2012). Instrumental evidence of normal mode rock slope vibration. *Geophys. J. Int.*, 188(2), 559-569.

[20] Calderoni, G, & Rovelli, A. Di Giovambattista R. ((2010). Large amplitude variations recorded by an on-fault seismological station during the L'Aquila earthquakes: evi-dence for a complex fault induced site effect. *Geophys. Res. Lett.*, 37, L24305, doi: 10.1029/2010GL045697.

[21] Cara, F. Di Giulio G., and Rovelli A. ((2003). A study on seismic noise variations at Colfiorito, Central Italy: implications for the use of H/V spectral ratios. *Geophysical Research Letters*, 30(18), 1972-1976.

[22] Carbone, S, Branca, S, Lentini, F, Barbano, M. S, & Corsaro, M. A. Di Stefano A., Fer-rara V., Monaco C., Longhitano S., Platania I., Zanini A., De Beni E. e Ferlito C. ((2009). Note Illustrative della Carta Geologica d'Italia alla scala 1:50000, foglio 634 Catania. *ISPRA, Servizio Geologico d'Italia, Organo Cartografico dello Stato, S.EL.CA. s.r.l., Firenze.*

[23] Castellaro, S, & Mulargia, F. (2009). The Effect of Velocity Inversions on H/V, *Pure appl. Geophys.* dois00024-009-0474-5., 166, 567-592.

[24] Catchings, R. D, & Lee, W. H. K. (1996). Shallow velocity structure and Poisson's ra-tio at the Tarzana, California, strong-motion accelerometer site. *Bull. Seism. Soc. Am.*, 86, 1704-1713.

[25] Chavez-garcia, F. J, Sanchez, L. R, & Hatzfeld, D. (1996). Topographic site effects and HVSR. A comparison between observations and theory. *Bull. Seism. Soc. Am.*, , 86, 1559-1573.

[26] Cultrera, G, Rovelli, A, Mele, G, Azzara, R, Caserta, A, & Marra, F. (2003). Azimuth-dependent amplification of weak and strong ground motions within a fault zone (Nocera Umbra, central Italy). *J. Geophys. Res.*, 108 (B3), 2156 doi:JB001929.

[27] Del Gaudio VCoccia S., Wasowski J., Gallipoli M. R., Mucciarelli M. ((2008). Detec-tion of directivity in seismic site response from microtremor spectral analysis. *Natu-ral Hazards and Earthquake System Science*, 8, 751-762.

[28] Del Gaudio VWasowski J. ((2011). Advances and problems in understanding the seis-mic response of potentially unstable slopes. *Engineering geology*, 122, 1-2, 73-83.

[29] Di Giacomo DGallipoli M.R., Mucciarelli M., Parolai S., Richwalski S.M. ((2005). Analysis and modeling of HVSR in the presence of a velocity inversion: the case of Venosa, Italy. *Bull. Seism. Soc. Am.*, 95, 2364-2372.

[30] Di Giulio GCara F., Rovelli A., Lombardo G., Rigano R. ((2009). Evidences for strong directional resonances in intensely deformed zones of the Pernicana fault, Mount Etna, Italy. *J. Geophys. Res.*, 114, B10308, doi:10.1029/2009JB006393.

[31] Di Grazia GFalsaperla S., Langer H. ((2006). Volcanic tremor location during the 2004 Mount Etna lava effusion. *Geophys. Res. Lett.*, 33, L04304, doi:10.1029/2005GL025177.

[32] Gallipoli, M. R, Mucciarelli, M, & Vona, M. (2009). Empirical estimate of fundamental frequencies and damping for Italian buildings. *Earthquake Engineering and Structural Dynamics* doi:eqe.878.

[33] Geli, L, Bard, P. Y, & Jullien, B. (1988). The effect of topography on earthquake ground motion: a review and new results. *Bull. Seism. Soc. Am.*, , 78, 42-63.

[34] Grasso, M, & Lentini, F. (1982). Sedimentary and tectonic evolution of the eastern hyblean plateau (South-Eastern Sicily) during Late Cretaceous to Quaternary time. *Paleoecology* , 39, 261-280.

[35] Herrmann, R. B. (2002). Computer programs in seismology. St. Louis University,, 4

[36] Hudson, M, Idriss, I. M, & Beikae, M. a computer program to evaluate the seismic response of soil structures using finite element procedures and incorporating a compliant base. *Center for Geotechnical Modeling, Department of Civil and Environmental Engineering*, University of California Davis, Davis California.

[37] Ibs-Von Seth MWohlenberg J. ((1999). Microtremor measurements used to map thickness of soft sediments. *Bull. Seism. Soc. Am.*, , 89, 250-259.

[38] Igel, H, Jahnke, G, & Ben-zion, Y. (2002). Numerical simulation of fault zone trapped waves: Accuracy and 3-D effects. *Pure Appl. Geophys.*, , 159, 2067-2083.

[39] Illies, J. H. (1981). Graben formation: the Maltese Islands, a case history, Tectonophysics, , 73, 151-168.

[40] Irikura, K, & Kawanaka, T. (1980). Characteristics of microtremors on ground with discontinuous underground structure. *Bull. Disaster Prev. Inst. Kyoto Univ.*, 30(3), 81-96.

[41] Jurkevics, A. (1988). Polarization analysis of three component array data, *Bull. Seism. Soc. Am.*, , 78, 1725-1743.

[42] Kolesnikov, Y. I, Nemirovich-danchenko, M. M, Goldin, S. V, & Seleznev, V. S. (2003). Slope stability monitoring from microseismic field using polarization methodology. *Natural Hazards and Earth System Sciences*, , 3, 515-521.

[43] Kramer, S. L. (1996). Geotechnical Earthquake Engineering. Prentice Hall, New Jersey.

[44] Lachet, C, & Bard, P. Y. (1994). Numerical and theoretical investigations on the possibilities and limitations of Nakamura's techniques. *Journal of Physics of the Earth* , 302(42), 377-397.

[45] LeBrun BHatzfeld D., Bard P.Y., Bouchon M. ((1999). Experimental study of the ground motion on a large scale topographic hill at Kitherion (Greece). *Journal of Seismology* , 3, 1-15.

[46] Lee, M. K. W, & Finn, W. D. L. (1978). DESDRA-2 Dynamic effective stress response analysis of soil deposits with energy transmitting boundary including assessment of liquefaction potential. *Soil Mechanics Series n. 38,* University of British Colombia, Vancouver.

[47] Lentini, F, Carbone, S, & Guarnieri, P. (2006). Collisional and postcollisional tectonics of the Apenninic-Maghrebian orogen (southern Italy). In: Dilek Y. & Pavlides S. (eds.), «Postcollisional tectonics and magmatism in the Mediterranean region and Asia». GSA Special Paper, , 409, 57-81.

[48] Lermo, J, & Chavez-garcia, F. J. (1993). Site effect evaluation using spectral ratio with only one station. *Bull. Seism. Soc. Am.,* , 83, 1574-1594.

[49] Lewis, M. A, Peng, Z, Ben-zion, Y, & Vernon, F. L. (2005). Shallow seismic trapping structure in the San Jacinto fault zone near Anza, California. *Geophys. J. Int.,* , 162, 867-881.

[50] Li, Y. G, Aki, K, Adams, D, & Hasemi, A. (1994). Seismic guided waves in the fault zone of the Landers, California, earthquake of 1992. *J. Geophys. Res.,* , 99, 11705-11722.

[51] Li, Y. G, Vidale, J. E, Aki, K, & Xu, F. (2000). Depth-dependent structure of the Landers fault zone using fault zone trapped waves generated by aftershocks. *J. Geophys. Res.,* , 105, 6237-6254.

[52] Lombardo, G, & Rigano, R. (2006). Amplification of ground motion in fault and fracture zones: observations from the Tremestieri fault, Mt. Etna (Italy). *J. Volcanol. Geotherm. Res.,* , 153, 167-176.

[53] Lombardo, G, & Rigano, R. (2007). Local seismic response in Catania (Italy): A test area in the northern part of the town. *Engineering Geology* , 94, 38-49.

[54] Lombardo, G, & Rigano, R. (2010). Local seismic response evaluation in natural and artificial cavities. *Proceedings 3rd International Symposium "Karst Evolution in the South Mediterranean Area", 29-31 May 2009, Ragusa, Italy,* Speleologia Iblea C.I.R.S. ISNN 1123-9875., 14

[55] Mantovani, M, Devoto, S, Forte, E, Mocnik, A, Pasuto, A, Piacentini, D, & Soldati, M. (2012). A multidisciplinary approach for rock spreading and block sliding investigation in the north-western coast of Malta. *Landslides,* DOIs10346-012-0347-3

[56] MEPA (Malta Environment and Planning AuthorityMapping Unit) ((2004). Xemxjia bay area map 1:2500.

[57] Mizuno, T, & Nishigami, K. (2006). Deep structure of the Nojima fault, southwest Japan, estimated from borehole observation of fault-zone trapped waves. *Tectonophysics*, , 417, 231-247.

[58] Mcguire, W. J, & Pullen, A. D. (1989). Location and orientation of eruptive fissures and feeder-dykes at Mount Etna: influence of gravitational and regional stress regimes. *J. Volcanol. Geotherm. Res.* , 38, 325-344.

[59] Monaco, C, Catalano, S, De Guidi, G, Gresta, S, Langer, H, & Tortorici, L. (2000). The geological map of the urban area of Catania (Eastern Sicily): morphotectonic and seismotectonic implications. *Mem. Soc. Geol. It.*, , 55, 425-438.

[60] Monaco, C, Tapponnier, P, Tortorici, L, & Gillot, P. Y. (1997). Late Quaternary slip rates on the Acireale-Piedimonte normal faults and tectonic origin of Mt. Etna (Sicily). *Earth Planet. Sci. Letters*, , 147, 125-139.

[61] Nakamura, Y. (1989). A method for dynamic characteristics estimation of subsurface using microtremor on the ground surface. *Q.R. Railway Tech. Res. Inst. Rept.*, , 30, 25-33.

[62] Neri, M, Acocella, V, & Behncke, B. (2004). The role of the Pernicana Fault System in the spreading of Mt. Etna (Italy) during the 2002-2002 eruption. *Bullettin of Volcanology* doi:10.1007/S00445-003-0322-X., 66, 417-430.

[63] Neri, M, Guglielmino, F, & Rust, D. (2007). Flank instability on Mount Etna: Radon, radar interferometry, and geodetic data from the southwestern boundary of the unstable sector, *J. Geophys. Res.*, 112, B04410, doi:10.1029/2006JB0047

[64] Nogoshi, M, & Igarashi, T. (1970). On the propagation characteristics of microtremors. *Journal of the Seismological Society of Japan* , 23, 264-280.

[65] Nunziata, C, Natale, M, & Panza, G. F. (1999). b. Estimation of cavity effects on response spectra for the 1980 earthquake in the historical centre of Naples. In: *Earthquake Geotechnical Engineering*, Sêco & Pinto (Eds), Balkema, Rotterdam, , 9-14.

[66] Pagliaroli, A, Pitilakis, K, Chávez-garcía, F. J, Raptakis, D, Apostolidis, P, Ktenidou, O. J, Manakou, M, & Lanzo, G. (2007). Experimental study of topographic effects using explosions and microtremors recordings, paper presented at 4th International Conference on Earthquake Geotechnical Engineering, Thessaloniki, Greece.

[67] Paolucci, R. (2002). Amplification of earthquake ground motion by steep topographic irregularities. *Earthquake Engineering and Structural Dynamics 31*, 1831-1853.

[68] Panzera, F, Pace, S, D'Amico, , Galea, S, & Lombardo, P. G. ((2011a). Preliminary results on the seismic properties of main lithotypes outcropping on Malta. *GNGTS 30° Convegno Nazionale, Mosetti tecniche grafiche, Trieste*, , 306-308.

[69] Panzera, F, Rigano, R, Lombardo, G, & Cara, F. Di Giulio G., Rovelli A. ((2011b). The role of alternating outcrops of sediments and basaltic lavas on seismic urban scenario: the study case of Catania, Italy. *Bulletin of Earthquake Engineering* dois10518-010-9202-x., 9, 411-439.

[70] Pedley, H. M, Clark, M, & Galea, P. (2002). Limestone isles in a cristal sea: the geology of the Maltese islands. *P.E.G. Ltd*, 9-99090-318-2

[71] Pedley, H. M, House, M. R, & Waugh, B. (1978). The geology of the Pelagian block: the Maltese Islands. In: *Narin, A. E. M., Kanes, W. H., and Stehli, F. G. (eds), The Ocean Basin and Margins, 4B: The Western Mediterranean, Plenum Press*, London, , 417-433.

[72] Peng, Z, & Ben-zion, Y. (2006). Temporal changes of shallow seismic velocity around the Karadere-Duzce Branch of the North Anatolian Fault and strong ground motion. *Pure Appl. Geophys.*, 163, 567-600.

[73] Pischiutta, M, Salvini, F, Fletcher, J, Rovelli, A, & Ben-zion, Y. (2012). Horizontal polarization of ground motion in the Hayward fault zone at Fremont, California: dominant fault-high-angle polarization and fault-induced cracks. *Geophys. J. Int.* 188(3), 1255-1272.

[74] Pitilakis, K. (2004). Chapter 5: Site Effects. In: *Recent advances in Earthquake Geothecnical Enginnering and Microzonation*, A. Ansal ed., Kluwer Academic Publishers, , 139-198.

[75] Poppeliers, C, & Pavlis, G. L. (2002). The seismic response of a steep slope: high-resolution observations with a dense, three-component seismic array. *Bulletin of the Seismological Society of America 92(8)*, 3102-3115.

[76] Raptakis, D, Chavez-garcia, F. J, Makra, K, & Pitilakis, K. (2000). Site effects at Euroseistest- I. Determination of the valley structure and confrontation of observations with 1D analysis. *Soil Dynamics and Earthquake Engineering*, 19, 1-22.

[77] Reuther, C, & Eisbacher, D. a. n. d G. H. (1985). Pantelleria Rift- crustal extension in a convergent intraplate setting. Geol. Rndsch., , 74, 585-597.

[78] Riedel, W. (1929). Zur Mechanik Geologischer Brucherscheinungen. Zentral-blatt fur Mineralogie, Geologie und Paleontologie B, , 354-368.

[79] Rigano, R, Cara, F, Lombardo, G, & Rovelli, A. (2008). Evidence of ground motion polarization on fault zones of Mount Etna volcano. *J. Geophys. Res.*,113, B10306, doi: 10.1029/2007JB005574.

[80] Rigano, R, & Lombardo, G. (2005). Effects of lithological features and tectonic structures in the evaluation of local seismic response: an example from Hyblean plateau (Eastern Sicily). *Geologica Carpathica, 564297306*

[81] Roten, D, Cornou, C, Fäh, D, & Giardini, D. (2006). D resonances in Alpine valleys identified from ambient vibtation wavefields. *Geophysical Journal International* , 165, 889-905.

[82] Rovelli, A, Caserta, A, Marra, F, & Ruggiero, V. (2002). Can seismic waves be trapped inside an inactive fault zone? The case study of Nocera Umbra, central Italy. *Bull. Seismol. Soc. Am.,* , 92, 2217-2232.

[83] Rust, D, & Neri, M. (1996). The boundaries of large-scale collapse on the flanks of Mount Etna, Sicily, In: *Volcano Instability on the Earth and Other Planets, edited by W.M. McGuire, A.P. Jones and J. Neuberg,* Spec. Pub. Geol. Soc. London, , 110, 193-208.

[84] Scherbaum, F, Hinzen, K. G, & Ohrnberger, M. (2003). Determination of shallow shear wave velocity profiles in the Cologne, Germany area using ambient vibrations. *Geophys. J. Int.,* , 152, 597-612.

[85] Schnabel, B, Lysmer, J, & Seed, H. (1972). SHAKE: a computer program for earthquake response analysis of horizontally layered sites. *Rep. E.E.R.C. 7212Earthq. Eng. Research Center,* Univ. Califormia, Berkeley.

[86] SESAME ((2004). Guidelines for the implementation of the H/V spectral ratio technique on ambient vibrations: Measurements, processing and interpretation. SESAME European Research Project WP12, deliverable D23.12, http://sesame-fp5.obs.ujf-grenoble.fr/Deliverables.

[87] Seeber, L, Armbruster, J. G, Ozer, N, Aktar, M, Baris, S, Okaya, D, Ben-zion, Y, & Field, E. (2000). The 1999 Earthquake Sequence along the North Anatolia Transform at the Juncture between the Two Main Ruptures, In: *The 1999 Izmit & Duzce Earthquakes: preliminary results, edit. Barka A., O. Kazaci, S. Akyuz & E. Altunel,* Istanbul technical university, , 209-223.

[88] Sgarlato, G, Lombardo, G, & Rigano, R. (2011). Evaluation of seismic site response nearby underground cavities using earthquake and ambient noise recordings: a case study in Catania area, Italy. *Engineering Geology.* doij.enggeo.2011.06.002., 122, 281-291.

[89] Spudich, P, Hellweg, M, & Lee, W. H. K. (1996). Directional topographic site response at Tarzana observed in aftershocks of the 1994 Northridge, California, earthquake: implications for mainshock motions. *Bull. Seism. Soc. Am.,* 86(1B), SS208., 193.

[90] Tchalenko, J. S. (1968). The evolution of kink-bands and the development of compression textures in sheared clays, Tectonophysics, , 6, 159-174.

[91] Various Authors(1993). Geological Map of the Maltese Island. Sheet Malta- Scale 1:25,000. Oil Exploration Directorate, Office of the Prime Minister, Malta,., 1.

[92] Wood, H. O. (1908). Distribution of apparent intensity in San Francisco, in the California earthquake of April 18, 1906. *Report of the State Earthquake Investigation Commission, Canergie Institute of Washington, Washington, D.C.,* , 1

Chapter 6

Seismic Ambient Noise and
Its Applicability to Monitor Cryospheric Environment

Won Sang Lee, Joohan Lee and Sinae Han

Additional information is available at the end of the chapter

1. Introduction

In seismology, most of researches have been investigated by analyzing major seismic 'signals', for instance, body waves and surface waves. Aki [1] first introduced 'coda' waves, which had long been recognized as 'noise', consisting of scattered S-waves during propagation through the heterogeneous Earth media. Since then, a number of studies have been conducted to measure medium heterogeneity using coda waves over the world (e.g. [2]). Another revolutionary research dealing with 'noise' in seismology has been reported by [3]. The authors introduced a remarkable method to determine surface wave velocity examining long sequences of seismic ambient noise. It gives us a great opportunity to explore velocity structures underneath by nothing but listening to noise.

An additional interesting feature in terms of 'noise' shown up in the broadband seismic records is microseisms having two predominant peaks in a frequency domain such as primary and secondary microseisms, which have been believed to be originated by long-period ocean waves. The most widely accepted mechanisms for the generation of microseisms are as follows: (1) When ocean waves impact the coast, a part of acoustic energy is transferred into the crust. The directly converted seismic energy (Primary Microseisms, PM) from ocean waves propagates mostly as Rayleigh waves having a predominant period near 8-20 s which is the same period as the ocean waves even P-waves have been observed [4]. (2) The most energetic ambient noise is referred to as the secondary microseisms, or Double-Frequency (DF) microseisms, with 4-10 s of predominant period and the generation mechanism is more complex than that of PM. As ocean waves travel toward and strike the coast, reflected waves are generated and nonlinearly interact with incident waves in shallow regions, which results in a frequency doubling of a standard ocean wave [5,6]. The pressure amplitude of propagating incident waves decays exponentially as water depth becomes deeper, whereas the amplitude of standing waves keeps nearly constant with depth. Such powerful DF microseisms could be efficiently excited by significant reflection of wave energy at steep coastlines.

There have been several research efforts (e.g. [7,8]) to identify the source regions of DF microseisms utilizing array analysis, which report that the most DF microseisms are excited near the coastal regions where the swell reaches steep coasts with normal incidence, in good agreement with the Longuet-Higgins model for the generation of DF microseisms [7]. In addition to the determination of source location, scientists have investigated if the DF microseisms vary seasonally in amplitude. According to the spectral analysis for seismic ambient noise by McNamara and Buland [9], the power of DF noise levels in winter is higher than the summer season, which could be observable over the northern hemisphere. Since the DF microseism has been originated by ocean waves, it tends to show seasonal variability, reflecting the vigor of ocean activities [10]. This phenomenon allows us to monitor the Earth's near-surface environment in the ocean using the ambient noise analysis. In Polar Regions, seasonal variation in the DF energy occurs inconsistently compared to the characteristics shown in the lower latitude regions. Recent observations (e.g. [11,12]) suggested that there is a possible relation between the seasonal variability of DF microseisms and sea-ice variability. Tsai and McNamara [13] has theoretically explained that the sea-ice concentration is responsible for the seasonal change of DF microseisms in power with a simple attenuation model.

Even though many literatures have reported fascinating results for seismic ambient noise in both the northern and southern hemispheres, there are only a few studies regarding the DF microseisms in Antarctica [11,12,14,15] due to a dearth of broadband seismic stations. In this chapter, we combine more 3-year data from 2009-2011 with the previous results [11] and present a seismic ambient noise study that could provide more reliable evidence to present strong association between the seasonal change of DF and the variability of sea-ice concentration, in turn we are able to monitor regional cryospheric environment.

2. Data and spectral analysis

Korea Polar Research Institute (KOPRI) has been operating a permanent broadband seismic station (KSJ1, 62.22°S/58.78°W; Figure 1) at the King George Island (KGI) in the South Shetland Islands, Antarctica, since 2001. The seismic station is mainly equipped with a three-component broadband Streckeisen seismometer (STS-2) and a 24-bit high-resolution data logger (Q4124). The sensor responses in amplitude and phase guarantee that signals recorded within the frequency range between 120 sec and 20 Hz are reliable to be used without severe distortion. We have collected seismic data with 1 (LHX) and 20 Hz (BHX) sampling rates, and used the BHZ data for the spectral analysis.

In order to examine the spectral characteristics of seismic ambient noise, we calculate the Power Spectral Density (PSD) of the seismic noise following the rigorous method by McNamara and Buland [9]. The method requires first parsing continuous time series into 1-hour time series segments, overlapping by 50 % and distributed continuously throughout the day, week, and month. The PSD estimate should be converted into decibels with respect to acceleration

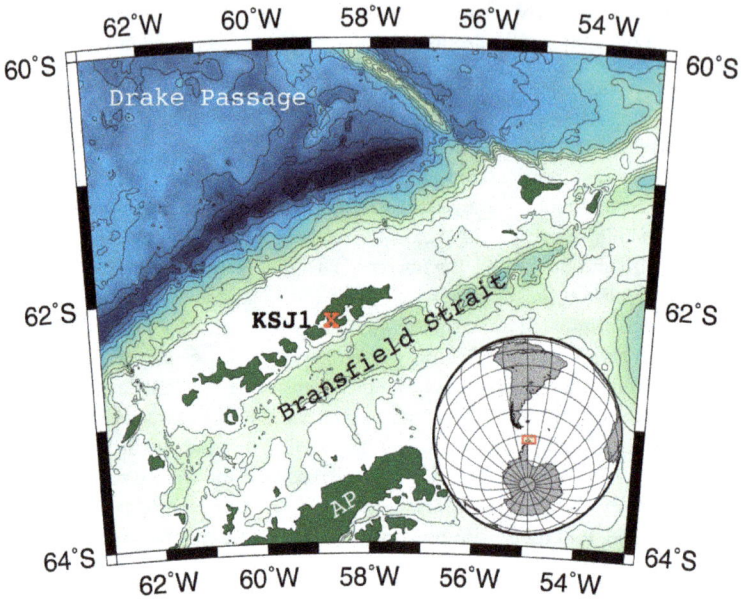

Figure 1. Location of a seismic station (KSJ1) at the King Sejong Korean Antarctic Base in the King George Island (KGI) marked by a red X. The Island is situated in between the Drake Passage and the Bransfield Strait near the Antarctic Peninsula (AP).

(meters/second^2)^2/Hertz for direct comparison to the standard noise model [16] in this study. The PSD technique provides stable spectra estimates over a broad range of periods (0.05–100 sec); however, it suffers from poor time resolution due to the long transforms (3600 sec) and requires many hours of data to compile reliable statistics. For better resolution at shorter periods, a larger number of shorter records should be analyzed [9]. From more than 90,000 PSDs for the period of 2006-2011, we could estimate Probability Density Functions (PDFs) to investigate the highest probability noise level (mode) for each channel as a function of period. As the method utilizes modes rather than higher energy level, we could obtain more reliable information to understand the characteristics of ambient noise, since even damaging earthquakes occurred near the station it is just a small portion out of background noise in terms of occurrence [9]. Moreover, it has a distinctive feature that we do not need to screen continuous quiet time window. At present, it has been known as the most common and robust technique to measure seismic ambient noise and evaluate the performance of seismic stations. More details regarding statistics and spectral analysis should be referred to [9].

Figure 2 demonstrates a statistical view of broadband PDFs of a vertical component for the period of 2006-2011. Two prominent peaks show up around 5 and 10 s in period, which correspond to secondary and primary microseisms, respectively. HNM and LNM indicate (gray curves in Fig. 2) the standard high and low noise model [16], respectively. Although

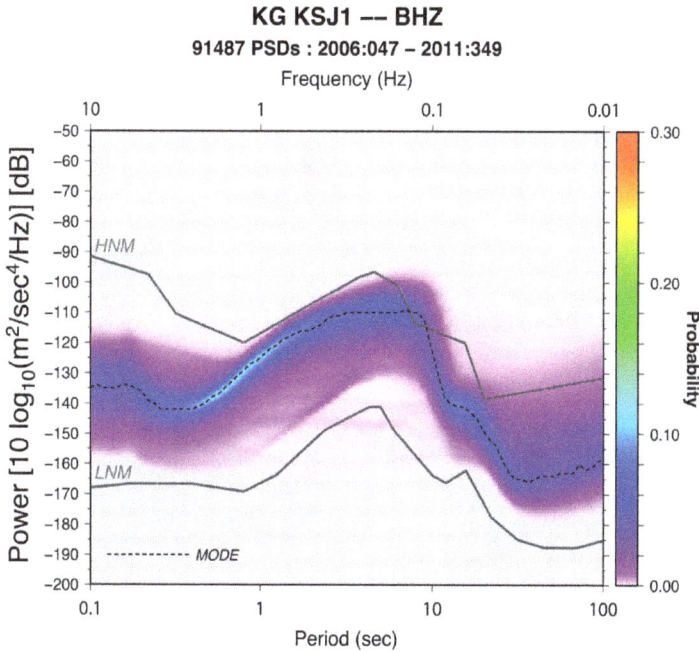

KG KSJ1 –– BHZ
91487 PSDs : 2006:047 – 2011:349

Figure 2. A Probability Density Function (PDF) plot of BHZ for KSJ1 during 2006-2011. Two predominant peaks appear around 5 and 10 s in period, corresponding to secondary (or DF) and primary microseisms, respectively. HNM and LNM indicate (gray curves) the standard high and low noise model (Peterson, 1993), respectively. The most probable energy with respect to frequency is presented by a dashed curve (mode).

several earthquakes occurred near the station during the operation period, they do not affect the overall PDFs as we mentioned earlier.

3. Seasonal variability of DF microseism and its association with sea-ice concentration

We investigated spectral characteristics for KSJ1 through the estimation of PSDs and PDFs of seismic background noise (Fig. 2), and found that the primary and secondary microseisms appear distinctly on the plot. Comparing to the HNM and LNM, we may evaluate that KSJ1 has been well operated in terms of system performance except slightly noisier (or might be higher energy level of DF microseisms) than the HNM around 10 s. In general, a plot of PDFs could provide helpful information on spectral signature of a station; however, temporal patterns of the microseisms are barely identifiable from it. In an attempt to examine temporal variation of the noise level of KSJ1, we obtain the statistical mode for the corresponding periods from daily PSDs so that we could construct a power spectrum with respect to time (Fig. 3). Empty spaces in the figure show data missing due to most likely system malfunctioning. There is nothing noticeable in the period of longer than 10 s throughout the operation time. Having interests in the feature near 4-10 s in period, i.e. DF microseisms, it happens that the DF energy comes to be weaker from July through September (austral winter). The behavior is apparent in 2007, 2009, and 2011, whereas it becomes ephemeral in 2006, 2008, and 2010, but rather weaker in power compared to other seasons in a year. This observation contrasts with the seasonal variability of seismic noises in the northern hemisphere; for instance, the amplitude of the Earth's hum reaches its seasonal maximum in winter season [17,18] revealed from an array analysis. The power of DF microseisms in the northern hemisphere shows a similar pattern (e.g. [19]) as the Earth's hum. Most literatures suggest that these characteristics are attributed to seasonal variation of the intensity of infragravity wave depending on swell amplitudes.

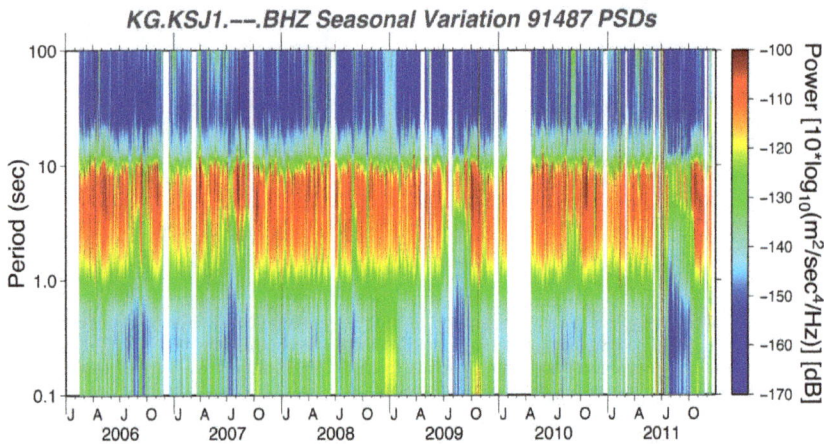

Figure 3. Spectral amplitude variation in seismic noise for BHZ (broadband vertical component) during the period of 2006-2011. Note that the seismic energy at the frequency range of DF microseisms (4-8 s) becomes weaker during July to September annually, which is a different behavior from that of the northern hemisphere except the Arctic region. Empty spaces in the plot indicate data missing or a period of malfunctioning.

Ringdal and Bungum [20] reported a pure sinusoidal pattern in long period noise level, i.e. seasonal maximum in winter and minimum in summer, from a spectral analysis of NORSAR data for three years. It does not, however, necessarily occur in the Polar Regions, especially Antarctica, and might be due to a regional difference between the northern and southern hemispheres. More specifically, [15] similarly observed weaker energy of DF microseisms in austral winter at the station DRV, Antarctica, and explained that the acoustic energy from ocean swell tends to be severely attenuated by sea ice and reflecting waves along the coast suffers as well causing fewer DF microseisms generated by sources. We refer to it as 'sea-ice damping effect' in this study. Recently, numerical modeling approach to figure out this phenomena has been made by Tsai and McNamara [13], which shows that 75-90 % of the variability in microseism power in the Bering Sea can be predicted using a simple model of microseism damping by

sea ice. Moreover, they argued that we could use the microseism as a good indicator to monitor the strength of sea ice that is not easily measured by through other means.

In order to carefully study the direct relation between the energy of DF microseisms and the
sea ice condition, we extract and integrate the DF power ranging 4-10 s in period out of the
power spectrum and collect Sea-Ice Concentration (SIC) information. To create time series of
SIC, we used data based on brightness temperature observations at 89 GHz obtained from the
AMSR-E (Advanced Microwave Scanning Radiometer for EOS Aqua) on board NASA's Aqua
satellite. The brightness of each image pixel is converted to the SIC using the ASI (ARTIST Sea
Ice) algorithm [21]. The data offer SICs on a grid with 6.25 km resolution with a complete daily
coverage of the Polar Regions. A domain for calculating SIC covers a part of the Drake Passage
allowing us to compare to the DF energy in this study. The calculated percentage of SIC is the
percentage of grid cells containing more than 15 % sea ice. This is mainly attributed to the fact
that the accuracy of the SIC is ±15 % in regions of first-year ice [22].

As shown in Figure 4, there is clear seasonal variation found in both the power of DF microseisms (red curve) and the SIC (black curve) from 2006-2011. To quantify how they are closely
related, we apply cross correlation that is a standard method of estimating the degree to which
two series are correlated. The bin size of each time series is chosen to be 1-day. The resultant
cross-correlation coefficient is given by -0.70 that is a strong negative correlation. The result
implies that as the SIC becomes higher, i.e. more sea-ice in the ocean, the DF power decreases,
which is coincident with the hypothesis of 'sea-ice damping effect'. We also determined the
lag time as almost zero from the cross correlation, which indicates that the DF energy responses
immediately to the sea-ice condition nearby. When one may take a closer look at the period of
May through September in a year, it becomes more prominent.

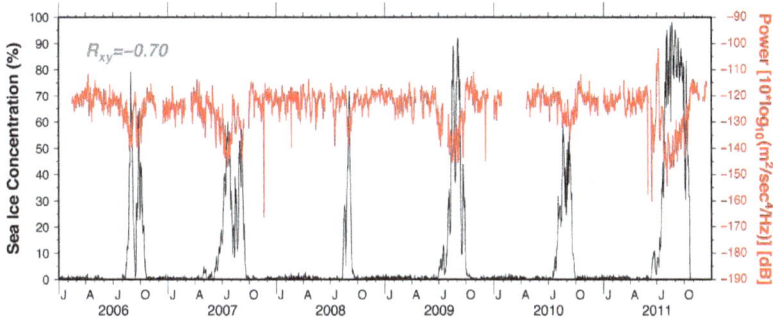

Figure 4. A comparative plot of Sea Ice Concentration (SIC, black curve) sampled near the KGI toward the Drake Passage vs. seismic energy of DF microseisms (red curve) observed at KSJ1 during 2006-2011. When either the SIC increases or decreases, the DF power responses immediately (negatively correlated), suggesting the DF energy is a relevant
seismic proxy to monitor cryospheric environment especially the sea ice condition nearby. The strong correlation
(-0.70) between them supports the hypothesis.

Remote sensing using satellites allows us to extensively improve our knowledge over various
scientific issues, especially in Polar Regions. For instance, we can measure surface melt extent

on ice, moving speed of glaciers, ice mass balance by means of AMSR-E, InSAR satellites, and GRACE, respectively. Even though all these methods give us great opportunities to monitor dramatic changes in the cryospheric environment, we should still conduct in-situ measurements to obtain physical and mechanical properties of ice such as stiffness. From the recent development on theoretical approach to predict the variability in DF microseisms [13] and the obvious evidence linking the variation of the DF energy to the SIC in this study, we anticipate that a long-term observation of the DF microseisms could be a good tool to monitor local climate change in Polar Regions. Most of literatures in seismology have dealt with several major issues such as determination of velocity and attenuation structures in the Earth, precise locating techniques, and investigation of earthquake sources. In this study, beyond that horizon, we find out that a seismological method could play a key role in understanding physical interaction between climate change and the cryosphere.

4. Conclusions

Substantial advances in seismograph technology allow us to consistently observe the Earth's continuous oscillation everywhere in the world. The DF microseism has been known that it is excited by ocean waves, thus it is likely to show seasonal variations [11]. Examining the ambient seismic noise level at KSJ1 during the period of from 2006-2011, we found a distinct seasonal pattern in the period of 4-10 s; the DF energy comes to be weaker from July through September (austral winter) in every year. Cross correlation results tell us that as the SIC becomes higher, the DF power decreases, and confirm that the DF energy responses immediately to the sea-ice condition. Consequently, we propose that a long-term observation of the DF microseisms should be necessary to monitor local climate change in Polar Regions, which contributes extra benefits to the satellite remote sensing.

Acknowledgements

This research has been supported by KOPRI research grants PN12040 (CATER 2012-8080) and PE13050.

Author details

Won Sang Lee[1], Joohan Lee[1] and Sinae Han[1,2]

*Address all correspondence to: wonsang@kopri.re.kr

1 Korea Polar Research Institute, Republic of Korea

2 Kangwon National University, Republic of Korea

References

[1] Aki K. Analysis of seismic coda of local earthquakes as scattering waves. Journal of Geophysical Research 1969; 74: 615-631.

[2] Sato H, Fehler MC. Seismic Wave Propagation and Scattering in the Heterogeneous Earth. AIP Press/Springer Verlag; 1998.

[3] Shapiro NM, Campillo M, Stehly L, Ritzwoller MH. High-Resolution surface wave tomography from ambient seismic noise. Science 2005; 307: 1615- 1618.

[4] Gerstoft P, Shearer PM, Harmon N, Zhang J. Global P, PP, and PKP wave microseisms observed from distant storms. Geophysical Research Letters 2008; 35 L23306: doi:10.1029/2008GL036111.

[5] Longuet-Higgins MS. A theory of the origin of microseisms. Philosophical Transactions of the Royal Society A 1950; 243: 1 –35.

[6] Hasselmann K. A statistical analysis of the generation of microseisms. Reviews of Geophysics 1963; 1: 177– 210.

[7] Chevrot S, Sylvander M, Benahmed S, Ponsolles C, Lefèvre JM, Paradis D. Source locations of secondary microseisms in western Europe: Evidence for both coastal and pelagic sources. Journal of Geophysical Research 2007; B11301: doi: 10.1029/2007JB005059.

[8] Traer J, Gerstoft P, Bromirski PD, Hodgkiss WS, Brooks LA. Shallow-water seismo-acoustic noise generated by tropical storms Ernesto and Florence. Journal of the Acoustical Society of America Express Letters 2009; doi:10.1121/1.2968296.

[9] McNamara DE, Buland RP. Ambient noise levels in the continental United States. Bulletin of the Seismological Society of America 2004; 94: 1517– 1527.

[10] Tanimoto T. Excitation of microseisms. Geophysical Research Letters 2005; 34 L05308: doi:10.1029/2006GL029046.

[11] Lee WS, Sheen D-H, Yun S, Seo K-W. The origin of double-frequency microseism and its seasonal variability at King Sejong Station, Antarctica. Bulletin of the Seismological Society of America 2011; 101: 1446-1451: doi:10.1785/20100143.

[12] Grob M, Maggi A, Stutzmann E. Observations of the seasonality of the Antarctic microseismic signal, and its association to sea ice variability. Geophysical Research Letters 2011; 38 L11302: doi:10.1029/2011GL047525.

[13] Tsai VC, McNamara DE. Quantifying the influence of sea ice on ocean microseism using observations from the Bering Sea, Alaska, Geophysical Research Letters 2011; 38 L22502: doi:10.1029/2011GL049791.

[14] Hatherton T. Microseisms at Scott Base. Geophysical Journal of the Royal Astronomical Society 1960; 3: 381-405: doi:10.1111/j.1365-246X.1960.tb01713.x.

[15] Stutzmann E, Roult G, Astiz L. GEOSCOPE station noise levels. Bulletin of the Seismological Society of America 2000; 90: 690– 701.

[16] Peterson J. Observation and modeling of seismic background noise. U.S. Geological Survey Technical Report 1993; 93(322): 1-95.

[17] Rhie J, Romanowicz B. Excitation of Earth's continuous free oscillations by atmosphere-ocean-seafloor coupling. Nature 2004; 431: 552–556.

[18] Bromirski PD, Gerstoft P. Dominant source regions of the Earth's "hum" are coastal. Geophysical Research Letters 2009; 36 L13303: doi:10.1029/2009GL038903.

[19] Sheen D-H, Shin JS, Kang T-S. Seismic noise level variation in South Korea. Geosciences Journal 2009; 13: 183–190.

[20] Ringdal F, Bungum H. Noise level variation at NORSAR and its effect on detectability. Bulletin of the Seismological Society of America 1997; 67: 479-492.

[21] Spreen G, Kaleschke L, Heygster G. Sea ice remote sensing using AMSR-E 89-GHz channels. Journal of Geophysical Research 2008; 113 C02S03: doi: 10.1029/2005JC003384.

[22] Parkinson CL, Coniso JC, Zwally HJ, Cavalieri DJ, Gloersen P, Campbell WJ. Arctic Sea Ice. 1973-1976: Satellite Passive Microwave Observations. National Aeronautics and Space Administration SP-489; 1987.

Chapter 7

Seismic Behaviour of Monolithic Objects: A 3D Approach

Alessandro Contento, Daniele Zulli and
Angelo Di Egidio

Additional information is available at the end of the chapter

1. Introduction

Despite the many progresses done in the modelling of rigid blocks, the grounding work for most of the research in this field remains [1], where a 2D model of the rigid block is obtained and the rocking and slide-rocking approximated conditions are written. Following papers on the dynamics of rigid bodies can be divided in two main groups, according to the kind of excitation used: earthquake excitation or sine-type pulse excitation (mainly one-sine). To the first group belong [2-5], in the second one, [6-10] can be found. In time, models of rigid blocks, very useful in many research fields, have been increased in complexity. Recently, for instance, sliding phenomena and the eccentricity of the center of mass have been considered (see [3, 11]). Some papers have been focused to specific problems, for example in [12] the behavior of two stacked rigid blocks has been considered, whereas in [13, 14] the attention has been pointed to blocks on flexible foundations. The dynamics and control of 2D blocks have also been analyzed in the framework of the bifurcation theory in [15, 16, 17].

The effects of the simultaneously presence of horizontal and vertical base excitations have been considered in some papers. For example, in [12, 18, 19] different problems related to the overturning of bi-dimensional rigid blocks have been studied in details.

Lately a large interest has been given to models of rigid bodies with base isolation systems, in order to improve the safety of art objects (see [20-22]). It has been proved that, in certain ranges of geometrical parameters of the rigid block, the effectiveness of base isolation can be amplified when coupling the isolating systems with devices able to limit the displacement of the oscillating base, so as to prevent the falling of the base of the body (see [23, 24]).

Recently, 3D models, mostly circular based, have been used in particular research fields, more precisely to study motions of a disk of finite thickness ([25, 26]), the wobbling of a frustum ([27]) or the sloshing in a tank ([28]).

A three-dimensional model of rigid body with a rectangular base, able to rock around a side or a vertex of the base, already presented by the authors in [29], is used herein to further study the dynamic behavior of rigid blocks. In particular the effects of a vertical one-sine excitation, acting concurrently to the horizontal one, and the seismic response of rigid bodies are considered. The body can experience only rocking motion since it is herein assumed that it possesses a slenderness for which bouncing is not triggered (see [22, 23]). Eccentricity of the center of mass, evaluated with respect to the geometrical center of the parallelepiped that ideally circumscribes the body, is also considered. The equations of motion of the body are obtained making use of the balance of moments. Impacts between the base and the ground are treated by imposing the conservation of the angular momentum before and after the impact. Starting conditions of rocking motion around a side or a vertex of the base are obtained by balancing the overturning moments and the resisting moments. Results are obtained by numerical integration of the equations of motion by using a IMSL routine developed in Fortran [30].

Rocking and overturning curves that furnish the amplitude of the one-sine pulses able to uplift or to overturn the body, versus the angular direction of the horizontal excitation, are obtained. The role of the period of the excitations, the geometrical characteristics of the body and the eccentricity of the center of mass are also highlighted. Particular attention is focused to the relative phase between the horizontal and vertical excitations. The presence of the vertical pulse can strongly change the behavior of the system with respect to the case where only the horizontal excitation is considered.

Regarding the seismic excitation, three different registered Italian earthquakes, with different spectrum characteristics, are used in the analyses. Two type of analyses are performed in the paper: the first is conducted by varying the direction of the seismic input to point out if, for some directions, the 3D model of rigid block furnishes more accurate results than the classical 2D models; the second is performed by fixing the direction of the input with the aim to highlight the role of the mechanical and the geometrical characteristics of the rigid block in the seismic response. Also in this case, rocking and overturning curves, that furnish the amplitude of the seismic excitation able to uplift or to overturn the body versus the angular direction of the excitation, are obtained. The role of the type of spectrum of the seismic excitation, the geometrical and mechanical characteristics of the body and the eccentricity of the mass center are also highlighted.

Finally, almost all the figures in the paper refer to a well-known statue. It is taken as example of the use of the model here discussed, but there are many other possible applications.

2. Description of the considered mechanical system

The base of the rigid block is supposed to be rectangular, with the four vertices indicated as A, B, C, D (see Fig. 1). The block is circumscribed by an ideal parallelepiped with upper vertices E, F, H, I, and lower vertices coinciding with A, B, C, D. The point M is the the centroid of the parallelepiped, while G is the centroid of the block.

When the body is at rest, at the $t = t_0$ initial time, the position vector of a generic point with respect the i-th vertex ($i = A$, B, C, D) is indicated as $\hat{\mathbf{r}}_i$ and the position vector of the other three vertices with respect to the vertex i is indicated as $\hat{\mathbf{r}}_{ij}(t)$ $(i = A, B, C, D, j \neq i)$. The body is allowed to rotate alternatively around one of the vertices, being this vertex in contact with the coordinate plane $z = 0$. If the body is rocking about the i-th vertex (see Fig. 1_e for $i = C$), the position vector rotates about the vertex; its time evolution is described by

$$\mathbf{r}_i(t) = \mathbf{R}(t)\hat{\mathbf{r}}_i \qquad (1)$$

where $\mathbf{R}(t)$ is the 3D finite rotation matrix which can be written in terms of three time-depending angles $\vartheta_1(t)$, $\vartheta_2(t)$, $\vartheta_3(t)$ (see Appendix for a representation of the matrix $\mathbf{R}$). Therefore the total acceleration of the generic point with respect to a fixed frame is written as

$$\mathbf{a} = \mathbf{a}_g - \mathbf{g} + \ddot{\mathbf{R}}\hat{\mathbf{r}}_i \qquad (2)$$

where $\mathbf{a}_g$ is the ground acceleration, $\mathbf{g}$ is the gravity acceleration and the dot stands for time differentiation. If the mass per volume of the block is indicated as ρ, the total volume force acting on the generic point of the block during the rocking motion around the vertex i is $f = \rho\mathbf{a}$, which becomes, using Eq. (2),

$$\mathbf{f} = \rho(\mathbf{a}_g - \mathbf{g} + \ddot{\mathbf{R}}\hat{\mathbf{r}}_i) \qquad (3)$$

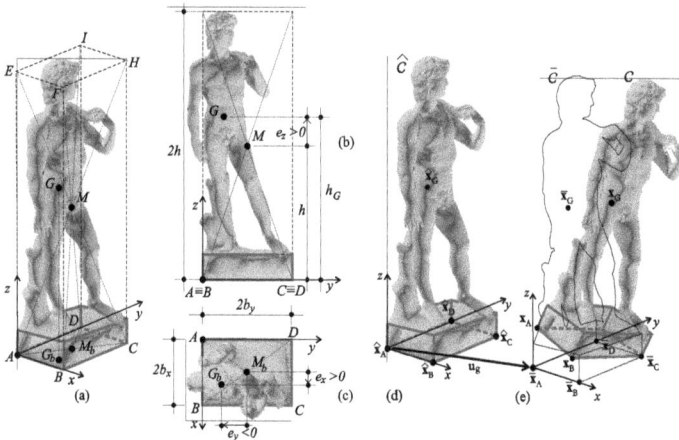

Figure 1. Geometrical characteristics of the rigid block: (a) three-dimensional view; (b) x-z plane projection; (c) x-y plane projection; (d-e) displacements of the rigid block: 3D rocking.

3. General formulation

3.1. Equations of motion

The equations of motions have been already presented in [29], where they are obtained imposing the balance of the moments acting on the body. In particular, when the body is rocking around the i-th vertex, which is assumed to lie on the horizontal, coordinate plane $z=0$, and is identified by its initial position $\hat{\mathbf{x}}_i(t)$, $i = A,\ B,\ C,\ D$, the equations of motion read

$$\text{skw}\left(\hat{\mathbf{s}}_i \otimes (\mathbf{a}_g(t) - \ddot{\mathbf{R}}(t)(\hat{\mathbf{x}}_i - \hat{\mathbf{x}}_A) - \mathbf{g})\mathbf{R}^T(t) - \ddot{\mathbf{R}}(t)\hat{\mathbf{J}}_{iA}\mathbf{R}^T(t)\right) = \mathbf{O} \tag{4}$$

where $\hat{\mathbf{s}}_i := \int_C \rho(\hat{\mathbf{x}} - \hat{\mathbf{x}}_i)dV$ is the vector of (initial) static moment of the body with respect to the vertex i ($\hat{\mathbf{x}}$ is the initial position of a generic point of the block); $\otimes$ is the tensor product; $\mathbf{a}_g(t)$ is the imposed base acceleration vector; $\mathbf{R}(t)$ is the 3D rotation matrix, which in turn depends on three angles $\vartheta_1(t)$, $\vartheta_2(t)$, $\vartheta_3(t)$; $\hat{\mathbf{J}}_{iA} := \int_C \rho(\hat{\mathbf{x}} - bh_i) \otimes (\hat{\mathbf{x}} - \hat{\mathbf{x}}_A)dV$ is the (initial) Euler tensor with respect to the vertices i and A; $\mathbf{g}$ is the gravity acceleration vector; the superscript T stands for transpose and the dot for time derivative; skw() is the skew part of the tensor in argument (see Fig. 1 and, for details, [29]). In Appendix A all the tensor and the vector quantities appearing in Eq. (4) are explicitly written, to make possible the reproduction of the numerical simulations reported in the following sections. Equations (4) reduce to the special case of 2D motion of the block around a side of the base (the same equation in [22]), when $\mathbf{R}(t)$ describes a planar rotation around one of the coordinate axes x or y.

3.2. Starting and ending conditions

The rigid block is assumed to be in a full-contact condition with the ideal horizontal support at the beginning of the base excitation. The rocking phase begins when the rigid block uplifts. An uplift occurs when the resisting moment M_r due to the weight of the body and to the vertical external acceleration is smaller than the overturning moment M_o due to the horizontal inertial forces. The uplift can occur around a side of the rectangular base (2D rocking motion) or around one of the four lower vertices of the base (3D rocking motion).

The eccentricity of the mass center with respect to the geometrical center of the parallelepiped, when considered, is obtained by introducing a concentrated mass $m_E = \beta m$ and, however, always keeping the total mass m of the body constant. Referring to Fig. 1, the following nondimensional eccentricities are introduced to characterize the system:

$$\varepsilon_x = \frac{e_x}{b_x}; \varepsilon_y = \frac{e_y}{b_y}; \varepsilon_z = \frac{e_z}{h} \tag{5}$$

and the slendernesses:

$$\lambda_x := \frac{h_G}{b_y} = \frac{h}{b_y}(1+\varepsilon_z); \lambda_y := \frac{h_G}{b_x} = \frac{h}{b_x}(1+\varepsilon_z) \qquad (6)$$

where $2b_x$, $2b_y$, $2h$ are the lengths of the three edges of the parallelepiped, respectively, e_x, e_y, e_z are the components of the distance vector between the center of mass of the body and the center of the parallelepiped, and $h_G = h + e_z$.

3.2.1. Starting condition of 2D rocking

An initial uplift around a side of the rectangular base leads to a 2D motion. In this case a rocking motion takes place when the overturning moment is equal or greater than the resisting moment, due to the vertical component of the acceleration $\ddot{u}_z$ of the center of mass. An initial uplift around a side parallel to the x direction (AB in Fig. 2_a) is considered. Similar conditions can be found for the orthogonal directions, since the mechanical system is symmetric. Thus the uplift occurs when:

$$M_o = ma_{g_y} h_G = M_r = m\ddot{u}_z(b_y + e_y) \qquad (7)$$

where a_{g_y} is the component of the ground acceleration along y. Using nondimensional quantities, Eq. (7) reads:

$$\frac{\ddot{u}_y}{\ddot{u}_z} = \frac{1+\varepsilon_y}{\lambda_x}. \qquad (8)$$

3.2.2. Starting condition of 3D rocking

An uplift on a vertex can occur either during a 2D rocking motion or directly from the full contact phase. In both cases a 3D motion is obtained.

In order to uplift the body directly from the rest, the rocking conditions around the two adjacent sides of the base have to be simultaneously satisfied. For example, to uplift the body and, consequently, to get 3D motion around the vertex A, rocking conditions around AB and AD have to be satisfied.

When the body is rocking around a side of the base, AB as example, the uplift condition on the vertex A is similar to the one for the 2D rocking motion on the side AD. The overturning moment M_o has to be at least equal to the resisting moment M_r:

$$M_o = ma_{g_x} h'_G = M_r = m\ddot{u}_z(b_x + e_x). \qquad (9)$$

In this case, h'_G is the actual height of the center G of the body during a rocking around the x-axis (see Fig. 2$_b$). The quantity h'_G can be evaluated as in [22] by the relation $h'_G = h_G \cos\vartheta_x + (b_y + e_y)\sin\vartheta_x$. By using the nondimesional quantities of Eq. (5), Eq. (9) reads:

$$\frac{\ddot{u}_x}{\ddot{u}_z} = \frac{1+\varepsilon_x}{\lambda_y} \frac{1}{\left(\frac{1+\varepsilon_y}{\lambda_x}\sin\vartheta_x + \cos\vartheta_x\right)} \tag{10}$$

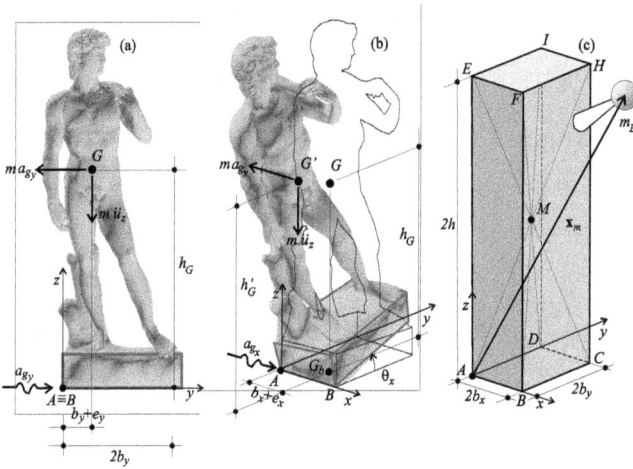

Figure 2. rigid block: (a) forces acting during the full-contact phase; (b) forces acting during a 2D rocking around the AB side; (c) shape of the body used in the numerical simulations.

The natural symmetry of the mechanical system leads to similar conditions for an uplift on one of the other three vertices of the base.

For a square based body with no eccentricity and under an excitation directed along the diagonal, the 3D motion around one of the vertices along the diagonal is directly triggered. This fact highlights how plausible is the occurrence of a 3D rocking motion for a square or near-square based body. An uplift on a vertex could easily manifest itself also in the case of excitation close to the diagonal, just after the occurrence of an uplift on a side and, therefore, during a 2D rocking motion. In fact the vertical position of the center of mass can increase enough ($h'_G > h_G$, see Fig. 2b) to satisfy Eq. (9).

On the contrary, for bodies with rectangular base and a side significantly larger than the other, an overturn around the largest side of the base is much more likely to occur before the uplift on a vertex.

3.2.3. Rocking termination and collapse conditions

No particular conditions are assumed to describe the termination of the motion and the return to the full-contact phase. This means that the rocking motion finishes when the energy associated to this phase is completely dissipated. A collapse event occurs when the body overturns. In the analyses, this condition conventionally manifests itself when one of the four upper vertices of the parallelepiped containing the body hits the ground.

3.3. Impact conditions

The impact conditions are taken from [29] too. They model the process of changing of the vertex around which the rocking occurs, and take place when the base of the block hits the ideal, horizontal, coordinate plane $z = 0$. Positions of the body underneath this plane are not allowed. For instance, if the body is rocking around the vertex i ($i = A, B, C, D$) and, at some special time t, another vertex, say j ($j \neq i$), hits the horizontal plane, then the impact process happens and j becomes the new center or rotation. The angles just after the impact (instant t^+) are exactly the same than those just before the impact (instant t^-): $\vartheta_1^+(t) = \vartheta_1^-(t)$, $\vartheta_2^+(t) = \vartheta_2^-(t)$, $\vartheta_3^+(t) = \vartheta_3^-(t)$, and therefore $\mathbf{R}^+(t) = \mathbf{R}^-(t)$. On the other hand, the evaluation of the time-derivatives of the three angles after the impact is made by equating the angular momentum around the (new) center of rotation just after and before the impact (see [29], [31]). In particular, the impact conditions read

$$\mathbf{R}^+(t)\hat{\mathbf{J}}_j\mathbf{R}^T(t) = \mathbf{R}^-(t)\hat{\mathbf{J}}_{ji}\mathbf{R}^T(t) \tag{11}$$

where $\hat{\mathbf{J}}_j := \int_C \rho(\hat{\mathbf{x}} - \hat{\mathbf{x}}_j) \otimes (\hat{\mathbf{x}} - \hat{\mathbf{x}}_j) dV$ and $\hat{\mathbf{J}}_{ji} := \int_C \rho(\hat{\mathbf{x}} - \hat{\mathbf{x}}_j) \otimes (\hat{\mathbf{x}} - \hat{\mathbf{x}}_i) dV$ are (initial) Euler tensors. Equations (11) provide a linear non-homogeneous algebraic system in the unknowns $\vartheta_1^+(t)$, $\vartheta_2^+(t)$ and $\vartheta_3^+(t)$, in terms of $\vartheta_1^-(t)$, $\vartheta_2^-(t)$ and $\vartheta_3^-(t)$.

4. Description of the excitations

4.1. One-sine excitation

The three-dimensional rigid body is excited by a one-sine pulse acceleration applied to the base of the body and acting along the horizontal direction ($a_h(t)$) or both along the horizontal and the vertical directions ($a_h(t)$, $a_v(t)$). The analyses are performed by varying the direction of the horizontal excitation, the period of the sine-pulses and their amplitudes. The direction is measured by a counterclockwise angle starting from the x-axis. The pulse-type acceleration used in the analyses are

$$
\begin{cases}
a_h(t) = A_h \sin\left(\dfrac{2\pi}{T_h} t\right) & 0 \le t \le T_h \\[2mm]
a_h(t) = 0 & T_h < t \le t_{max} \\[4mm]
a_v(t) = -A_v \sin\left(\dfrac{2\pi}{T_v} t + \phi\right) & 0 \le t \le T_v \\[2mm]
a_v(t) = 0 & T_v < t \le t_{max}
\end{cases}
\tag{12}
$$

where h and v stand for horizontal and vertical, respectively; T_h and T_v are the periods of the two one-sine pulses, A_h and A_v are their amplitudes and T_{max} is the maximum time used in the numerical integrations. This is always taken at least five times the period T_h. The phase of the one-sine horizontal excitation is assumed always equal to zero in this paper, although this parameter, in principle, can affect the behavior of the system. Only the phase of the vertical excitation will be taken into account, since the objective of the analysis is to point out the role of the difference of phase between the horizontal and vertical sine pulses.

4.2. Seismic excitation

In the seismic analysis, only the horizontal effects of the seismic source $\ddot{u}_g = \gamma f(t)$ are considered, where $f(t)$ is the registered seismic acceleration normalized to a $PGA = 1\,g$ (PGA stands for Peak Groung Acceleration) and γ is a variable coefficient used to scale the amplitude of the seismic accelerations. The time-histories and the elastic response spectrums of the three normalized seismic inputs used in the analyses are shown in Fig. 3. Brienza, Buia and Calitri earthquakes are choosen with the aim to perform a simplified analysis, able to evaluate the influence of the spectral characteristic of the earthquake in the dynamics of the three-dimensional rigid block.

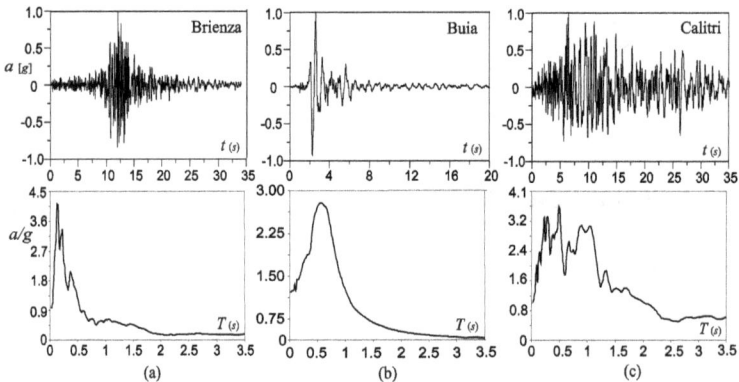

Figure 3. Normalized time-history and spectrum of the registered Italian earthquakes used in the numerical simulations: (a) Brienza earthquake $(PGA = 2.2\,m/s^2,\ lenght = 34.012\,s)$; (b) Buia earthquake $(PGA = 2.3\,m/s^2,\ lenght = 20.252\,s)$; (c) Calitri earthquake $(PGA = 1.2\,m/s^2,\ lenght = 35.019\,s)$.

4.3. Description of the simulation

Results have been obtained by the numerical integration of the equations of motion. A Fortran code has been implemented by using the IMSL Math libraries [30]. In particular the DASPG routine, able to numerically integrate the equations of motion in implicit form, has been chosen. It uses the well known Gear's Backward-Differentiation-Formulas method. Special care has been devoted to the detection of impacts. The integration time step has been fixed for all the simulation to $1/2^{16}$ sec. At each integration step, checks are made in order to verify if, under vertical excitation, the conditions of sliding or free-flight occur. Consequently the results of the evaluation have not been taken into account, since the model is not able to describe them.

For the one-sine excitation, the analyses are conducted by varying continuously the direction of the horizontal excitation and by evaluating the amplitudes of the horizontal or vertical one-sine pulse at which an uplift or an overturning collapse event manifests itself. This type of analysis is performed for several values of other parameters, such as period of the excitations, phase between the horizontal and vertical pulses, eccentricity and geometrical characteristics of the body.

The seismic analyses are performed by exciting rigid blocks with different mechanical and geometrical characteristics, by three different Italian registered earthquakes acting along different directions. Two type of analyses are performed in this study: the first is conducted with the aim to point out if for some directions of the escitation the 3D model of rigid block furnishes more accurate results than the classical 2D models; the other is performed by fixing the direction of the input, with the aim to highlight the role of the mechanical and the geometrical characteristics of the rigid block in the seismic response.

In the following analyses, a rigid body in the shape of a parallelepiped with a volume equal to $V = 8(b_x b_y h)$ is always assumed. The eccentricity of the mass center with respect to the geometrical center of the parallelepiped, when considered, is obtained by introducing a concentrated mass $m_E = \beta m$ and, however, always keeping the total mass m of the body constant (Fig. 2c). This means that the mass of the body with eccentricity $m = \bar{\rho}V(1 + \beta)$ is taken equal to the mass of the body without eccentricity $m = \rho V$. As a consequence the mass density $\bar{\rho}$ of a body when an eccentricity is considered leads to $\bar{\rho} = \frac{\rho}{1+\beta}$. The value of $\beta = 0.20$ and $\rho = 2000 kg / m^3$ are always taken in the following analyses.

5. Description of the results

Results are shown by using polar diagrams where, along the angle-axis (external circle), the angle that measures the direction of the horizontal excitation with respect to the x-axis, positive if counterclockwise, is reported. Along the radial-axis, the amplitude of the horizontal or vertical excitation able to uplift or to overturn the body is reported. These diagrams have been obtained by a massive use of calculator. Increments equal to 1.0^o for the direction of the pulse and equal to $0.01g$ for the amplitudes have been adopted to obtain the following diagrams. To give an idea of the calculus time needed to get all the results, from which those shown in this paper have been selected, a calculator with 12 Gb ofRAM and a Intel-I7 quad-core CPU with 2.0 GHz clock has been running for about two months.

In the following figures (Figs. 4, 5, 7-13), the same line styles are used. Solid thick lines refer to overturning events (they furnish the amplitude of the excitation at which the first occurrence of an overturning event manifests itself for a specific direction of the pulse). Dashed thick lines refer to the uplift of the body on a side of the rectangular base. In particular, for a specific direction of the excitation, below this curves, the body remains in perfect contact with the ground (full-contact) whereas, above them, a 2D rocking motion occurs. Dotted curves furnish the amplitude of the excitation for which an uplift on a vertex of the base occurs. Above this amplitude, a 3D rocking motion takes place. Directions of the excitation where the body manifests an uplift directly on a vertex (where dashed and dotted curves touch each others) always exist. The analyses performed in this paper do not permit to obtain the so-called survival regions, that could exist also in 3D rocking motions above the first overturning occurrence, as found for 2D rocking motions (see [6, 11]).

5.1. Rocking motion due to one-sine excitation

5.1.1. Rigid block with square base and no eccentricity

The first analysis, shown in Fig. 4, is conducted with the aim to check the influence of the phase ϕ of the one-sine vertical excitation. These results refer to a body with a square base and without any eccentricity of the center of mass, excited by a vertical pulse with fixed amplitude, when considered, and by a horizontal excitation with variable amplitude. The curves reported in the diagrams represent the value of the horizontal amplitude that causes the uplift or the overturning of the body. The angular sectors where a 3D rocking motion manifests itself are marked along the angular circle with solid thick lines. The diagrams shown in Fig. 4_a refer to the case in absence of vertical excitation. It is a case already reported and discussed in [29]. When a vertical one-sine pulse is added to the system, the uplift and overturning curves change as shown in Figs. 4_{b-d}.

In particular when $\phi = 90^o$ (Fig. 4b) the sectors where a 3D rocking motion occurs are wider, the system uplifts for smaller horizontal amplitudes and overturns for higher amplitudes (gray curve represents the results reported in Fig. 4_a in absence of vertical pulse). Therefore, with respect to the overturning collapse events, in this case, the presence of the one-sine pulse with this specific phase acts in favor of safety. A change of the phase ($\phi = 180^o$, see Fig. 4_c) produces a worsening of the situation in terms of overturning events. Many angular sectors, where a smaller horizontal amplitude is required to overturn the body, appear; they are contained in 2D or 3D rocking regions. When the phase is $\phi = 270^o$ (Fig. 4_d) the amplitudes necessary to overturn the body become smaller than those obtained in absence of vertical excitation (gray curve) in almost all the angular space.

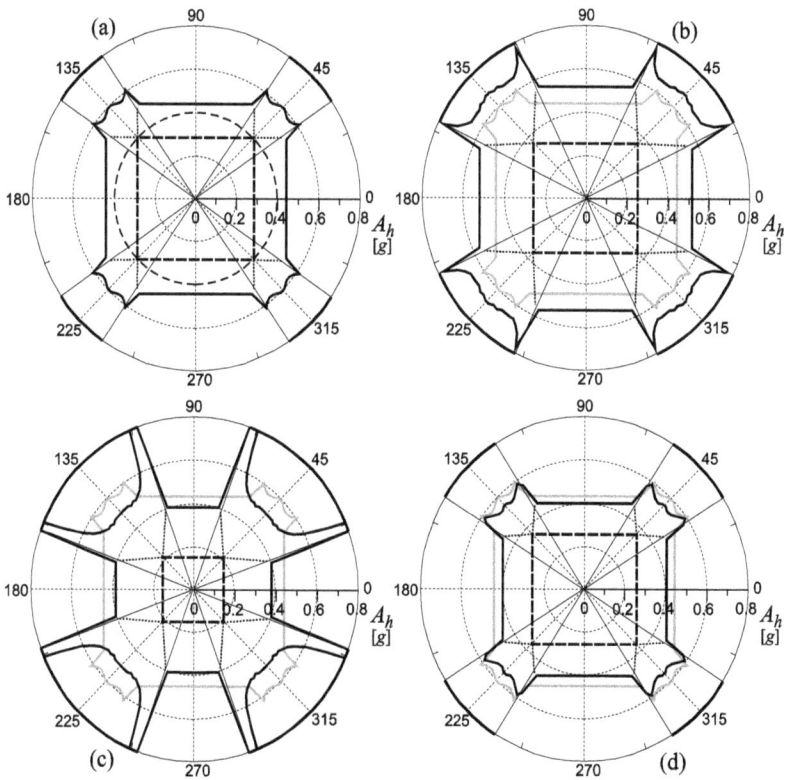

Figure 4. Direction vs Horizontal Amplitude of the excitation: (a) A_v =0;(b) A_v =0.5g, ϕ =90°; (c) A_v =0.5g, ϕ = 180°; (d) A_v =0.5g, ϕ =270°; (T_h =0.75s, T_v =0.75s, b_x =b_y =0.3m, h =1.0m, ε_x =ε_y =0).

With respect to the previous case, results shown in Fig. 5 refer to different periods of the excitations. In Fig. 5$_a$, the case in absence of vertical excitation is shown ([29]). When a vertical excitation with fixed amplitude and ϕ =0 is considered (Fig. 5$_b$), the uplift and the overturning curves strongly change. In particular, in many angular sectors, the horizontal amplitude able to overturn the body becomes smaller than the one in absence of vertical pulse (gray curve). Also a very critical condition takes place: in the gray angular sectors, contained into the 3D rocking regions, the horizontal amplitudes able to overturn the body become smaller than those obtained along the directions 0, 90, 180, 270 degrees, where the excitation is orthogonal to one of the four sides of the base (dash-dot circle). Since, in order to obtain the amplitudes along the directions 0, 90, 180, 270 degrees, a bi-dimensional model of rigid block is sufficient, the necessity to use a 3D model of rigid block to remain in favour of safety, is confirmed

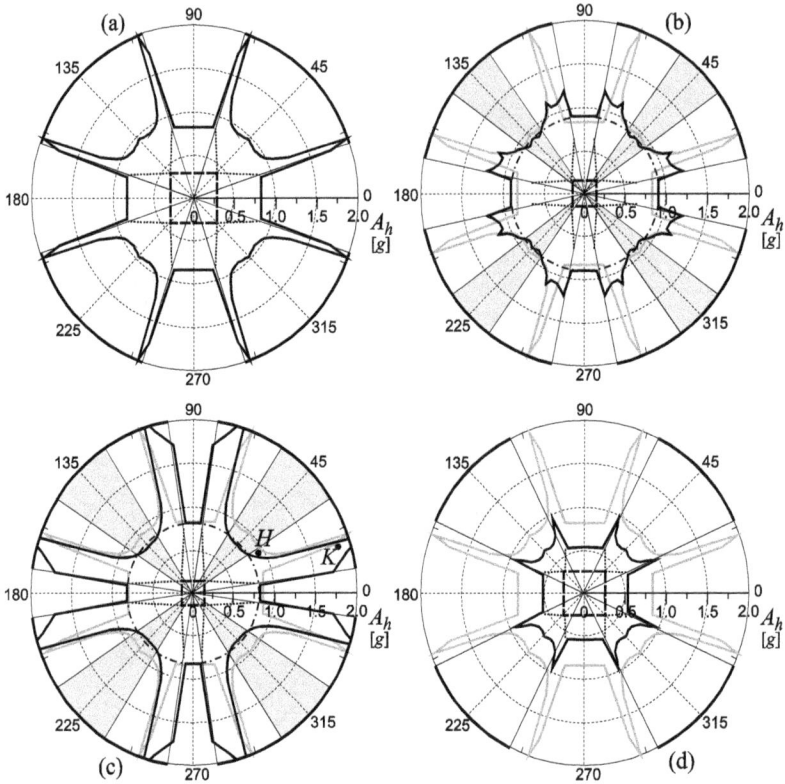

Figure 5. Direction vs Horizontal Amplitude of the excitation: (a) $A_v=0$;(b) $A_v=0.5g$, $\phi=0°$; (c) $A_v=0.5g$, $\phi=180°$; (d) $A_v=0.5g$, $\phi=270°$; ($T_h=0.5s$, $T_v=0.5s$, $b_x=b_y=0.3m$, $h=1.0m$, $\varepsilon_x=\varepsilon_y=0$).

It is interesting to note that the case in absence of vertical excitation (Fig. 5_a) does not manifest the necessity of the use of a 3D model. Therefore, it is possible to assert that this critical situation is only caused by the presence of vertical excitation. A change of the phase of the vertical pulse ($\phi=180°$, see Fig. 5_c) causes the enlargement of the critical sector where the amplitudes obtained by a 3D model are smaller than the amplitude of a 2D model (dash-dot circle). A further change of the phase ($\phi=270°$, see Fig. 5_d) strongly changes the scenario. The critical gray sectors disappear whereas, in all the angular plane, the overturning amplitudes become smaller than the case in absence of vertical excitation (gray curve).

In Fig. 6 the time-histories of vertical position zA of the base point A and of the angle ϑ_y, referring to the cases labeled with H and K in Fig. 5_c, are shown. Point H refers to a case where, in absence of vertical pulse, the body does not overturn whereas, when the vertical excitation is considered, the body does overturn (Fig. 6_a); point K refers to a case where, in absence of vertical pulse, the body overturns whereas, when the vertical excitation is considered, the body does not overturn (Fig. 6_b). Solid curves represent the case with vertical excitation, whereas dashed curves the case without vertical excitation. In both cases, where the body does not overturn, the time-histories zA touch and remain on the $zA = 0$ axis in several time ranges. This happens when, during the 3D rocking motion, the point A hits the ground and becomes the instantaneous rotation center.

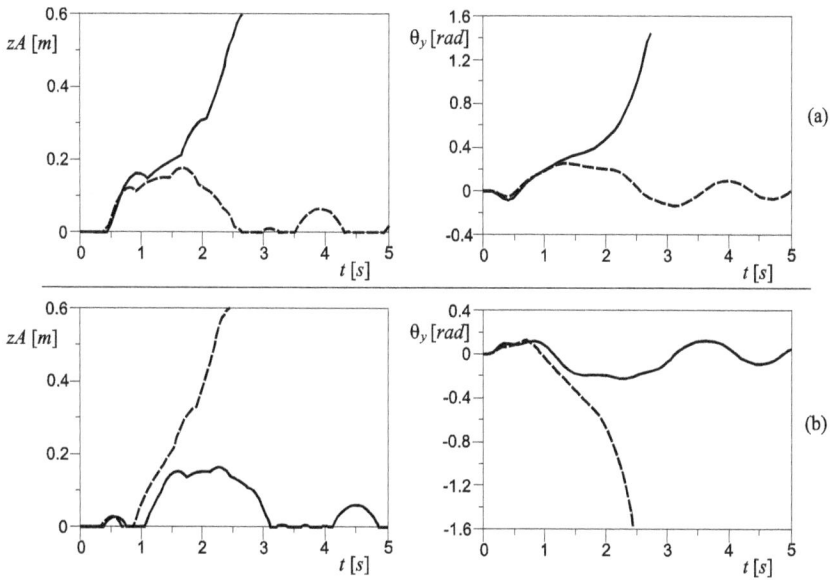

Figure 6. Time-histories: (a) Vertical position of the vertex A and angle $_y$ vs time calculated at point H labeled in Fig. 5c $(A_h = 0, 875g, A_v = 0.5g, a = 30°, \phi = 180°)$; (b) Vertical position of the vertex A and angle $_y$vs time calculated at point K labeled in Fig. 5c $(A_h = 1.74g, A_v = 0.5g, a = 18°, \phi = 180°)$; $(T_h = 0.5s, T_v = 0.5s, b_x = b_y = 0.3m, h = 1.0m, \varepsilon_x = \varepsilon_y = 0$;solid lines: with vertical excitation, dashed lines: without vertical excitation).

Very interesting is the case shown in Fig. 7. The results refer to a value of the period of the horizontal one-sine pulse for which, in absence of vertical excitation, many sectors where the 3D model furnishes more accurate results with respect to the classical 2D model exist (gray sectors in Fig. 7_a, see [29] for more details). The vertical excitation, also in this case, strongly changes the scenario. In particular, when the $\phi = 0$ (Fig. 7_b), the critical gray sectors reduce but, in all the angular plane, the overturning amplitudes become smaller than those obtained without vertical pulse (gray curve). For $\phi = 90°$ (Fig. 7_c) in several angular sectors the amplitude able to overturn the body becomes higher than the one obtained without vertical excitation, but the critical gray sectors, where it is necessary the use of a 3D model to better evaluate the overturning of the body, completely disappear. These critical regions appear again for $\phi = 180°$

(Fig. 7_d) but the overturning amplitudes become higher than those obtained in absence of vertical pulse, in all the angular plane.

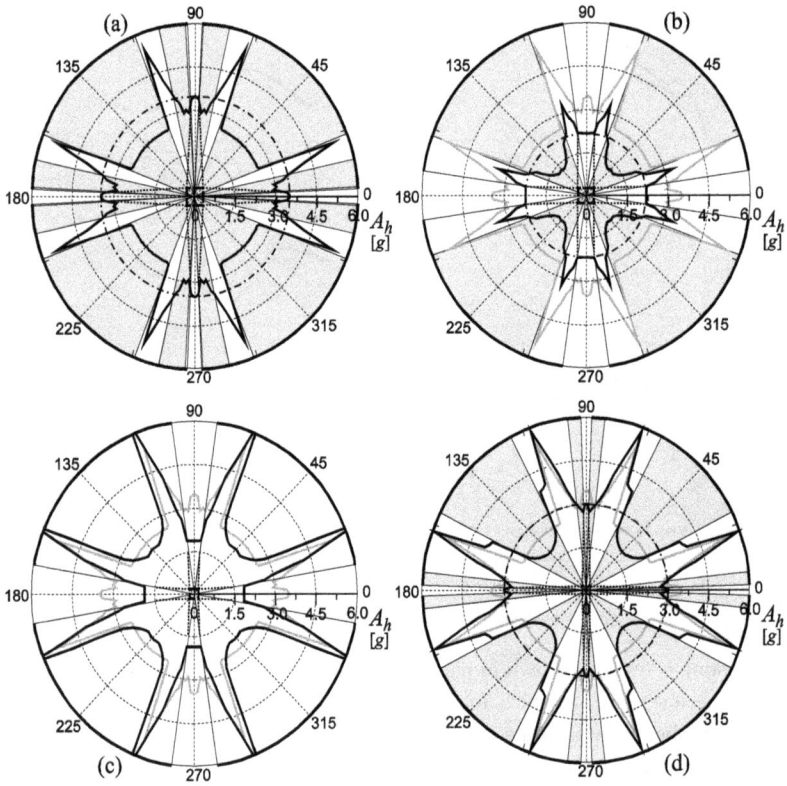

Figure 7. Direction vs Horizontal Amplitude of the excitation: (a) $A_v=0$;(b) $A_v=0.5g$, $\phi=0°$; (c) $A_v=0.5g$, $\phi=90°$; (d) $A_v=0.5g$, $\phi=180°$; ($T_h=0.35s$, $T_v=0.5s$, $b_x=b_y=0.3m$, $h=1.0m$, $\varepsilon_x=\varepsilon_y=0$).

5.1.2. Rigid block with square base and with eccentricity

In Fig. 8 the role of a fixed vertical excitation applied to a square based, eccentric, rigid body is investigated. The case of absence of vertical excitation, horizontal pulse with period $T_h = 0.75s$ and eccentricities $\varepsilon_x = \varepsilon_y = 0.25$, is shown in Fig. 8_a (see [29] for more details). Two critical regions (gray sectors) where a 3D model is necessary to better evaluate the overturning collapse events manifest themselves. When a vertical fixed excitation is considered, a change of the previous scenario occurs. In particular, for $\phi = 0$ and $\phi = 270^o$ (Fig. 8_b and 8_d respectively), a slightly modification of the critical sectors takes place: A diminution of the horizontal overturning amplitude with respect to the case in absence of vertical pulse (gray curve) manifests itself. For $\phi = 90^o$ (Fig. 8_c), the critical regions disappear.

5.1.3. Rigid block with near-square base

Finally, in Fig. 9, the role of a fixed vertical excitation applied to a near-square based, eccentric, rigid body is investigated. The case of absence of vertical excitation, without and with eccentricity ($\varepsilon_x = 0$, $\varepsilon = 0.5$) are shown in Fig. 9_a and Fig. 9_b, respectively (see [29] for more details). The eccentricity of the mass center of the body is the cause of the apparition of the critical regions (gray sectors, Fig. 9_b) where the overturning amplitude obtained during a 3D rocking motion is smaller than the overturning amplitude furnished by a 2D model (dash-dot circle). The vertical excitation, in addition to changing the overturning curves, acts also on the critical regions. In particular, for $\phi = 0$ (Fig. 9_c), two new critical regions appear whereas, for $\phi = 90^o$ (Fig. 9_d), these critical regions change their position and amplitude with respect to the case where the vertical excitation is null (Fig. 9_b).

5.2. Rocking motion due to seismic excitation

5.2.1. The role of the direction of the input

In the first analysis, square based block with constant height are excited by the three different earthquakes. In particular results reported in Fig. $10_{a,b}$ refer to two different blocks ($30 \times 30 \times 200$ cm^3 and $50 \times 50 \times 200 cm^3$, respectively) subject to the Brienza earthquake. It is possible to observe a general increase of the PGA able to overturn the body when the dimension of the base increases. Also a slight increase of the sectors of 3D rocking motion manifests itself for higher bases of the block. The overturning amplitude in 3D regions is always larger than the amplitude able to overturn the body during a 2D rocking motion (the value observed along the 0°, 90°, 180°, 270° directions) and marked in the graphs with a dash-dotted circle. However, for larger bases of the body, the 3D overturning amplitude becomes very close to the 2D overturning PGA, along some specific directions of the excitation. A similar behavior can be observed when the same previous blocks are excited by the Buia earthquake (Fig. $10_{c,d}$). Very different are the results obtained by exciting the body by the Calitri earthquake (Fig. $10_{e,f}$). The amplitude of the sectors where a 3D rocking motion manifests itself are a lot smaller than the previous cases and the 3D overturning amplitude remains always far enough from the 2D overturning PGA (dash-dotted circle). However, smaller values of the PGA than the previous earthquakes are requested to cause a collapse event. Considering the results of these first analyses, it does not seem necessary the use of a 3D model to study the seismic behavior of a rigid block, since the most dangerous situations manifest themselves during a 2D rocking motion.

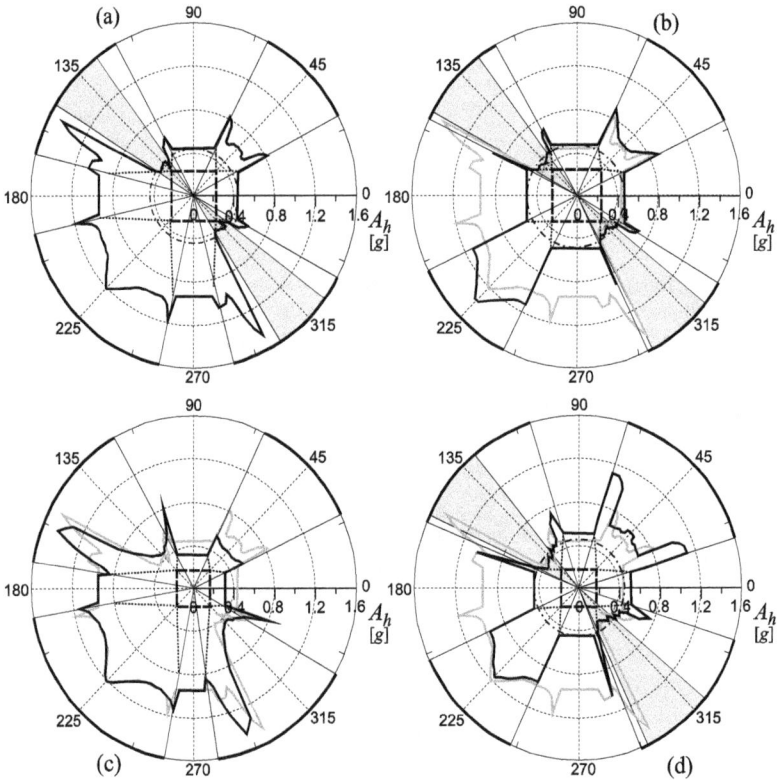

Figure 8. Direction vs Horizontal Amplitude of the excitation: (a)$A_v=0$; (b) $A_v=0.5g$, $\phi=0°$; (c) $A_v=0.5g$, $\phi=90°$; (d) $A_v=0.5g$, $\phi=270°$; ($T_h=0.75s$, $T_v=0.5s$, $b_x=b_y=0.3m$, $h=1.0m$,$\varepsilon_x=\varepsilon_y=0.25$).

In Fig. 11 the effect of the eccentricity of the mass center is outlined. In particular, a block of dimensions $40\times40\times200cm^3$, with and without eccentricity, is excited by the three different earthquakes. Results shown in Fig. 12$_{a,b}$ refer to the case without eccentricity and the case with eccentricity along the y-axis ($\varepsilon_y=0.25$, $e_y=5$ cm) respectively. The presence of an eccentricity sensibly changes the overturning curve, that loses one symmetry axis and becomes more irregular than the case without eccentricity.

However, a very interesting phenomenon occurs in presence of the eccentricity: in some directions inside the 3D rocking regions (marked with thick lines along the external circle), overturnig PGA's smaller than the minimum required during the 2D rocking motion (dash-dotted circle) manifest themselves. In other words, inside the gray sectors, during a 3D rocking motion, an overturning collapse event occurs for a PGA smaller than the minimum obtained by using a 2D model of rigid block. Results shown in Fig. $11_{c,d}$, that refer to Buia earthquake, confirm what previously said. The situation changes if the block is excited by the Calitri earthquake (Fig. $11_{e,f}$). In this case the 3D overturning amplitudes remain always far enough from the 2D overturning PGA's, also when an eccentricity is considered.

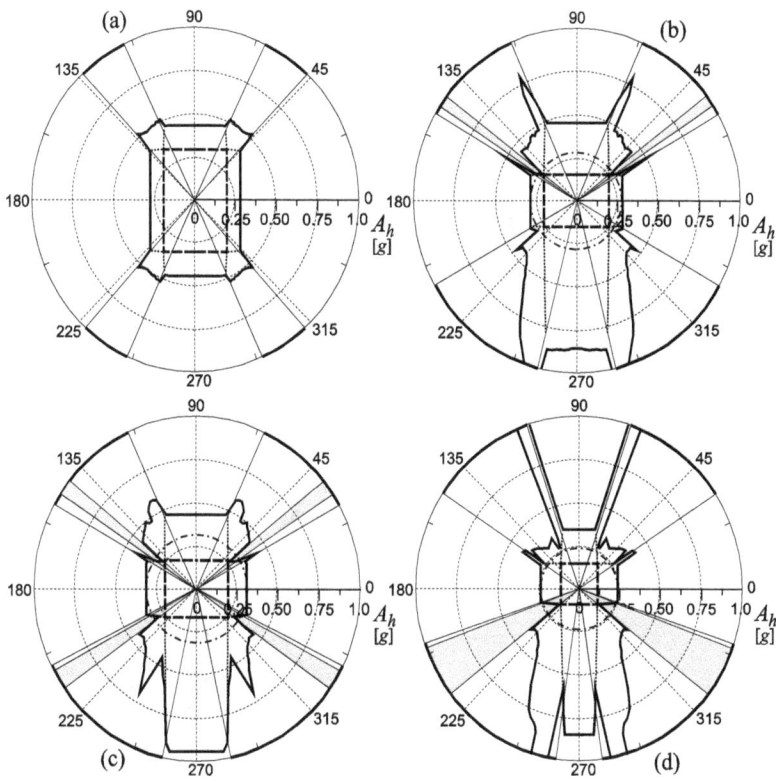

Figure 9. Direction vs Horizontal Amplitude of the excitation: (a) $A_v = 0$, $\varepsilon_x = \varepsilon_y = 0$; (b) $A_v = 0$, $\varepsilon_x = 0$, $\varepsilon_y = 0.5$; (c) $A_v = 0.5g$, $\varepsilon_x = 0$, $\varepsilon_y = 0.5$, $\phi = 0°$; (d) $A_v = 0.5g$, $\varepsilon_x = 0$, $\varepsilon_y = 0.5$, $\phi = 0°$; ($T_h = 0.75s$, $T_v = 0.5s$, $b_x = b_y = 0.3m$, $h = 1.0m$).

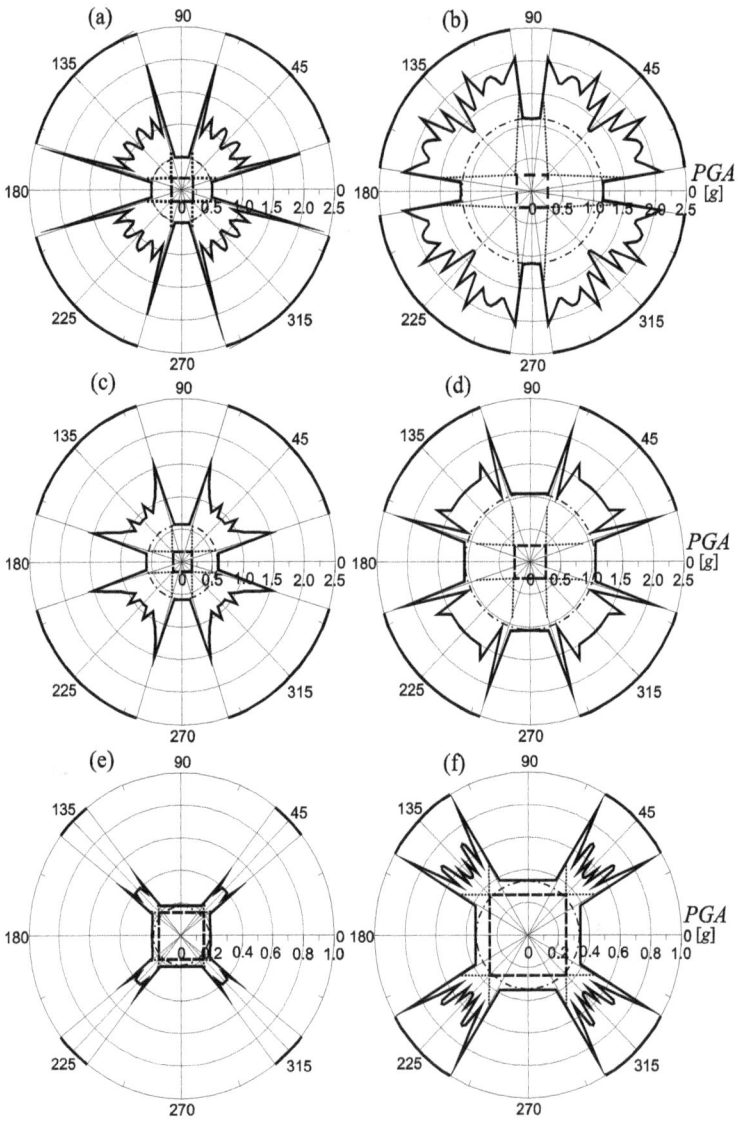

Figure 10. Direction vs Amplitude of the excitation. Brienza earthquake: (a)$30 \times 30 \times 200$; (b) $50 \times 50 \times 200 cm^3$. Buia earthquake: (c)$30 \times 30 \times 200$; (d) $50 \times 50 \times 200 cm^3$. Calitri earthquake: (e)$30 \times 30 \times 200$; (f) $50 \times 50 \times 200 cm^3$,($\varepsilon_x = \varepsilon_y = 0$).

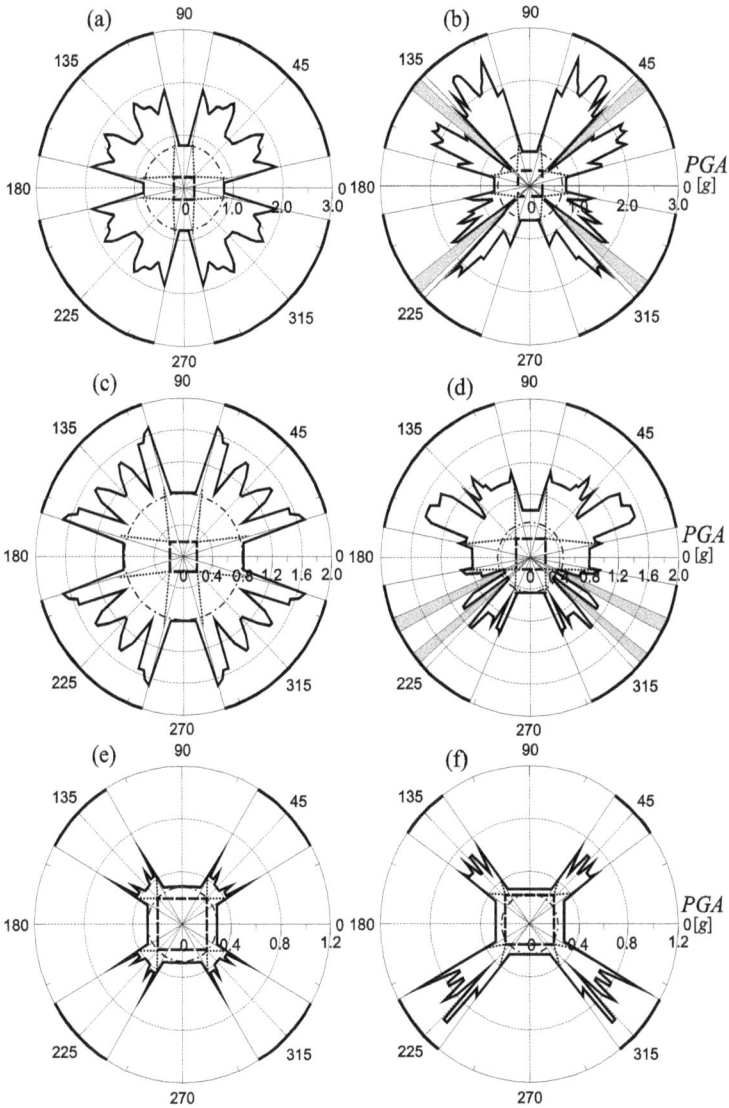

Figure 11. Direction vs Amplitude of the excitation. Brienza earthquake: (a)$\varepsilon_x=\varepsilon_y=0$; (b)$\varepsilon_x=0$, $\varepsilon_y=0.25$. Buia earthquake: (c)$\varepsilon_x=\varepsilon_y=0$; (d)$\varepsilon_x=0$, $\varepsilon_y=0.25$. Calitri earthquake: (e)$\varepsilon_x=\varepsilon_y=0$; (f) $\varepsilon_x=0$, $\varepsilon_y=0.25$ $(40\times40\times200)cm^3$.

To summarize, when the block is excited by an earthquake with narrow spectrum (Brienza and Buia earthquakes, see Fig. $3_{a,b}$), the presence of a small eccentricity makes possible the existence of angular sectors inside the 3D rocking regions, where the use of the 3D model of rigid block is necessary to obtain results in favour of safety. On the contrary, a wide spectrum earthquake (Calitri earthquake, see Fig. 3_c) does not require the use of the 3D model of rigid block, since the eccentricity of the mass center never causes the existence of these critical sectors inside the 3D rocking regions.

Finally, the case of blocks with a rectangular base is considered. Results shown in Fig. $12_{a,b}$ refer to rectangular based block ($30 \times 40 \times 200cm^3$), subject to Brienza earthquake, without and with eccentricity ($\varepsilon_x = 0$, $\varepsilon_y = 0.25$), respectively.

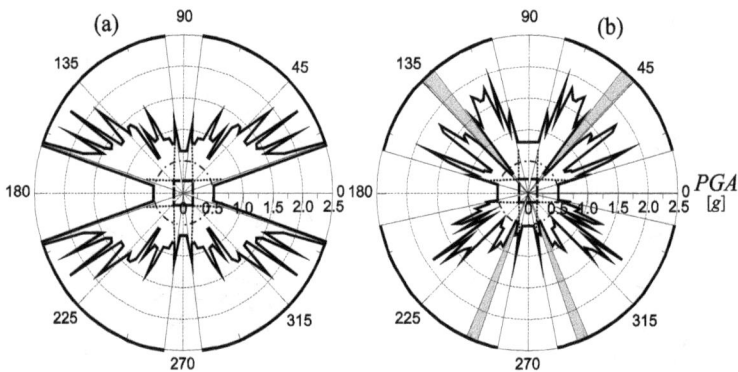

Figure 12. Direction vs Amplitude of the excitation. Brienza earthquake: (a) $\varepsilon_x = \varepsilon_y = 0$; (b) $\varepsilon_x = 0$, $\varepsilon_y = 0.25$ $(30 \times 40 \times 200)cm^3$.

As it is possible to observe, also in blocks with rectangular base, subject to a narrow spectrum earthquake (Brienza), sectors inside 3D rocking regions where the overturnig PGA's are smaller than the minimum required during the 2D rocking appears. This case is very interesting since, when a rectangular base of a rigid body occurs, the rocking motion is usually analysed by a 2D model in the plane of the smaller dimension of the base. As a consequence, the eccentricity of the mass center in the direction orthogonal to the plane of the analyzed motion is not taken into account. These results highlight the fact that also in this case, to evaluate the behavior of the system in favour of safety, it is necessary the use a 3D model of rigid body.

5.2.2. The role of the mechanical and geometrical characteristics of the body

In this section the analyses are performed by fixing the direction of the excitation and varying other geometrical and mechanical characteristic of the block, to point out their role in the seismic response of a square based body. In particular a first analysis is conducted by fixing the slenderness of the body and varying the base of the body (i.e. varying the volume of the body); the second analysis is performed by fixing the mass of the body (i.e. fixing the volume of the body) and varying its slenderness. Results of these analyses are reported in Fig. 13. In all the graphs, solid curves refer to PGA able to overturn the body when the excitation angle is $\alpha = 40°$, while dashed curves refers to overturning events when the excitation angle is $\alpha = 0$. Dotted curves, when reported, refer to overturning amplitude when the direction of the excitation is $\alpha = 40°$ and in presence of a very small eccentricity ($\varepsilon_x = 0$, $\varepsilon_y = 0.1$). The chosen directions permit to compare the overturning seismic response of the body during the 2D rocking motion ($\alpha = 0$) and the 3D rocking motion ($\alpha = 40°$).

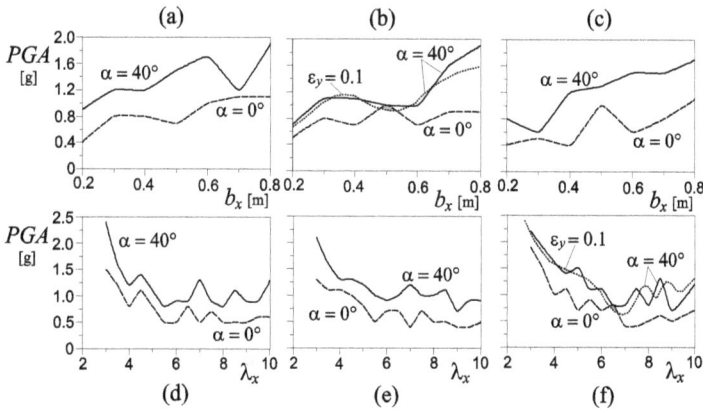

Figure 13. Overturning curves under Brienza earthquake: (a)$\lambda_x = \lambda_y = 5$; (b)$\lambda_x = \lambda_y = 6$; (c)$\lambda_x = \lambda_y = 7$; (d)$m = 500\,kg$; (e) $m = 640\,kg$; (f)$m = 800\,kg$.

The sequence of results shown in Fig. 13_{a-c} refers to cases with fixed slenderness $\lambda_x = \lambda_y = 5$, 6, 7). As it is possible to observe the two curves (solid and dashed curves) approach to each other at different values of the dimension of the base, depending on the slenderness used in the analysis. An increasing slenderness requires a decreasing base dimension to make possible the approach of the two curves. Very interesting is the case shown in Fig. 13_b ($a_x = \lambda_y = 6$) where the two curves touch each other. For this particular dimension of the base of the block, the same PGA is required to overturn the body during the 2D rocking motion and the 3D rocking motion, when the direction of the excitation is $\alpha = 40°$. The presence of a small eccentricity changes the 3D overturning curve (dotted curve) by making possible the existence of a region where the overturning PGA during the 3D motion is smaller than the one during the 2D motion. On the contrary, the sequence of results shown in Fig. 13_{d-f} refers to cases with fixed mass ($m = 500$, 640, $800\,kg$). The two curves (solid and dashed curves) approach to each other at different values of the slenderness, depending on the mass of the body used in the analysis. Very interesting is the case shown in Fig. 13_f ($m = 800\,kg$) where the two curves touch each other. Also in this case the presence of a small eccentricity (dotted curve) makes possible the existence of a region where the overturning PGA during the 3D motion is smaller than the one during the 2D motion.

Finally, when the body is excited by a narrow spectrum earthquake, it is always possible to find cases where a 3D model of rigid block is necessary to evaluate the seimic responce in favour of safety.

6. Conclusion

The rocking motion around a side or a vertex of a rectangular based rigid body has been deeply studied, making use of a three-dimensional model already proposed by the same authors. Starting conditions of motion have been found by means of equilibrium between overturning and resisting moments, whereas the impact has been described considering the conservation of the angular momentum.

The dynamics of the rigid body excited by one-sine pulse horizontal and vertical excitations and horizontal seismic excitation has been analyzed. Rocking and overturning curves versus the angular direction of the horizontal pulse have been obtained. The influence on the motion of several parameters, such as the period of the excitations, the geometrical characteristics of the body and of the eccentricity of the mass center has been pointed out.

The vertical one-sine pulse strongly modifies the behavior of the system with respect to the case where only horizontal excitation acts on the body. Results show that, in presence of vertical excitation and in significant ranges of the parameters, as happens when just horizontal base acceleration is considered, bi-dimensional models are not enough accurate to correctly evaluate the occurrence of the overturning and, therefore, a three-dimensional model is needed.

The seismic response of the rigid body excited by three different Italian registered earthquakes has been analyzed, reporting rocking and overturning curves. Results show that, for narrow spectrum earthquakes, bi-dimensional models are not enough accurate to correctly evaluate the occurrence of the overturning since, in significant sectors inside the 3D rocking regions, the overturning amplitudes are smaller than the ones given by the 2D models. Hence a 3D model of rigid block is necessary to evaluate the seismic response of a rigid block in favour of safety.

Appendix A. Vector and tensor quantities

The rotation $\mathbf{R}$ is the composition of three planar rotations: if $\{\mathbf{e}_x, \mathbf{e}_y, \mathbf{e}_z\}$ is the canonical basis, the first rotation, indicated as $\mathbf{R}_1$, of angle ϑ_1, is around the axis $\mathbf{e}_x$; the second, indicated as $\mathbf{R}_2$, of angle ϑ_2, around the axis $\mathbf{R}_1 \mathbf{e}_y$; the third, indicated as $\mathbf{R}_3$, of angle ϑ_3, around the axis $\mathbf{R}_2 \mathbf{R}_1 \mathbf{e}_z$. The representation of $\mathbf{R}$ on the canonical basis is

$$\left[\mathbf{R}(t)\right]_{\mathbf{e}_{x,y,z}} = \begin{pmatrix} c_2 c_3 & s_1 s_2 c_3 - c_1 s_2 & c_1 s_2 c_3 + s_1 s_3 \\ c_2 s_3 & c_1 c_3 + s_1 s_2 s_3 & c_1 s_2 s_3 + s_1 c_3 \\ -s_2 & s_1 c_2 & c_1 c_2 \end{pmatrix} \tag{13}$$

where, for $k = 1, 2, 3,$

$$\begin{aligned} c_k &:= \cos(\vartheta_k(t)) \\ s_k &:= \sin(\vartheta_k(t)) \end{aligned} \tag{14}$$

When the block is a parallelepiped of uniform mass density, with sides of length $2b_x,\ 2b_y,\ 2h$, respectively, the positions of the base vertices are

$$\begin{aligned} \hat{\mathbf{x}}_A &= 0 \\ \hat{\mathbf{x}}_B &= \hat{\mathbf{x}}_A + 2b_x \mathbf{e}_x \\ \hat{\mathbf{x}}_C &= \hat{\mathbf{x}}_A + 2b_x \mathbf{e}_x + 2b_y \mathbf{e}_y \\ \hat{\mathbf{x}}_D &= \hat{\mathbf{x}}_A + 2b_y \mathbf{e}_y \end{aligned} \tag{15}$$

The mass is $m = 8\rho b_x b_y h$. The static moment with respect to the point A is

$$\hat{\mathbf{s}}_A = m(b_x \mathbf{e}_x + b_y \mathbf{e}_y + h \mathbf{e}_y) \tag{16}$$

The representation of the Euler tensor with respect to the point A is

$$[\hat{\mathbf{J}}_A]_{\mathbf{e}_{x,y,z}} = m \begin{bmatrix} \dfrac{4}{3}b_x^2 & b_x b_y & b_x h \\ b_x b_y & \dfrac{4}{3}b_y^2 & b_y h \\ b_x h & b_y h & \dfrac{4}{3}h^2 \end{bmatrix} \tag{17}$$

To get the generic static moment $\hat{s}_i$ and the generic Euler tensor $\hat{J}_{ji}$, the transport rules read:

$$\begin{aligned} \hat{\mathbf{s}}_i &= \hat{\mathbf{s}}_A + m(\hat{\mathbf{x}}_A - \hat{\mathbf{x}}_i) \\ \hat{\mathbf{J}}_{ji} &= \hat{\mathbf{J}}_A + \hat{\mathbf{s}}_A \otimes (\hat{\mathbf{x}}_A - \hat{\mathbf{x}}_i) + (\hat{\mathbf{x}}_A - \hat{\mathbf{x}}_j) \otimes \hat{\mathbf{s}}_A \\ &\quad + m(\hat{\mathbf{x}}_A - \hat{\mathbf{x}}_j) \otimes (\hat{\mathbf{x}}_A - \hat{\mathbf{x}}_i) \end{aligned} \tag{18}$$

where the tensor product $\otimes$ is defined such that $(u \otimes v)w = (u \cdot w)v$ for any vectors u, v, w of the same vector space.

Author details

Alessandro Contento, Daniele Zulli and Angelo Di Egidio

Dipartimento di Ingegneria Civile,Edile-Architettura, Ambientale University of L'Aquila, Italy

References

[1] Shenton H, Jones N. Base excitation of rigid bodies. I: Formulation. Journal of Engineering Mechanics 1991; 117: 2286–2306.

[2] Taniguchi T. Non-linear response analyses of rectangular rigid bodies subjected to horizontal and vertical ground motion. Earthquake Engineering and Structural Dynamics 2002; 31: 1481–1500.

[3] Boroscheck R, Romo D. Overturning criteria for non-anchored non-symmetric rigid bodies. In: Proceeding of the 13th World Conference on Earthquake Engineering, 1-6 August 2004, Vancouver, Canada.

[4] Agbabian M, Masri F, Nigbor R, Ginel W. Seismic damage mitigation concepts for art objects in museum. In: Proceeding of the 9th World Conference on Earthquake Engineering, 1998, Tokyo-Kyoto, Japan.

[5] Zhu Z, Soong T. Toppling fragility of unrestrained equipment. Earthquake Spectra 1998; 14: 695–712.

[6] Zhang J, Makris N. Rocking response of free-standing blocks under cycloidal pulses. Journal of Engineering Mechanics 2001; 127: 473–483.

[7] Makris N, Roussos Y. Rocking response of rigid blocks under near-source ground motions, Géotechnique 2000; 50: 243–262.

[8] Spanos P, Koh A. Rocking of rigid blocks due to harmonic shaking. Journal of Engineering Mechanics 1984; 110: 1627–1642.

[9] Makris N, Zhang J. Rocking response of anchored blocks under pulse-type motions. Journal of Engineering Mechanics 2001; 127: 484–493.

[10] Kounadis A. On the overturning instability of a rectangular rigid block under ground excitation. The Open Mechanics Journa 2010; l4: 43–57.

[11] Purvance M, Anooshehpoor A, Brune J. Freestanding block overturning fragilities: numerical simulation and experimental validation. Earthquake Engineering and Structural Dynamics 2008; 37(5): 791-808.

[12] Spanos P,Roussis P, P N P A. Dynamic analysis of stacked rigid blocks. Soil Dynamics and Earthquake Engineering 2000; 21: 559–578.

[13] Spanos P, Koh A. Harmonic rocking of rigid block on fexible foundation. Journal of Engineering Mechanics 1986; 112: 1165–1181.

[14] Spanos P, Koh A. Analysis of block random rocking. Soil Dynamics and Earthquake Engineering 1986; 5: 178–183.

[15] Lenci S, Rega G. Heteroclinic bifurcations and optimal control in the non linear rocking dynamics of generic and slender rigid blocks. International Journal of Bifurcation and Chaos 2005; 5: 1901–1918.

[16] Lenci S, Rega G. A dynamical systems approach to the overturning of rocking blocks. Chaos, Solitons and Fractals 2006; 28: 527–542.

[17] Lenci S, Rega G. Optimal control and anti-control of the nonlinear dynamics of a rigid block. Philosophical Transactions of the Royal Society A 2006; 364: 2353–2381.

[18] ujita K, Yoshitomi S, Tsuji M, Takewaki I. Critical cross-correlation function of horizontal and vertical ground motions for uplift of rigid block. Engineering Structures 2008; 30: 1199–1213.

[19] Iyengar R, Manohar C. Rocking response of rectangular rigid blocks under random noise base excitations. International Journal of Non-Linear Mechanics 1991; 26: 885–892.

[20] Vestroni F, Di Cintio S.,Base isolation for seismic protection of statues. In: Twelfth World Conference on Earthquake Engineering, 2000, New Zealand.

[21] Caliò I, Marletta M. Passive control of the seismic response of art objects. Engineering Structures 2003; 25: 1009–1018.

[22] Contento A, Di Egidio A. Investigations into the benefts of base isolation for non-symmetric rigid blocks. Earthquake Engineering and Structural Dynamics 2009; 38: 849–866.

[23] Di Egidio A,Contento A. Base isolation of sliding-rocking non-symmetyric rigid blocks subjected to impulsive and seismic excitations. Engineering Structures 2009; 31: 2723–2734.

[24] Di Egidio A,Contento A. Seismic response of a non-symmetric rigid block on a constrained oscillating base. Engineering Structures 2010; 32: 3028–3039.

[25] Koh A, Mustafa G. Free rocking of cylindrical structures. Journal of Engineering Mechanics 1990; 116: 34–54.

[26] Batista M. Steady motion of a rigid disc of finite thickness on a horizontal plane. Journal of Nonlinear Mechanics 2006; 41: 850–859.

[27] Stefanou I, Vardoulakis I, Mavraganis A. Dynamic motion of a conical frustum over a rough horizontal plane. International Journal of Nonlinear Mechanics 2011; 46: 114–124.

[28] Taniguchi T. Rocking behavior of unanchored fat-bottom cylindrical shell tanks under action of horizontal base excitation. Engineering Structures 2004; 26: 415–426.

[29] Zulli D, Contento A, Di Egidio A. Three-dimensional model of rigid block with a rectangular base subject to pulse-type excitation. International Journal of Non-Linear Mechanics 2012; 47(6): 679-687.

[30] IMSL Fortran Library User's Guide, Visual Numerics, 2003.

[31] Housner G.The behaviour of inverted pendulum structures during earthquakes. Bulletin of the Seismological Society of America 1963; 53: 404–417.

Engineering

Chapter 8

Pushover Analysis of Long Span Bridge Bents

Vitaly Yurtaev and Reza Shafiei

Additional information is available at the end of the chapter

1. Introduction

It has been observed that most of the bridges damaged in earthquakes were constructed before 1971 and had little or no design consideration to seismic resistance. Since the 1971 San Fernando earthquake in California, the standards for earthquake design have been strengthened considerably, and bridge structural behavior has been more accurately evaluated. Since then, structural ductility, a crucial element for the survival of bridges under severe earthquakes has become a key consideration in structural analysis and design.

However, bridges that were constructed prior to 1971 are still in use and play important roles in our transportation systems, which may be susceptible to failure due to their structural deficiencies. To ensure safety and performance of these bridges, a seismic retrofit and strengthening program has been one of the major efforts of the Washington Department of Transportation and the Federal Highway Administration, aiming at improving seismic performance of older bridges. Retrofitting methods such as restrainers and column jacketing have proven to be effective in recent earthquakes. Techniques to retrofit other bridge members have also been developed such as soil anchors, footing retrofit involving increased plan dimension and reinforced overlay, construction of link beams, and system isolation and damping device.

The goal of seismic retrofit is to minimize the likelihood of structural failure while meeting certain performance requirements. This allows engineers to design repair strategies based on performance needs. As a consequence, some level of damage may be acceptable during a design-level earthquake. The California Department of Transportation (Caltrans) has required that bridge retrofits provide survival limit-state protection at seismic intensities appropriate for new bridges. This makes possible the proposition of efficient and effective strengthening measures with optimized retrofitting schemes, and the adoption of the plan that is the most economical for the acceptable damage level. One of the ways of implementation the retrofit program for the structures is providing a nonlinear static analysis.

Nonlinear static analysis under monotonically increasing lateral loading is becoming an increasingly popular tool for seismic performance evaluation of existing and new structures. Pushover analysis can be viewed as a method for predicting seismic force and deformation demands, which accounts in an approximate manner for the redistribution of internal forces occurring within the inelastic range of structural behavior. It is expected to provide information on many response characteristics that cannot be obtained from an elastic static or dynamic analysis. Pushover analysis is based on the assumption that the response of the structure can be related to the response of an equivalent SDOF system. This implies that the response is controlled by a single mode, and that the shape of this mode remains constant through the time history response. These assumptions are likely to be reasonable if the structure response is not severely affected by higher mode effects, or the structure has only a single plastic mechanism that can be detected by an invariant load pattern. The use of at least two load patterns that are expected to bound inertia force distribution is recommended. For structures that vibrate primarily in the fundamental mode, pushover analysis will very likely provide good estimates of global as well as local inelastic deformation demands. It will also expose design weaknesses that may remain hidden in an elastic analysis. Such weaknesses include storey mechanisms, excessive deformation demands, strength irregularity, and overloads on potentially brittle elements, such as columns and connections. On the negative side, the most critical is the concern that the pushover analysis may detect only the first local mechanism that will form in an earthquake and may not expose other weaknesses that will be generated when the structure's dynamic characteristics change after the formation of the first local mechanism.

2. Objectives of the analysis

The purpose of this research is to evaluate the displacement capacity of bents from a long span bridge. A three-dimensional nonlinear finite-element model of the bridge bents were developed to determine the inelastic response by performing nonlinear pushover analysis. Modal pushover analyses were carried out in the transverse direction. Detailed data of performance was collected and interpreted to use as a baseline in a parametric study. Separate parametric study was carried out on a single column within this bridge in order to locate appropriate plastic hinge locations. These results were then transferred to individual bents, where multiple columns were modeled based on the results from the single column parametric study.

3. Description of the bridge

The bridge is located at the Primary State Highway No.1, Seattle Freeway Ravenna Boulevard Overcrossing North Bound. Figure 1 shows an aerial view of the bridge. The North Bound Bridge is the first bridge from below shown in the picture. The bridge length is 1310 ft back to back of pavement seats and consists of twenty spans. Plan and elevation views are shown in Figure 2 and Figure 3.

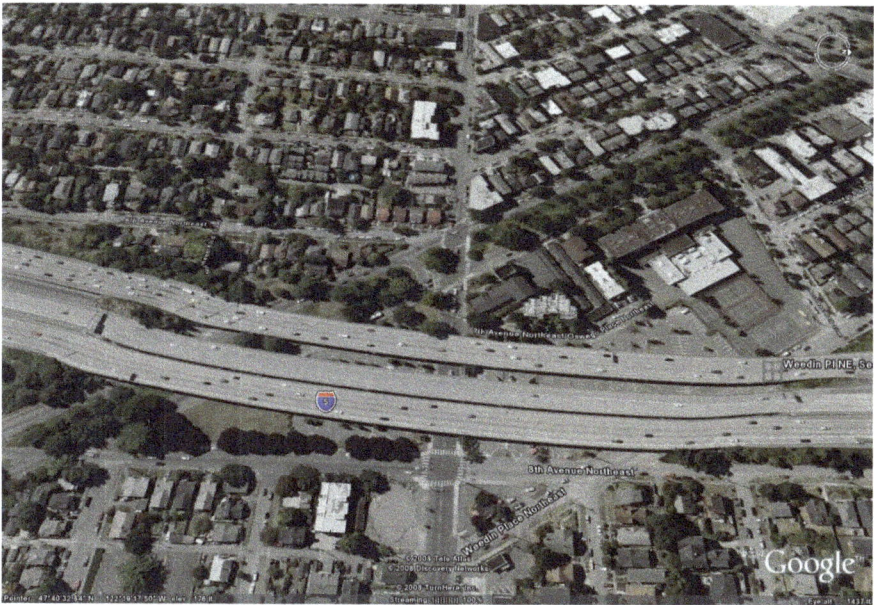

Figure 1. Aerial view of the North Bound bridge

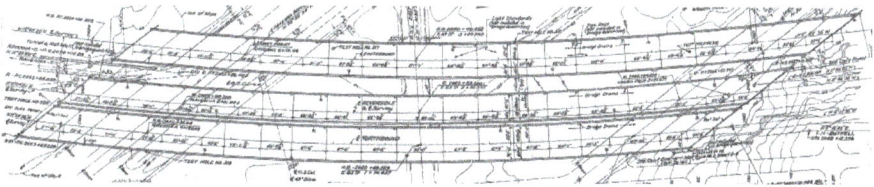

Figure 2. Plan view of the North Bound Bridge

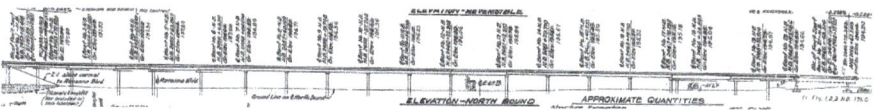

Figure 3. Elevation view of the North Bound Bridge

The superstructure is composed of pre-tensioned concrete beams. Each span includes twelve girders, and the general girder cross-section varies for each span. This complicates the calculation of the total mass of the superstructure. In order to simplify the procedure, individual span cross-sections were drawn in AutoCAD. The sections can be found is Figure 4 relating them to the spans they are assembled for. A table with calculated weight and length for each span can be found in the Appendix. Overlaid on top of the girders is a 5 in thick, approximately 60 ft wide reinforced concrete deck slab.

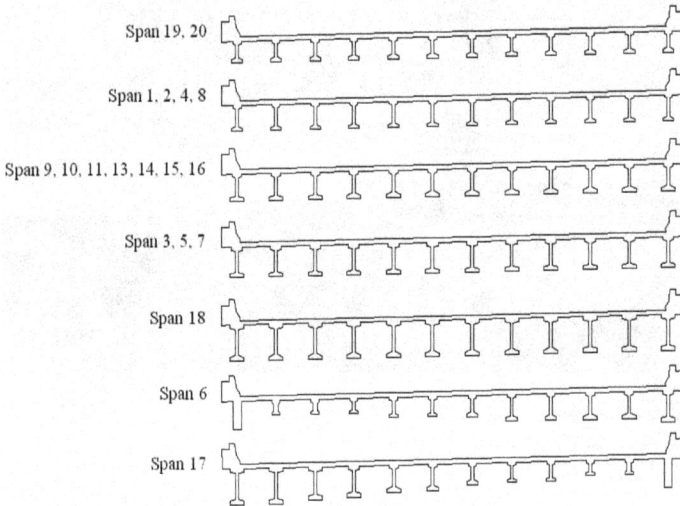

Figure 4. Superstructure Sections

There are a total of 19 bents in the bridge. Five are 6-column bents (#1-6), three are 7-column bents (#18-20) and ten are 4-column bents (#7-17). The cross-beam plans for the three types of bents are shown in Figure 5. Each bent has a unique elevation above the ground. Also, because of the curved shape of the bridge, each bent has a slight rotation in the vertical direction. Consequently, there is column height variation within each bent. The various column height values can be found in the Appendix.

At each bent, a 3x4.6 ft crossbeam transversely connects the columns. Figure 6 below shows the geometry and steel reinforcement. The length of the beams varies for each bent, which can be found in the Appendix. The steel reinforcement consists of nine No. 10 bars located at the top and at the bottom of each crossbeam. Two No. 5 bars are located at the side edges and run longitudinally along the crossbeam. For shear reinforcement, No. 5 stirrups are spaced evenly along each member.

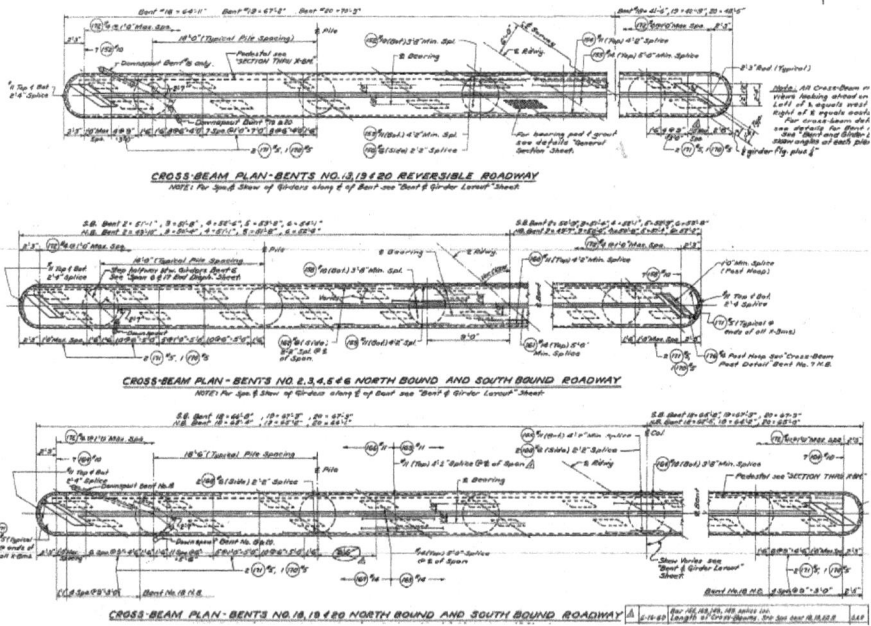

Figure 5. Cross-Beam Plan for Bents

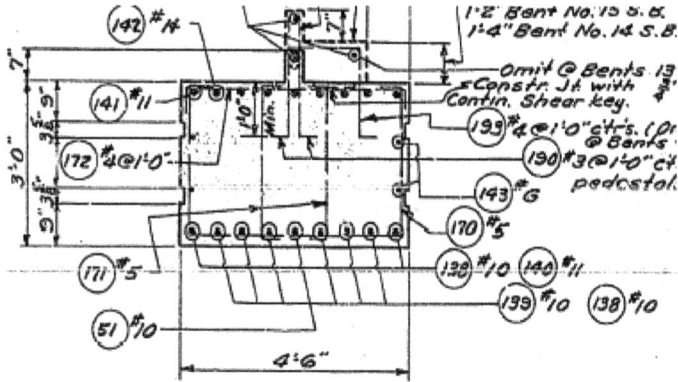

Figure 6. Section Thru Cross-Beam

The columns are spaced at 18 ft centerline to centerline. Each column is hollow with an outer diameter of 48 in and a wall thickness of 5 in. Twelve evenly spaced No. 5 bars provide the longitudinal reinforcement within each column. The columns also include twelve No. 3 steel cables each post-tensioned initially to 61 kips. Transverse reinforcement is provided by No. 2 spiral hooping spaced at 6 in on center. Figure 7 shows the plan column section. The columns are extended approximately 27 ft into the ground to act as piles.

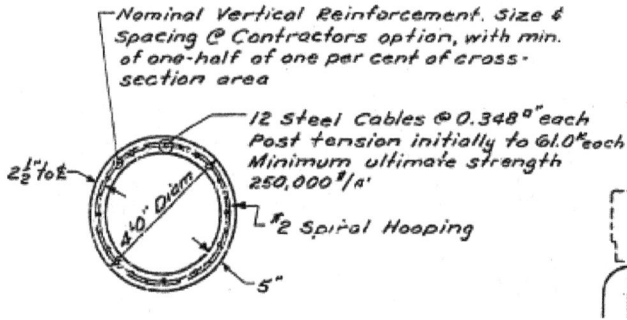

Figure 7. Plan Column Section

The columns and crossbeam were cast monolithically adding considerable rigidity to each bent. Figure 8 shows the elevation view of a typical bent. Further, the top 4 ft of each column is filled with class A concrete. This fill is further reinforced with sixteen No. 8 bars longitudinally, and No. 3 hoops spaced at 12 in transversely. In this section of the column, the hollow column is transversely reinforced with No. 2 spiral hooping spaced at 3 in over center. Figure 9 shows a typical pile.

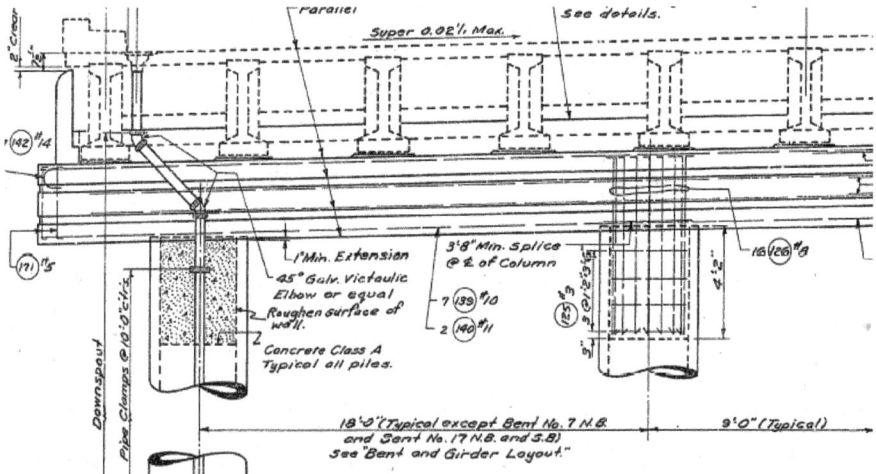

Figure 8. Elevation View of Bents

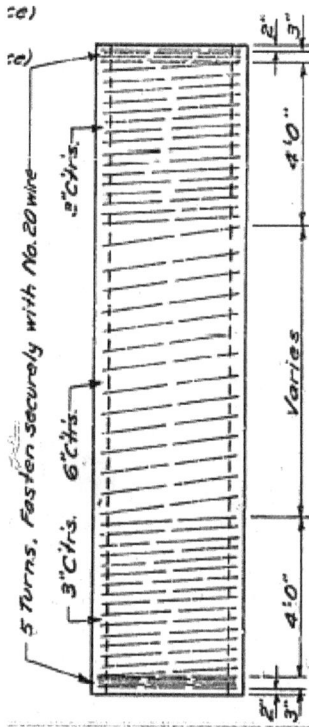

Figure 9. Typical Pile

4. Modeling of the bents

A spine model of each bent is created in the finite element program SAP2000. Line elements can behave three-dimensionally in the form of beam, beam-column elements and springs. The superstructure is represented as a distributed dead load which represents the dead weight of the superstructure based on tributary length of related spans for each bent. A table in the Appendix provides the distributed load values used in the analysis for each bent. The soil-structure interaction is represented by springs. In order to capture nonlinear behavior of the columns, plastic hinges were defined at maximum moment points. The general model is represented in 3D in Figure 10.

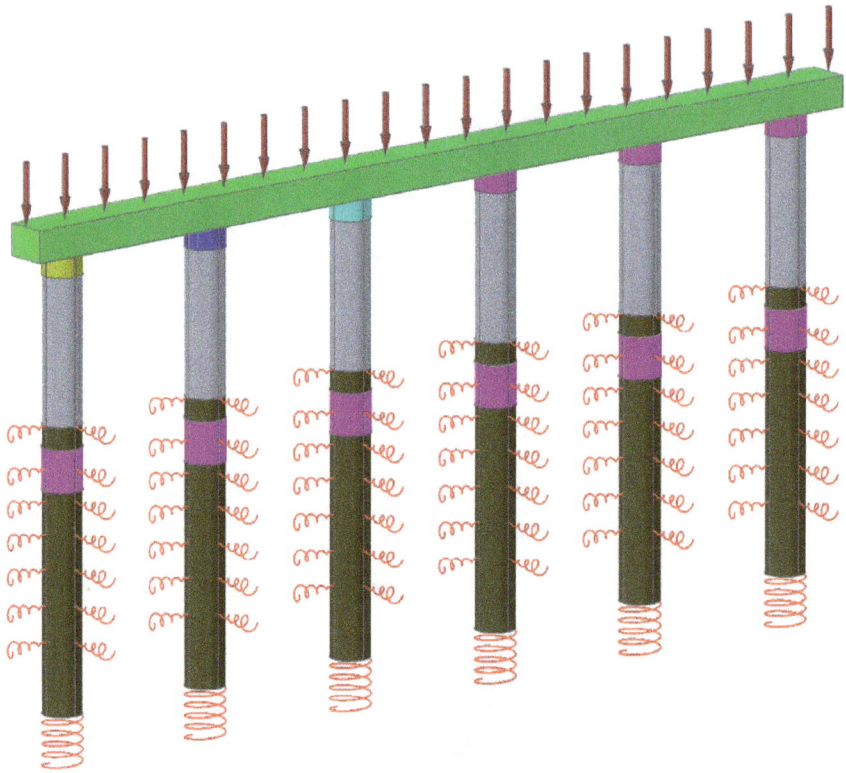

Figure 10. Model of Six-Column Bents

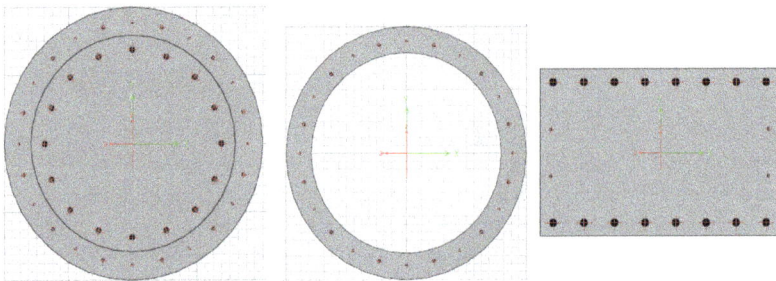

Figure 11. Bent Element Cross-Sections

The cross-sections of the cap beam and the column were accurately modeled by using the subprogram offered in SAP2000 called Section Designer. Section Designer lets the user draw the shape of the cross-section and also include the steel reinforcement. Figure 11 shows the drawn sections used in the analysis.

Piles

The columns were considered fixed in the cap beam. Nonlinear springs along the pile shafts were used to model the resistance provided by the surrounding soil. The L-Pile software (2002) was used to compute the P-Y curves, based on the stiff sand soil model with free water at 15 depths.

To build an exact computer model of a structure beard against underground elements-piles it is necessary to know how interaction between soil and a pile can be simulated, to get more precise result of the analysis. The p-y curves is a strait interpretation of the relation between deflection of an element and soil pressure on a particular depth. The pressure from the soil on the element is distributed within certain length which depends on the number of springs assigned to it Figure 12.

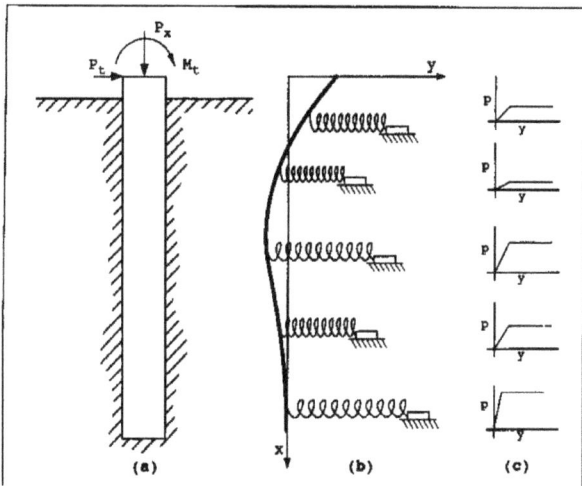

Figure 12. Model of laterally loaded pile

A physical definition of the soil resistance p is given in Figure 13. There was made an assumption that the pile has been installed without bending so the initial soil stresses at the depth x_i are uniformly distributed as shown in Figure 13b. If the pile is loaded laterally so that a pile deflection y_i occurs at the depth x_i the soil stresses will become unbalanced as shown in Figure 13c. Integration of the soil stresses yielding the soil resistance p_i with units F/L equation 1.

$$p_i = E_s y_i \tag{1}$$

where,

E_s – a parameter with the units F/L^2, relating pile deflection y and soil reaction p.

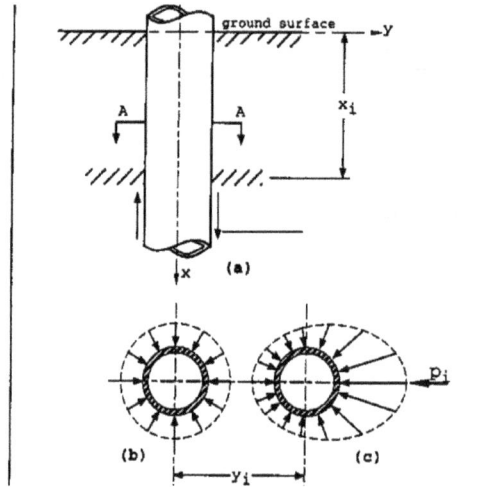

Figure 13. Definition of p and y as Related to Response of a Pile to Lateral Loading

Once the p-y curves at various depths of the pile have been obtained, a force-displacement relationship can be calculated by multiplying p with the tributary length of the pile between springs. Figure 14 shows a bilinearization of the force-displacement relationship at different depth based on the data retrieved from LPILE single pile analysis. These results were used to define multi-linear elastic links (springs) in SAP2000 in order to represent the SSI of the piles. The piles of all bents were assumed to extend 27 ft under the ground, so all bents had the same pile modeling.

5. Plastic hinge

It is well known that well-confined concrete structures can deform inelastically without significant strength loss through several cycles of response. Ductility describes such ability of structures, which is often defined as the ratio of deformation at a given response level to the deformation at yield response. Commonly used ductility ratios include displacement ductility, curvature ductility and rotation ductility. In the software of XTRACT, developed by Imbsen & Associates Company (2002) with the capability of analyzing structural cross sections, curvature ductility can be calculated for a given section and are defined in Equation 2 (Paulay and Priestley, 1992).

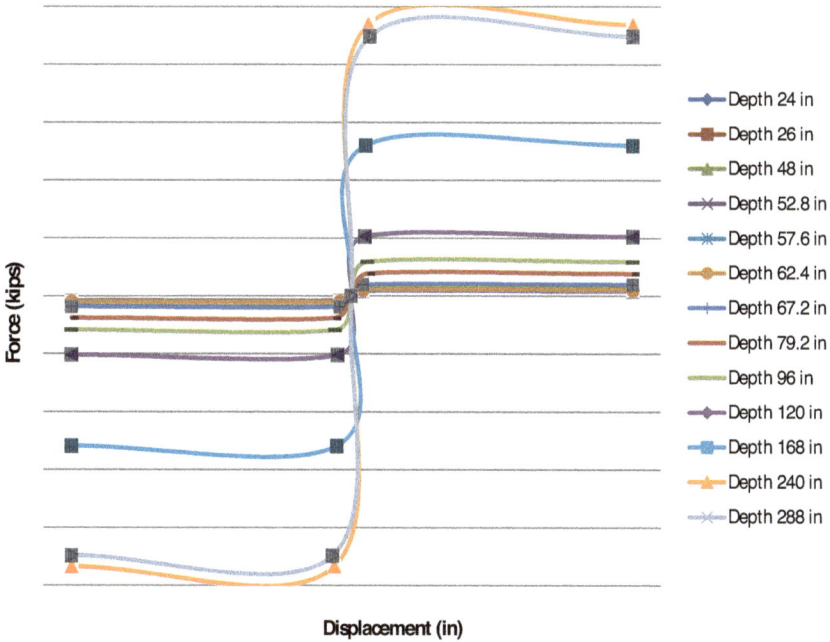

Figure 14. Bilinearized Force-Displacement of SSI at Different Depths

$$\mu_\varphi = \frac{\varphi_u}{\varphi_y} = \frac{\varphi_p + \varphi_y}{\varphi_y} \qquad (2)$$

in which ϕ_y is yield curvature, ϕ_p is plastic curvature, and ϕ_u is summation of yield curvature and plastic curvature that presents the ultimate curvature capacity of a section.

Figure 15 and Figure give a moment-curvature diagram for the column sections in the North Bound Bridge, calculated by the XTRACT. Curvature properties are section dependent and can be determined by numerical integration methods. Input data of a cross-section include nonlinear material properties of concrete and steel, and the detailed configuration of the section. For the North Bound bridge, all the columns have the identical section dimension, however, the moment-rotation relationships may not be the same because of the different axial loads.

Hinge length

The plastic hinge length for piles depends on whether the hinge is located at the pile/deck interface or is an in-ground hinge. For prestressed piles where the solid pile is embedded in the deck, the plastic hinge length at the pile/deck interface can be taken as (PIANC):

Moments about the X-Axis - kip-in

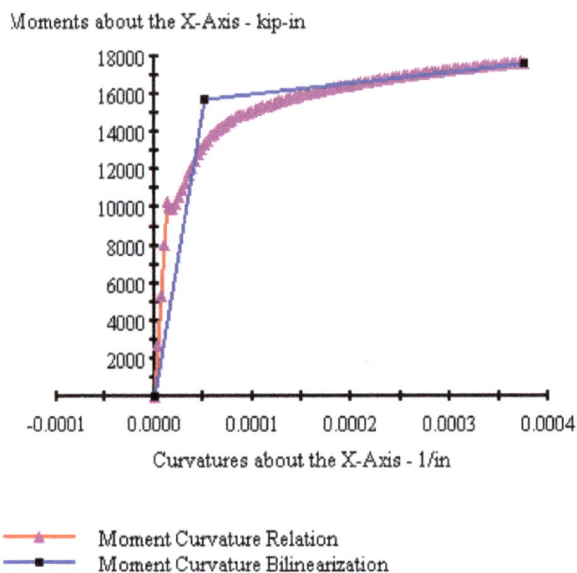

Curvatures about the X-Axis - 1/in

———▲——— Moment Curvature Relation
———■——— Moment Curvature Bilinearization

Figure 15. Bilinearization of the Moment–Curvature Curve for Hollow Column

Moments about the X-Axis - kip-in

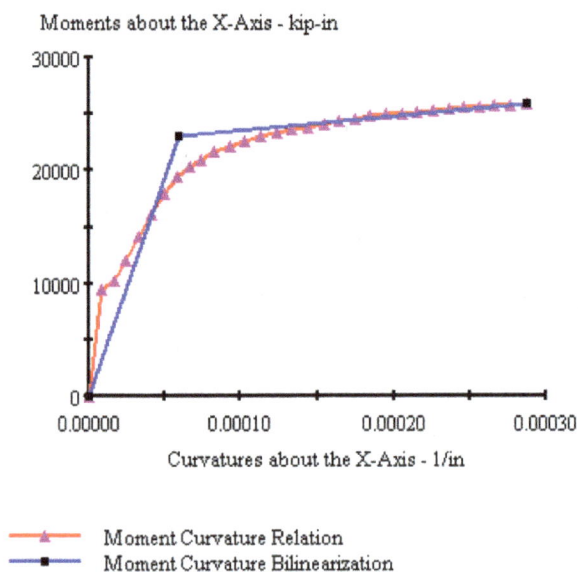

Curvatures about the X-Axis - 1/in

———▲——— Moment Curvature Relation
———■——— Moment Curvature Bilinearization

Figure 16. Bilinearization of the Moment–Curvature Curve for Filled Hollow Column

$$L_p = 0.5\, D_p \qquad\qquad (3)$$

where,

D_p – diameter of a pile

For in-ground hinges, the plastic hinge length depends on the relative stiffness of the pile and the foundation material. Because of the reduced moment gradient in the vicinity of the in-ground hinge, the plastic hinge length is significantly longer there. In this report CALTRANS interpretation of in-ground hinges for a non-cased pile shaft was used. Figure 17 describes the calculation steps provided by CALTRANS.

$$L_p := D_e + 0.6 \cdot L$$

L_p = Plastic hinge length (in)

L = Length of pile shaft from
            ground surface to point of
            contraflexure above ground (in)

D_e = Diameter or least cross sectional
            dimension (in)

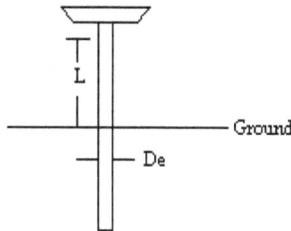

Figure 17. In-Ground Hinge Length

Hinge location

In order to locate the plastic hinge locations, a separate push over analysis was run on single column. Figure 18 shows the single column element modeled in SAP2000. Top of the column is restrained against rotation to represent the rigid connection between the column and the deck. The SSI is represented by links just as discussed for general bents. The pin connection at the bottom of the pile restricts the pile from vertical movement.

Figure 19 provides the moment diagram of the above column/pile under horizontal loading. The diagram has two points of maximum moment. The plastic hinge should be placed at these locations in order to represent the most conservative nonlinear behavior of the column/pile.

Figure 18. Single Column Finite Element Stick Model

The placement of the first hinge should be at the column/bent connection as expected before. The second hinge has to be place under the ground, but the location of maximum moment in that area changes in a pushover analysis. A parametric study was run in order to locate the worst location for an in-ground hinge. The placement of the in-ground hinge was varied for multiple pushover analysis. Figure 20 shows the results of this parametric study, where the hinge depth below ground level is compared to column top displacement capacity. The plot in Figure 20 shows that placing the hinge 20% of pile length under the ground would give a displacement capacity of 2.25 in, which is less than any other location. Figure 21 shows the placement of the plastic hinges in four column bent.

Figure 19. Moment Diagram of Single Column under Horizontal Load

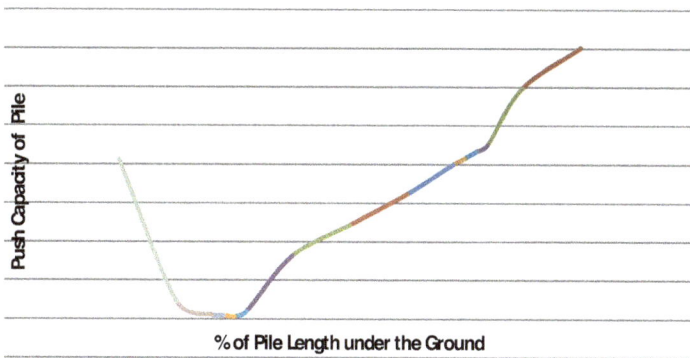

Figure 20. Single Column Parametric Study Results

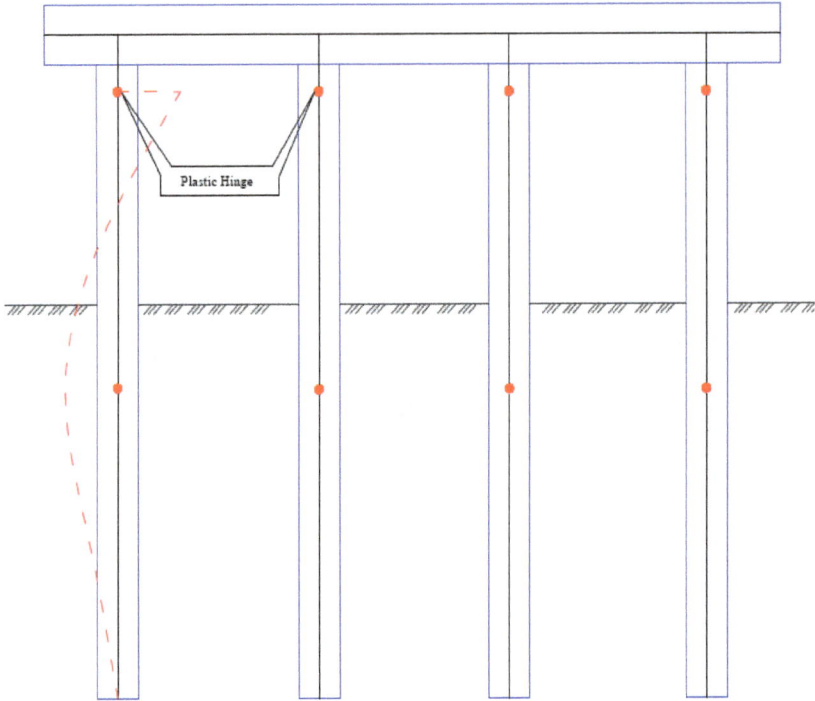

Figure 21. Location of the Plastic Hinge

Plastic hinge property

The Manual of SAP2000 recommends a distributed plastic hinge model assuming 0.1 of element length as the plastic hinge length, but information on how to define distributed plastic hinge properties is not provided. In this research, a concentrated plastic hinge model is used with the assumption that plastic rotation will occur and concentrate at mid-height of a plastic hinge. Input hinge properties consist of the section yield surface, plastic rotation capacity, and acceptance criteria.

A plastic rotation, θ_p, can be calculated by the plastic curvature given the equivalent plastic hinge length L_p as shown in Equation 4.

$$\theta_p = \varphi_p L_p = L_p(\varphi_u - \varphi_y) \tag{4}$$

The plastic rotation is an important indicator of the capacity of a section to sustain inelastic deformation and is used in SAP to define column plastic hinge properties. FEMA 356 provides a generalized force-deformation relation model shown in Figure 22 for the nonlinear static analysis procedure, which is the defaulted model in SAP for the Axial-Moment hinge.

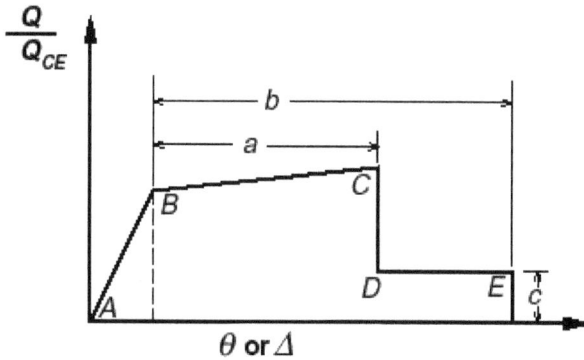

Figure 22. Generalized Force-Deformation Relations for Concrete Elements (FEMA-356)

Three parameters, a, b and c are defined numerically in FEMA-365, and are permitted to be determined directly by analytical procedures. The moment and rotation are normalized by yield moment and yield rotation respectively, i.e., $\frac{M}{M_y}$ and $\frac{\theta}{\theta_y}$. By default SAP will calculate the yield forces and the yield rotation based on reinforcement and section provided.

In Table 6-8 of FEMA 356, modeling parameters and numerical acceptance criteria are given for reinforced concrete columns in various categories. Columns investigated are all primary structural elements. A conforming transverse reinforcement is defined by hoops spaced in the flexural plastic hinge region less than or equal to $\frac{d}{3}$, and the strength provided by the hoops (Vs) being greater than three-fourths of the design shear. Thus, the category of the column is decided in Table 6-8 of FEMA 356, and values and relationship of the performance levels can be utilized.

In SAP, an absolute rotation value can overwrite the default value in defining a hinge property. The plastic rotation capacity angle, a, calculated with Equation 4-12 for a given column is at point C. The ultimate rotation angle, which is inputted as b in SAP, is taken as 1.5 times the plastic angle. It is indicated at point E, which defines a local failure at a plastic hinge. A larger value could be used to allow the structure to form a global failure due to instability.

The three discrete structural performance levels are Immediate Occupancy (IO), Life Safety (LS) and Collapse Prevention (CP) shown in Figure 23.

The ultimate plastic hinge angle calculated by the XTRACT was taken as the Collapse Prevention level. Its value was indicated as "a" in Figure 1. The permissible deformation for the Life Safety performance level is taken as three quarters of the plastic rotation capacity "a".

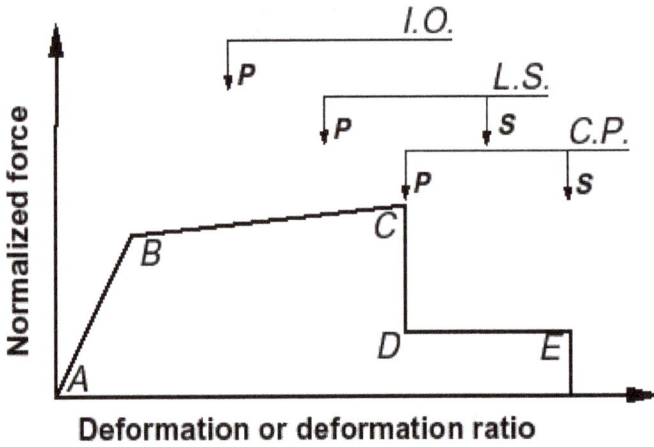

Figure 23. Performance Level on Generalized Force-Deformation Relations for Concrete Elements (FEMA-356)

The increase of moment strength at point C is taken as the over strength factor computed by XTRACT, ignoring the strength softening effect. The actual moment strength at point C is the product of the factor and the yielding moment. FEMA 356 defines a 0.2 residual strength ratio before plastic hinge eventually fails. Figure 24 presents moment-rotation curves for one of the columns.

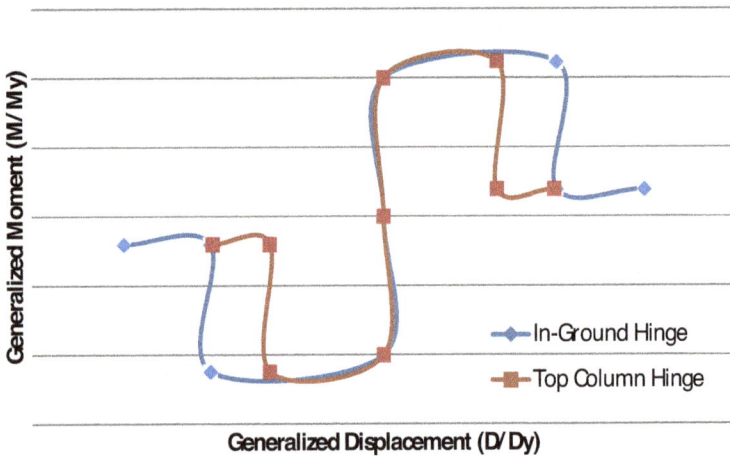

Figure 24. Moment-Rotation Relationship of the Columns

A concrete interaction surface was obtained from XTRACT for the frame hinges under combined bending and axial load. A generated interaction surface is shown in Figure 25.

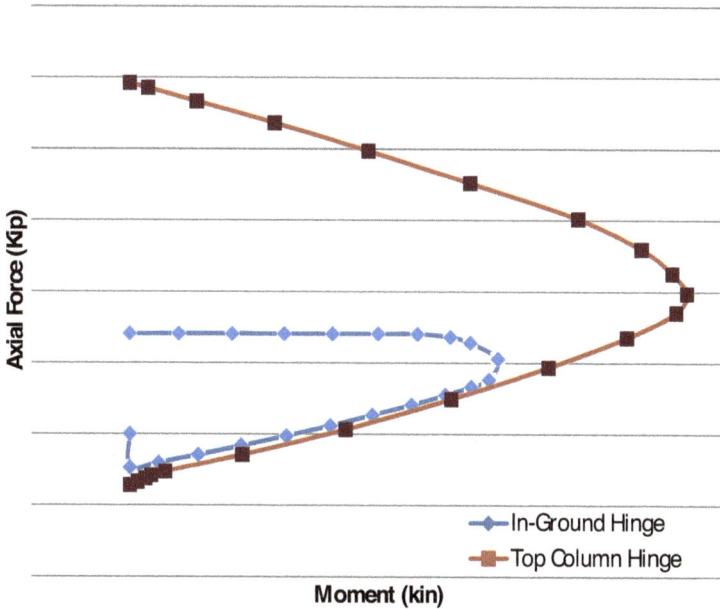

Figure 25. Axial Load-Moment Interaction Curve (Compression force is negative)

6. Results

The results of the pushover analysis are organized in the following pages. Each bent has two pages of results. The first page shows the general characteristics of the bent and the push over curve. The second page shows the results of the step by step push over analysis and the development of hinges. The first and the last bent on the plan are abutments, so they are not included in this report.

Maximum displacement capacity exhibited is 3.9 inches by the fifth bent. The lowest bent capacity is 2.08 inches by the second bent. The reason for this low displacement capacity in bent number two is because of its shorter columns compared to other bents.

Using two hinges per column creates a much lower capacity than using only one hinge at the top of the column. The shear force capacity stays almost the same, but the ductility reduces for all bents.

Author details

Vitaly Yurtaev[1,2*] and Reza Shafiei[1,2]

*Address all correspondence to: vitalijj@gmail.com

*Address all correspondence to: vitalijj@hotmail.com

1 Moscow State University of Railways, Department Of Civil Engineering, (MIIT), Moscow, Russia

2 Washington State University, Pullman, WA, USA

References

[1] FEMA 356(2000). table , 6-8.

[2] Priestley, M. J. N. Frieder Seible, Gian Michele. Seismic Design and Retrofit of Bridges, Wiley-IEEE, (1996). p., 1996, 200-250.

[3] Pui-shum, B. Shing, National Science Foundation (U. S.), Tadaaki Tanabe, Nihon Gakujutsu Shinkōkai, National Science Foundation (U.S.), Nihon Konkurīto Kōgaku Kyōkai, Nihon Gakujutsu Shinkōkai, Nihon Konkurīto Kōgaku Kyōkai, Modeling of inelastic behavior of RC structures under seismic loads, ASCE Publications, (2001). p., 2001, 425-450.

[4] Rainer, J. H, & Vanselst, A. Dynamic properties of Lions Gate Bridge', Proc. Conf. on Dynamic Response of Structures: Instrumentation,.esting Methods and System Identi,cation, ASCE Specialty Conf., University of California, (1976). , 243-252.

[5] Rainer, J. H, & Pernica, G. Dynamic.esting of a Modern Concrete Bridge, National Research Council of Canada, (1979). , 447.

[6] Nishimura, A, & Haya, H. Assessment of the structural integrity of bridge foundation by the impact vibration test', GEO-COAST, Yokohama, (1991). , 719-724.

[7] Shepherd, R, & Sidewell, G. K. Investigations of the dynamic properties of concrete bridges, 4th Australian Conf. on the Mechanics of Structures and Materials, University of Queensland, Brisbane, Australia, (1977). , 261.

[8] Dougla, B. M, Drown, C. D, & Gordon, M. L. Experimental dynamics of highway bridges', in G. Hart (ed.) Proc. Specialty Conf. on Dynamic Response of Structures, Experimentation, Observation, Prediction and Control, ASCE, New York, (1981). , 698.

Chapter 9

FE Based Vulnerability Assessment of Highway Bridges Exposed to Moderate Seismic Hazard

C. Mullen

Additional information is available at the end of the chapter

1. Introduction

The assessment of seismic vulnerability in regions where the risk from earthquake shaking is considered moderate poses special problems in terms of establishing critical conditions for failure and the importance and urgency for taking action. Research studies sponsored at the University of Mississippi (UM) over a period of about 10 years by the Mississippi Emergency Management Agency (MEMA) and Mississippi Department of Transportation (MDOT), respectively, have been aimed at identifying the vulnerability of select critical highway bridges subject to significant ground shaking from the New Madrid Seismic Zone (NMSZ).

The historical occurrence of multiple but infrequent major seismic events in the NMSZ exceeding seismic moment of $M\,7$ has been established by geophysicists and seismologists through numerous surveys of surface rupture features and paleoseismological excavations conducted throughout the region (for example, see [10, 18]). Planners in both state and federal agencies are concerned about the consequences of both physical and economic damage posed by the next major recurrence of a potentially catastrophic earthquake along the fault. The United States (US) Federal Emergency Management Agency (FEMA) sponsored a major research study [3] to investigate the multi-state regional consequences of a hypothetical event of $M\,7.7$ on both buildings and bridges. The bridges in Mississippi discussed in this chapter represent critical lifelines exposed to the earthquake threat that are located along the evacuation routes and economic supply chains for communities in the northern part of the state as well as the tri-state metropolitan area of the city of Memphis, Tennessee, having population of about 1.3 million.

A myriad of uncertainties exist for both the rare but potentially catastrophic seismic events and the multiple factors affecting the response of these soil-foundation-structure systems. In the absence of ground motion records for the severe historical events in the seismic zone under consideration, a simulation based approach is adopted to highlight the salient features of both the input and response at the site. The vulnerability assessment requires that reasonable behavioral response and multiple failure limit states be examined under a range of possible ground motion intensities. While a probabilistic approach is desirable overall, a deterministic approach enables the examination of the key response characteristics and the detailed information needed to establish relative importance of different limit states including soil capacity, pile/column axial and flexural strength, and member/system instability.

The bridge seismic vulnerability studies in this chapter highlight the challenges posed by the need to balance the level of sophistication of the finite element (FE) simulation with the:

1. state of knowledge of the bridge facilities, their seismic exposure, and local site conditions

2. project objectives in order to provide safe and economic decision making for hazard mitigation and emergency response and mobilization planning

Lessons learned and discussed herein are the result of over a decade of research at UM sponsored at the multidisciplinary Center for Community Earthquake Preparedness (CCEP) and graduate level studies by a number of students supported by the Department of Civil Engineering.

2. Seismic hazard and inventory characterization for the study region

In [10] the 1811, 1812 sequence of three distinct earthquakes corresponding to rupture along separate segments of the irregular shaped New Madrid fault is described. More recently, in the FEMA study [3], a scenario established for emergency planning purposes comprising a single M7.7 event consisting of sequential rupture along all three segments. Seismicity of smaller events recorded using a strong motion instrument array during an almost 30 year span is plotted in Figure 1 to which the approximate location of the study region has been added.

According to the 2012 data compiled for the National Bridge Inventory (NBI) [6] in the US by the Federal Highway Administration (FHWA), a total of 18,459 highway bridges are found in the 82 counties in the state of Mississippi (MS). The study region contains only a small subset of this inventory and may be approximately characterized as the counties located in north MS most likely to experience moderate ground shaking from a major event in the NMSZ. Based on default inventory data contained in the GIS-based software, Hazards US-Multihazard (HAZUS-MH) [5] created under sponsorship by FEMA for use in emergency management planning, 1133 bridges are exposed to the moderate seismic hazard.

The seismic vulnerability of all bridges in MS has been examined from a risk or loss estimation point of view in both [3] and [13]. In each study, the HAZUS-MH methodology has been implemented which depends on use of fragility curves assigned to bridge classes included in the NBI system. No study has yet been performed to assess seismic vulnerability using FE as the basis of the loss estimation. The present study provides a first step toward such a more comprehensive study and focuses on five bridges at a variety of sites in the study region

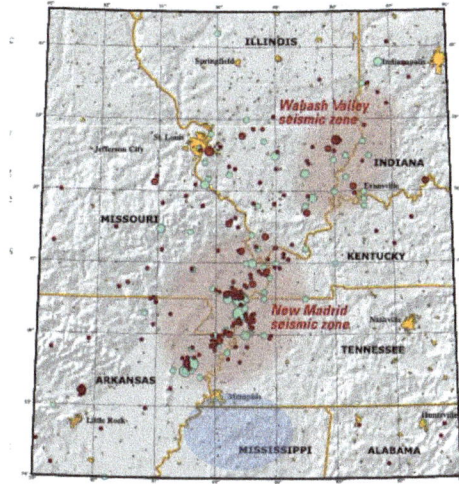

Figure 1. Recent seismicity in NMSZ and surrounding multi-state region exposed to risk of a repeat of historic catastrophic events (M7-M8); red circles give epicenters for events > M2.5 during the period, 1974-2002 [from 19]; the study region is represented by the blue shaded area

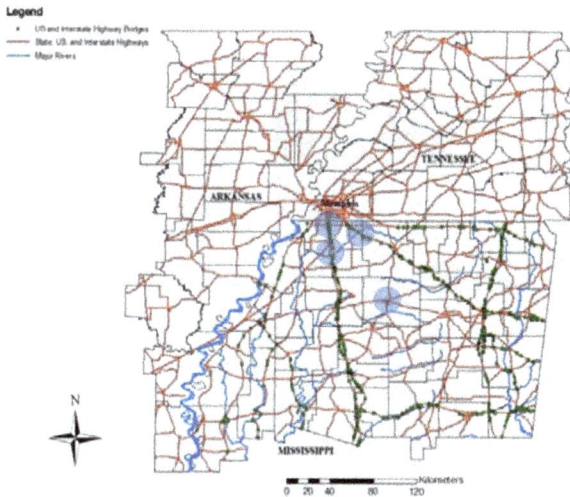

Figure 2. NBI bridge inventory in study region shown by open circles; green circles show bridges located in north Mississippi on the major access routes for the Memphis metropolitan area, those investigated using FE based seismic vulnerability analysis lie within shaded areas; red lines indicate highways on federal and state system; blue lines represent major rivers to show critical water crossings

investigated during three separate projects. Figure 2 shows the locations of the sites in relation to the NBI inventory supplied in HAZUS-MH and federal and state highway system.

The select bridges studied have been modeled to varying degrees of complexity with both two-dimensional (2D) and three-dimensional (3D) computational simulations including eigenvalue, linear dynamic, nonlinear static, and nonlinear dynamic. Earthquake time histories have been generated to capture a range of intensities from M6 to M8 and peak ground accelerations (PGA) in the approximate range, PGA=5-50% g, depending on the site, study objectives, and methodology. It is noteworthy that, over the period of the studies, significant changes have occurred in the understanding of the earthquake risk and level of ground shaking to be expected. Each study used the best available knowledge at the time.

The earliest study was performed for MEMA in the context of a broader study of the seismic vulnerability of facilities located on the UM campus [17]. The motivation for the study was the belief that the University was and remains a key to economic development in the state as well as a place of both historic value and a population center of relatively high density. The study included an approximately 70 year old bridge that serves as a major entrance as seen in Figure 3. The bridge was designed by MDOT prior to any recognition of a significant seismic threat in the applicable design code. This bridge served as the first attempt at a detailed 3D FE-based evaluation of seismic response and vulnerability assessment [12]. The evaluation was performed with both fixed base and soil-foundation-substructure interaction boundary conditions to capture the influence of high embankments on the response of the structural components with emphasis on the pier columns.

Figure 3. East Gate Bridge carrying traffic from University Avenue in the City of Oxford at the entrance to the UM main campus [17]; present day bridge, old private railroad tracks, and right-of-way have been replaced by a modern city roadway

A second study [14] was performed for MDOT for what the Bridge Division deemed a critical facility that provides access from a major interstate highway to a vital economic development region in the state. The region is located within the fastest growing county in the state and one of the fastest growing in the nation due to its proximity to the metropolitan area of the city of Memphis, Tennessee. Approximately 30 years old, this bridge shown in Figure 4 was built to low seismic standards. The code recognized by MDOT was and remains the one published by the American Association of State Highway and Transportation Officials (AASHTO). Even when the first edition of the AASHTO Bridge Load and Resistance Factor Design (LRFD)

Specifications appeared in 1994, the ground motion demand at the site was only about PGA=0.15g.

The third and most recent study was performed for MEMA to investigate findings of the FEMA sponsored NMSZ catastrophic earthquake study [3] on the impact of an M 7.7 event on bridges in MS. Using a HAZUS-MH fragility curve based analysis which estimated conditional probabilities of damage at four basic limit states (slight, moderate, extensive, and complete), the study found that only six bridges in the entire state would have a significant probability exceeding slight damage. The purpose of the FEMA study was to provide states affected by the NMSZ a basis for establishing earthquake components of their federally mandated mitigation plans. The MEMA study used an FE based approach to establish vulnerability considering more site and facility specific information. In consultation with MDOT personnel, three bridges shown in Figure 5 were identified for study. All are located on major evacuation/mobilization routes which crossed the Coldwater River. The bridges were deemed near the edge of significant ground shaking based on the FEMA study. The rationale was that if these showed evidence of significant vulnerability then bridges closer to the NMSZ would then be at similar or higher risk.

Figure 4. Bridge carrying *MS* 302 over Interstate highway *I* 55; (left) looking toward Southaven, MS, a fast growing city forming part of the metropolitan area of the City of Memphis; (right) view of the intermediate bents and girders of the two closely spaced bridges

Figure 5. Three bridges crossing the Coldwater River on lifelines serving the study area; (left) view of southbound *I* 55 bridge; (middle) nearby *US* 51 bridge showing piled bents carrying simple spans; (right) view of northbound *US* 78 bridge

3. Ground motion simulation for the study sites

The lack of seismic records of significant earthquake events in the NMSZ makes the task of selecting ground motion excitation for response analysis a challenge. The state of knowledge of the causative features of the fault and the expected attenuation of motions from the source has changed over time and remains an area of significant debate and research. Spectral physics-based parametric source and attenuation models have provided a rational basis for the case studies presented here.

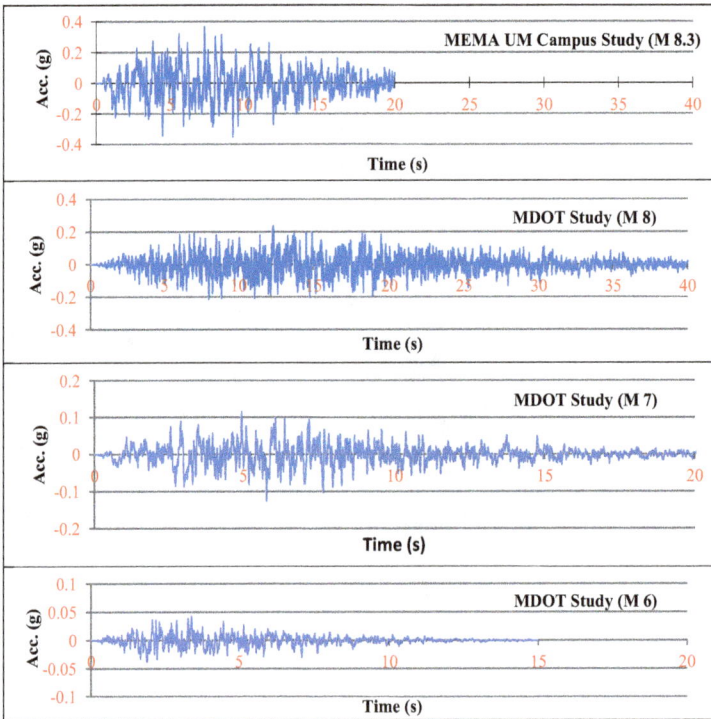

Figure 6. resultant horizontal ground acceleration time histories used in FE model analyses; MEMA UM campus study [17]; MDOT study [14]; MEMA Coldwater River bridges study [16]

Figure 6 shows resultant horizontal ground motion realizations generated for the various studies assuming 2D propagation from an assumed epicenter usually taken as Marked Tree, Arkansas, the town nearest to the southernmost position of the New Madrid fault. In the MEMA campus and MDOT bridge studies, orientation of the bridge was considered and component motions were then extracted for application to the FE models. In the absence of a 3D propagation model, requiring definition of layered media in a spherical coordinate system, vertical motion was obtained by uniformly scaling the resultant horizontal motion by the commonly assumed factor of 2/3.

In the MEMA UM campus study [17], the input horizontal motion realization for the UM campus was generated by others for M6.3 and M 8.3 events having source along the nearest (southernmost) segment of the NMSZ. In the MDOT bridge study [14], software was obtained from the US Geological Survey (USGS) and source model parameters and attenuation relations were identified in consultation with USGS and the University of Memphis enabling simulation of multiple realizations at arbitrary intensities. Events of nominal M 6, 7, and 8 were selected in order to capture different response levels. In the MEMA Coldwater River bridges study [16], a FEMA scenario of M 7.7 was adopted to be consistent with their results for the multi-state NMSZ region which were based on a distributed source model involving slip along the entire southern segment of the New Madrid fault. Since the study provided only PGA contours, not time histories, the MDOT study realization for the M 8 scenario case (Fig. 6) was scaled to achieve an input motion with PGA corresponding to that of the M 7.7 scenario at the bridge site locations (approximately PGA=0.25g).

Source spectral models for the very large intensity events were such that all ground motions have significant energy in the 1-2 Hz range coinciding with fundamental and low natural frequencies of the bridges.

4. FE modeling options

When using FE as the basis of vulnerability assessments, it is important to make several basic decisions regarding modeling approach including probabilistic versus deterministic and simple versus complex. These choices influence at the most general level, the software to be used, and at the most specific level, the key modeling assumptions such as system scope, boundary conditions, incorporation of soil-structure interaction (SSI), and focus on lumped parameter, 2D structural, or 3D continuum finite elements. Rather than propose a comprehensive view on the proper choices for all possible objectives, the select bridge study cases are offered as the possible range one might consider.

In the MEMA UM campus study [17], no prior knowledge existed. As a result of this uncertainty about what might be expected as well as a strong desire to ensure the safety of the many thousands of students, employees, and visitors to the campus and a major concern about the impact of significant losses to the future functioning of the university enterprise and consequential economic impacts on the state, the sponsors sought the most realistic view possible given the state of the art at the time. In response to this objective, the analysts committed to full 3D nonlinear dynamic FE simulation including SSI in cases where it might have a significant influence on the response. The project was initiated in the mid-1990s when the software ABAQUS [7] provided many desirable features including 3D nonlinear beam-column (structural) elements (B33) with user input moment-curvature relations and 3D continuum (solid) "infinite" elements (CIN3D8) with shape functions capturing radiation damping, in effect providing non-reflecting boundaries which allow dissipation of wave energy propagating radially away from the FE model.

There was little experience with the modeling approach at the time of the study and no experience with the nonlinear beam-column and radiation damping elements, so validation analyses were performed [9]. Detailed drawings were available from the bridge designer (MDOT), and a series of detailed models were developed to establish confidence in each subsequent level of complexity. Static self-weight analysis was first performed using a so-called fixed-based model (no soil stiffness included) to represent structural connectivity and weight and stiffness characteristics. Basic features of the fixed base model are shown in Fig. 7.

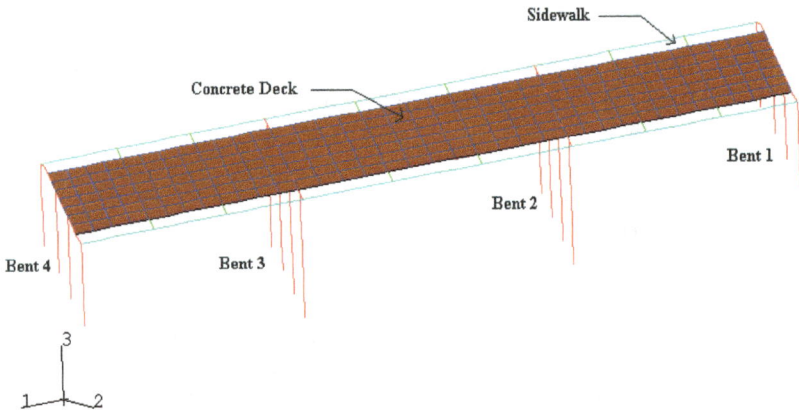

Figure 7. Fixed-base FE model of East Gate Bridge for MEMA UM campus study[17]; bents modeled with nonlinear beam-column elements; composite concrete deck-steel girder superstructure modeled using concrete plate elements for deck and linear beam elements for steel girders; no soil degrees-of-freedom

Once an acceptable result was obtained from the static analysis, an eigenvalue analysis was performed to estimate structural mass distribution characteristics and associated mode shapes and frequencies. Since the ground motions shown in Figure 6 accounted primarily for propagation through the earth's crust, modification and possible amplification as the seismic waves propagated through soil at the bridge site was not considered. To account for this limitation, a one-dimensional (1D) vertical wave propagation analysis [12, 17] was performed using a model of the top 100 ft of soil layers based on data obtained from soil borings. The analysis incorporated nonlinear softening of dynamic shear moduli at high strains and enabled generation of input motions to all fixed degrees-of-freedom (DOF) in the FE model regardless of elevation, in this case, at both the base of the columns of the intermediate bents and the level of the end abutment pile caps.

As Figure 3 shows, there is a significant difference (over 30 ft) in elevation between the abutments and the intermediate bents. Furthermore, the deck girders are built into concrete end walls where fill material is placed beneath the roadway. Between the abutments and what is now a roadway, steep embankments are found. To incorporate the interaction between the soil immediately below the footings of the intermediate piers, the embankments, and the structural system, the significantly more elaborate model shown in Figure 8 was developed [12,17].

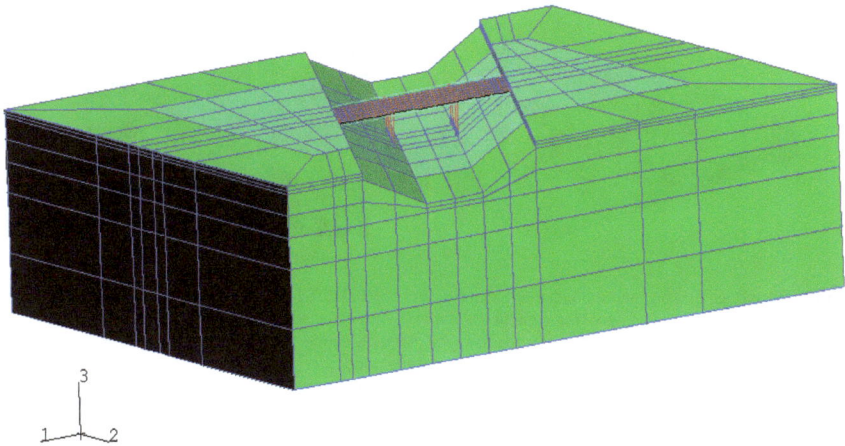

Figure 8. Subsurface geology and embankment interaction FE model of East Gate Bridge for MEMA UM campus study [12, 17]; end walls modeled with shell elements; active/passive soil pressure resistance modeled with nonlinear springs connecting end wall and back fill soil elements; embankment soil and subsurface geology modeled with elastic 3D solid elements; radiation damping at absorbing boundaries modeled with 3D solid infinite elements

The MDOT study was the first earthquake vulnerability study performed in the state for its Bridge Division. Again because of the many uncertainties, a 3D detailed FE based simulation approach [15] was adopted to provide the most accurate estimate of likely response. The bridge system was much larger than the one in the UM campus study due to the overcrossing of an interstate highway which now carries three lanes of traffic in each direction and the presence of two bridge frame substructures separated by a only a small gap between bents (see Fig. 4). The servicing of a large commercial center and a rapidly growing residential community required the bridge to carry a total of nine lanes of traffic, each substructure carrying traffic in one of the two directions. Embankments again created a significant difference in elevation of approximately 20 ft between soil beneath respective roadway pavements, but here the embankments were sloped to accommodate access to/from the interstate highway.

As shown in Figure 9, there were four continuous deck spans totaling approximately 350 ft. The substructures now included both piled footings at the end abutments and central intermediate bent and spread footings at the two other intermediate bents. A low-rise building SSI study [9] had demonstrated the importance of including a refined mesh locally around spread footings to account for soil softening under large seismic shaking. The detail view in Figure 9 shows the refinement pattern used around the bridge footings.

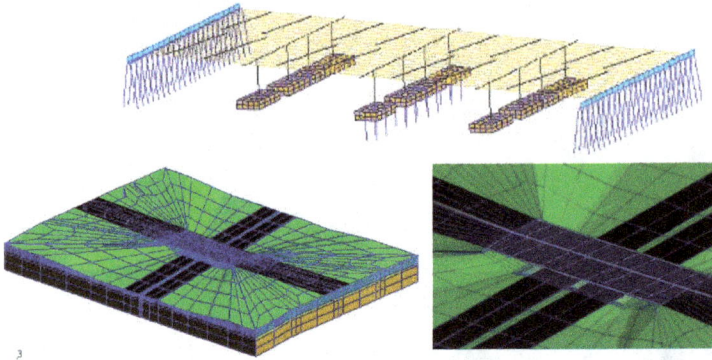

Figure 9. Subsurface geology and embankment SSI FE model of *I55/MS*302 Goodman Road Overcrossing for MDOT study [14]; concrete girders and bent frame members modeled with 3D nonlinear beam elements; concrete deck and footing modeled with shell elements (top figure shows soil elements connecting to footing shell elements); soil modeled using 3D solid elements with a Drucker-Prager cap material model for nonlinear response at high strains; radiation damping at absorbing boundaries modeled with 3D solid infinite elements

The MEMA Coldwater River bridges study [16] was originally intended to support a multi-state regional (National Level) earthquake Exercise (NLE) sponsored by FEMA with participation by MEMA. A major flood along the MS River threatened to overtop the levees protecting the farming communities in the Delta region, so MEMA personnel were called away from the exercise, and the input from the bridge study was not required as planned. The long term objective of the study to assess the bridge vulnerability was nonetheless pursued but without as much urgency.

The three Coldwater River bridges consisted of multiple intermediate bents (up to 42 in one case) supporting composite concrete deck slabs over short simple spans (40-50 ft) and a longer central span (100-120 ft) over the main navigable channel. The deck in the central span was usually continuous over several adjacent spans and consisted of a multi-cell concrete box girder or a composite concrete steel girder section. With a limited budget and time frame, a 3D model of the entire bridge with SSI was not attempted. A simpler approach was taken that focused on characterizing the main perceived sources of vulnerability.

Again, design drawings were available from MDOT along with soil borings and test pile logs. The drawings indicated the structures had been built in the 1950s and 1960s, and lacked any consideration of seismic loading in the design. The location of the bents in the flood plain of the river with, in several cases, soil in the top layer permanently saturated, allowed the possibility of weak lateral resistance of the soil and liquefaction under strong ground shaking.

The modeling approach thus focused on 2D representation of lateral resistance of the typical intermediate bents in each bridge and 3D representation of the continuous span box girders.

Figure 5 shows that the intermediate bents consist of 4-5 relatively short concrete piles with batters on the outer piles tied together by a concrete pile cap that support bearings for the deck girders. Figure 10 shows the representation of this structural system as modeled in the SAP2000 software [1]. The piles in this system were designed for vertical (deck weight and vehicle live) loads primarily, so the potential vulnerability is from lateral inertial load generated by seismic shaking. Under lateral forces, the piles have a tendency to bend under the lateral resistance from the soil. Furthermore, the overturning moment associated with the deck lateral load develops increased compressive axial loads in the outer (batter) piles far in excess of their design assumptions.

Key aspects of the modeling are the axial and bending capacity of the concrete pile section, the lateral stiffness of the soil, the unsupported length of the pile, and the depth of pile embedment. In keeping with the simplified assumptions, linear vertical and horizontal soil springs were used to represent the soil resistance. Surprisingly, standard geotechnical and bridge engineering textbooks and even some advanced earthquake engineering ones offer little on methods to determine the stiffness properties of soil, choosing to focus rather exclusively on capacity estimation. Results presented in a FEMA guidance document [4] were used to estimate the spring constants considering the projected area of the pile and the elastic modulus of the soil.

Isolation of the intermediate bents for lateral load analysis is valid to the extent that the deck moves uniformly so that no bending or torsional resistance is provided by adjacent bents. The simple deck spans help to minimize this effect through the discontinuity of the bearings. In the case of continuous main spans, however, the deck is supported on pile supported concrete piers with either one or two columns of significantly different heights and size, so significant resistance from adjacent bents is anticipated. Figure 11 shows a 3D model developed using another FE software [2] oriented toward bridge design analysis used to explore the effect of the interaction between bents in these spans.

5. FE Evaluation process – System behavior analysis

The previous section indicates that the goals of the vulnerability evaluation influences the selection of FE modeling options including software (structural or general purpose), level of analysis (2D or 3D), element selection (structural or continuum), connectivity (rigid connections or flexible bearings), boundary conditions (fixed, flexible, or absorbing). These choices not only influence the behavior and response details that may be estimated and Visualized, they also determine what output measures are available for estimating physical damage, performance characteristics, and vulnerability.

In the MEMA UM campus study [17], a basic analysis approach was established that was followed throughout all the studies. Before proceeding to the complex nonlinear dynamic time history analysis, linear static and eigenvalue preliminary analyses were first performed. The linear static gravity load analysis requires processing of all parameters and procedures involved in estimating the stiffness properties of the system. It is relatively fast computationally and enables visual and quantitative confirmation of element connectivity and effect of support fixity (fixed base models), support flexibility (soil springs), or absorbing boundary conditions (SSI models). The eigenvalue analysis requires processing of all parameters and procedures involved in estimating the mass properties of the system. The analysis is also relatively fast computationally and yields mode shapes and frequencies. These modal properties provide insight into the expected dynamic response characteristics under earthquake loading.

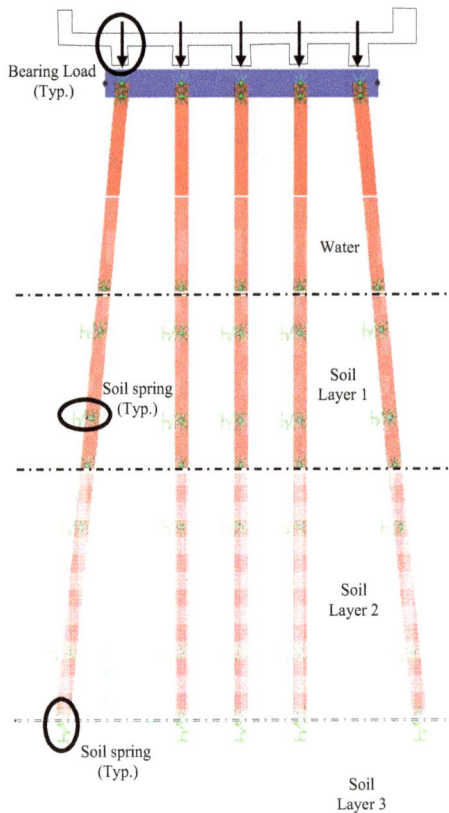

Figure 10. Model of typical intermediate bent for Coldwater River bridge [16] carrying two lanes of interstate highway traffic; concrete piles and cap modeled as frame elements (section behavior modeled with axial-bending interaction using fiber model); ground motion applied to soil springs

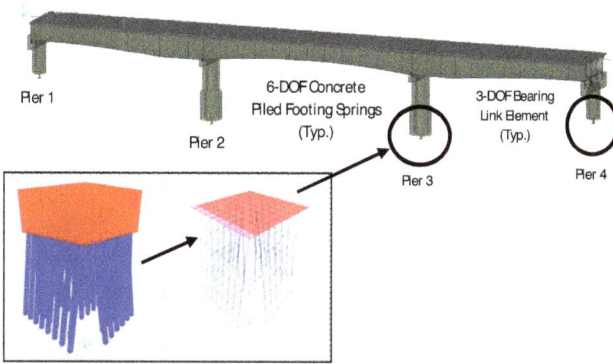

Figure 11. Model [16] of typical 3-span continuous concrete box girder bent for Coldwater River bridge carrying two lanes of interstate highway traffic; concrete pier columns and footing piles modeled with frame elements; box girder flanges and webs and footing pile cap modeled with shell elements; footings modeled with equivalent 6-DOF springs; ground motion applied to footing springs

Figure 12 illustrates some of the benefits of performing the preliminary analyses before proceeding to the nonlinear time history response analysis. The issues of stiffness and mass distribution become evident from the plotting and animation of the mode shapes associated with global movements of the system. These shapes may be broadly categorized as ones that involve significant net movement of the center of mass of the system and those that do not (sometimes called breathing modes). In the case of the campus bridge shown, it is seen that the mode involving transverse movement of the mass center becomes coupled with a rotational movement because of the skew of the deck necessitated by the angle between the centerlines of the street carried and the one crossed. Also visualized in the case shown in Figure 12 is the effect of the SSI, in this case the embankments and abutments interacting with the main span deck and intermediate bents.

Behavior similar to that observed for the MEMA UM campus study bridge is found in the case of the MDOT study bridge. Figure 13 shows the transverse mode shape for the fixed base model. The bridge proportions (both deck length to width and deck span to column height ratios) and skew angle are different in the two cases. The translational and rotational coupling is less pronounced, and the transverse column bending is more pronounced.

The eigenvalue analyses not only provide insight regarding the expected deformation patterns, they also provide the frequencies associated with these characteristic modes. These frequencies provide quantitative information which provide insight into the expected influence of the SSI effects as well as the dominance of deformation modes associated with specific earthquake events.

The influence of SSI was examined in detail in the MDOT study which included ambient vibration measurements using a portable array of accelerometers [11, 14]. Simultaneous readings were taken at each bent location under excitation of the bridge by truck traffic. Using

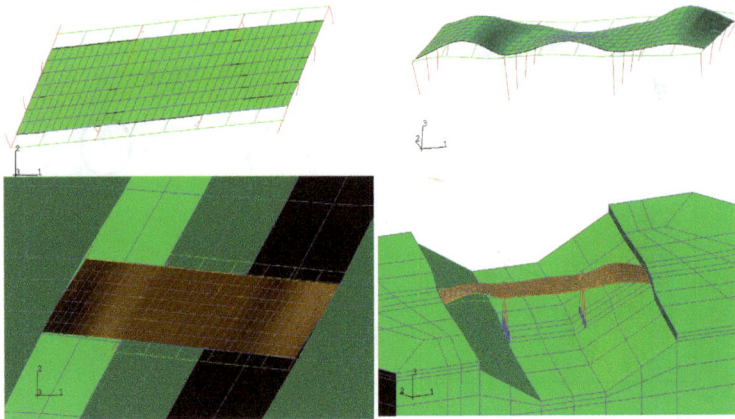

Figure 12. Eigenvalue analysis results for the MEMA UM campus study bridge models [17]; top figures show plan and isometric views of fixed based model transverse mode causing bent and abutment column deformation; bottom figures show comparable modes for SSI model; skew of roadway alignment introduces coupling of translation and rotation of the deck mass as well as bending and torsion of the deck; the resistance provided by the embankment is apparent from the contact developed during rotation

a point on the bridge deck as a reference point, frequency response functions were derived that eliminated the influence of the excitation, and system response frequencies were extracted corresponding with excellent correlation to the 3D model SSI case without any model parameter modification. Accelerometers were then moved to the abutments and frequency extraction performed [11] revealing evidence of the participation of the abutments in the transverse mode shape comparable to the one in Figure 13.

In the MEMA bridge study, the preliminary analyses were again performed prior to time history analysis. Figure 14 shows that the fundamental mode of vibration for a typical intermediate bent in the interstate highway river crossing is one involving net translation of the deck and corresponding bending of the piles which were designed as axially loaded members. Consideration of the eccentricity of the deck mass with respect to the center of resistance of the soil-pile system provides for expectation of an overturning moment. Such a moment would generate an increase of axial force in one of the batter piles which would combine with the bending action.

6. FE evaluation process — Seismic response analysis

The benefit of FE based evaluation is that a great bit of detail of the response of the system is made available through the analysis especially when the time history approach is taken. In essence all DOF selected in the modeling process are accessible over the full length of the simulated event. With further post-processing whether computational or graphical, additional response quantities and behavior can be accessed, plotted, and visualized.

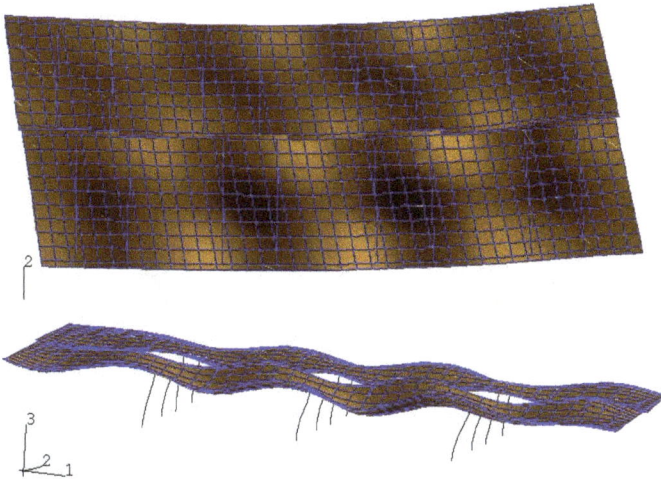

Figure 13. Eigenvalue analysis results for the MDOT bridge study; plan and isometric views of fixed base model transverse mode; lateral deck and column bending dominates response in this pair of adjacent bridge structures; some bending and torsional coupling in the deck is evident

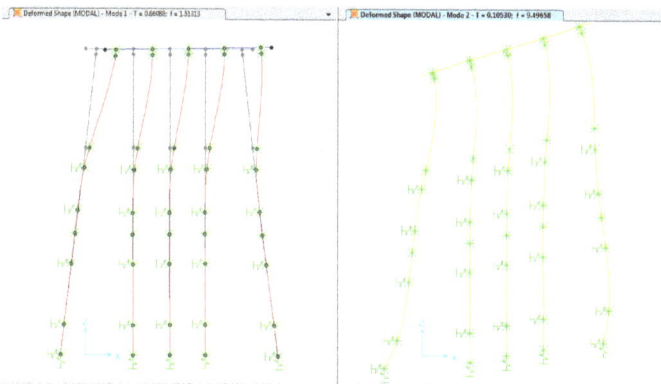

Figure 14. Eigenvalue analysis results for the MEMA bridge study [16]; elevation views of 2D intermediate bent model showing deck transverse displacement (left) and rotation (right) modes; pile bending dominates response although combined action of bending and axial force in the piles is implied

In the MEMA UM campus study [17], it became particularly useful to examine hysteresis of the column section in the plastic hinge region. Figure 15 shows a typical plot of simulated moment-curvature response during the severe (M 8.3) event case. The results demonstrate that the yield limit state is achieved in both directions for a corner column, and the ultimate limit state is achieved in one direction. The latter result provided clear evidence of vulnerability and the possibility of complete failure or collapse.

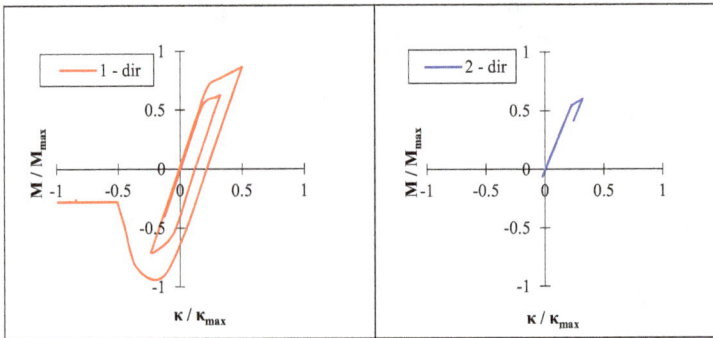

Figure 15. Normalized moment-curvature hysteresis plots for a corner column of the MEMA UM campus study bridge (see Figs. 3, 8, 12) subject to an extreme event (see Fig. 6); nonlinear user-input curves developed separately using a fiber section model [12, 17]

In the MDOT study [14], with the availability of the USGS simulation tool for developing random realizations of input ground acceleration time histories (Fig. 6) at different intensities, limit state determination was enabled for the columns and piles over the full range of damaging events. Comparison of the peak and characteristic responses enabled a performance evaluation of the system based on critical material, section, or member limit states such as first cracking, first yield, plastic hinge formation, and plastic collapse mechanism. In the case of the bridge studied, it was learned that the piles at the abutments and the columns of the central bent provided substantial energy absorption in the extreme event case through ductile hysteretic response in these members. It was also learned that the pile system at the abutments and central bent adequately distributed lateral forces so that the soil remained linear throughout the event. While nonlinear slip at the superstructure to abutment pile cap bearing connection was attempted, this proved too difficult for the software to resolve and convergence was never reached. Ultimately, rigid connections were assumed and the slip mechanism was interpreted as another potential energy absorption source.

In the MEMA Coldwater River bridges study [16], linear dynamic response was performed for most of the analysis runs. An example of the response motion time histories at the level of a typical intermediate bent pile cap is shown in Figure 16. Peak internal force (axial force, shear, and bending moment) responses in the piles were obtained and compared with design values and pile test data.

In the critical case, a nonlinear static pushover analysis was performed to estimate the capacity of the pile system. In the FE model, both geometric and material nonlinear options for the software were used. In the latter option, a fiber representation of the cross-section was used that accounted for 1D nonlinear normal stress-normal strain behavior in the concrete and the steel reinforcement, enabling computation of the force-displacement behavior shown in Figure 17. A nonlinear time history was also run for this critical case.

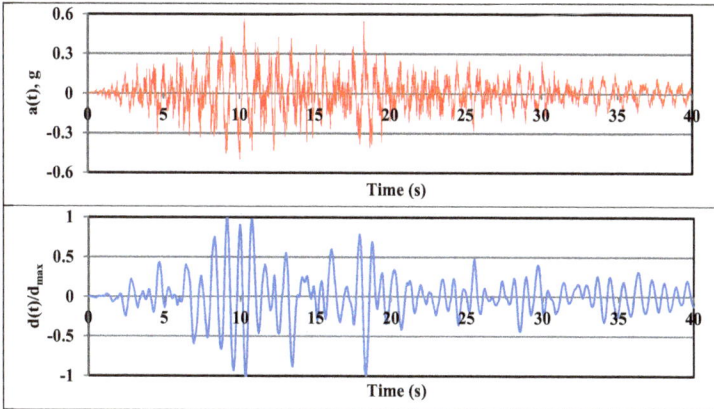

Figure 16. Normalized time history plots of response at center of intermediate bent pile cap for one of the Coldwater River bridges studied [16] (see Figs. 5 and 10) subject to an extreme event (see Fig. 6 scaled to PGA=0.25g); top plot shows acceleration; bottom plot shows displacement)

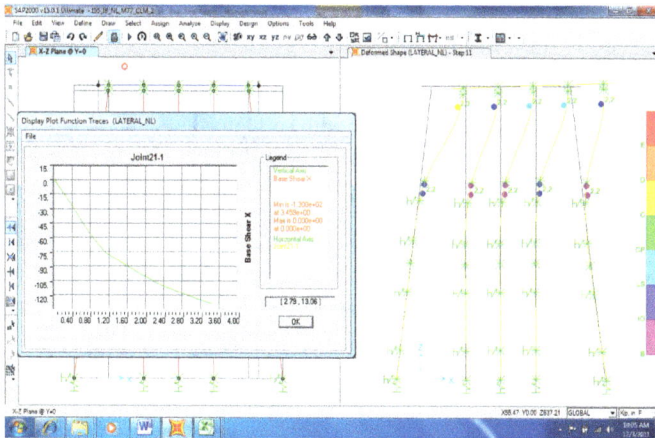

Figure 17. Plots of pushover response to lateral load at bearing positions of intermediate bent pile cap for one of the Coldwater River bridges studied [16] (see Figs. 5 and 10); left plot shows force-displacement response; right plot shows deformed shape and limit state condition for plastic hinge locations

In assessing the results of the time history analysis, a simplified analysis was performed using a single DOF equivalent model using a lumped stiffness based on a unit force lateral load analysis. When compared to the linear dynamic analysis of the full bent, this simplified analysis was able to demonstrate that the first mode of the system dominated the overall response and could have been used as a predictive tool.

A decision was made not to depend on nonlinear dynamic response analysis for all the bridges. This was in part due to the lack of confidence in the soil properties at the site from which reasonable assumptions could be made for simulation of nonlinear soil response and in part due to the scope of the work which was limited as has been mentioned previously. A limited attempt was made to verify at least the elastic properties of the soil adjacent to the sites using seismic refraction tests performed near the embankments which were accessible on dry land.

7. Conclusions

This chapter highlights objectives, modeling options, and analysis results for a range of FE based vulnerability assessments of highway bridges performed at the University of Mississippi by the author. To illustrate the range of conditions and considerations, three projects have been selected as case studies. Finite element software, and the operating systems and hardware, especially microprocessors which support the software, have advanced significantly over the time period since the first of these projects was conducted. The lessons learned, however, remain fundamental in a sense for vulnerability assessments that are premised on mechanics of highway bridge materials, elements, and structural systems that inevitably include soil and construction materials such as concrete and steel.

The objectives of the vulnerability analysis depend in part on the nature of the hazard, the inventory exposed to the hazard, and the agency concerned with the inventory. For the study cases the hazard for the region is characterized by a high consequence but low probability event. The inventory is not near enough to the hazard to be considered in a high seismic exposure but the potential ground shaking is significant enough to inflict severe damage especially to older bridges that pre-date seismic design provisions. Furthermore, and perhaps of most interest to the state emergency management and transportation officials that have sponsored the studies, the bridges selected for detailed evaluation are located on important lifelines between a major metropolitan area and the multiple surrounding communities that are growing rapidly. Many of these communities would become isolated in the event of the complete functional loss of the highway network. Both urban and rural stakeholders will depend on these bridges remaining serviceable not only for the densely populated area closer to the hazard needing to exit the concentrated region of potential seismic damage but also for incoming emergency responders and other personnel providing assistance. The understanding of both the hazard and the inventory in the study region is evolving even at the present time, and a research study is now underway. The study is exploring the short and long term impacts of potential damages on traffic flow in north Mississippi as well as the resulting economic losses.

The development of seismic ground motion records for the study cases is addressed only to the extent necessary to characterize the hazard and help interpret the results obtained from the FE simulation. In most situations the input motion is a major uncertainty in both the model analysis and the vulnerability assessment. A performance based approach has been adopted where consistent with the study objectives. A range of hazard and ground motion intensities has been considered in these studies and FE based time history response analysis has formed the basis of the performance evaluation. In one case, the objective was to validate results of a regional study that did not consider many of the key details of the structural system using FE based analysis. In this case, the hazard and ground motion intensity were selected to be consistent with that used in the regional study.

Examples of 2D and 3D structural models are presented that incorporate a wide variety of finite element and material types and consider the effects of soil-foundation-structure interaction which, in the view of the author, is an essential part of reliably establishing the performance of bridge structural systems. The incorporation of such interaction presents challenges to the analyst which are not well represented in standard textbooks on highway bridge design and even some of the FE literature. To complement such reference works, a portion of the chapter is devoted to discussion of preliminary analyses that are quite naturally performed while the more complex models for final evaluation are constructed. The discussion highlights the importance of first capturing behavioral aspects of the system revealed by static response analysis under gravity or idealized lateral loads and subsequent examination of vibration mode shapes and natural frequencies obtained by eigenvalue analysis. These preliminary analyses provide not only quality assurance but also insight that may guide expectations for the results of the more complex models. The information obtained from subsequent static, nonlinear, and dynamic response analysis is then maximized so that the most useful or telling information is extracted from the analysis under seismic excitation.

The FE based approach to vulnerability assessment ensures that quantitative data formulated on basic mechanics principles is generated for consideration during the assessment. Extracting the data and using it to establish measures of performance remains somewhat of an art. In the study cases, a range of measures has been adopted including peak dynamic response acceleration and displacement as well as maximum internal forces in critical members and damage distribution in major subsystems. It is hoped that an appreciation of the complexity of highway bridge systems has been provided through the description of the many details of the FE models and the results obtained from analysis of response to seismic excitation.

Application of the results of FE analysis to a specific vulnerability assessment requires consideration of the objectives and end-user needs. A range of complexity in successive models used in the evaluation may be appropriate depending on the sensitivity of the evaluation on the outcomes of the analysis. Furthermore, the availability of powerful analysis tools should not overshadow lack of confidence in data provided to the analysis. In particular, soil property and earthquake intensity and motion characteristics are often not known precisely.

In regions of moderate seismic hazard it may prove difficult to establish a sense of urgency for action on the basis of the results of a vulnerability analysis whether or not is based on FE modeling and considered highly accurate. In such a context it may be useful to incorporate the seismic vulnerability assessment in a broader one considering multiple hazards exhibiting comparable levels of risk.

Acknowledgements

The material in this chapter represents the accumulated effort of the author with others involved in the cited bridge seismic vulnerability studies. These include Dr. Robert Hackett, former chair of the Department of Civil Engineering at the University of Mississippi, and Mr. Charles Swann, now Associate Director of the Mississippi Mineral Resources Institute. I would also like to thank the many graduate research assistants in the Department of Civil Engineering at the University of Mississippi under my advisement who contributed to these studies and used them as the basis of master's theses and a doctoral dissertation. The academic works are cited separately in the relevant references given. Lastly, the financial support of the Department of Civil Engineering is appreciated as is that of the various sponsors of the cited studies. Sources of the latter include two FEMA mitigation grants administered by the Mitigation Bureau of the Mississippi Emergency Management Agency and a federal grant administered by the Research and Bridge Divisions of the Mississippi Department of Transportation. The specific sponsored project and grant details are described in more detail in the relevant references given.

Author details

C. Mullen

Department of Civil Engineering, University of Mississippi-Oxford, USA

References

[1] Computers and Structures, Inc. SAP2000 Users Manual-Version 15. CSI; 2012.

[2] Computers and Structures, Inc. CSIBridge Users Manual-Version 15. CSI; 2012.

[3] Elnashai AS, Jefferson, T, Fiedrich F, Cleveland LJ, Gress, T. Impact of New Madrid Seismic Zone Earthquakes on the Central USA:- Volume I. MAE Center Report No. 09-03. Mid-America Earthquake Center, University of Illinois; 2009.

[4] Federal Emergency Management Agency. FEMA 273: NEHRP Guidelines for the Seismic Rehabilitation of Buildings. FEMA; 1997.

[5] Federal Emergency Management Agency. HAZUS-MH (Hazards US-Multihazard)-Version MR5. FEMA; 2010.

[6] Federal Highway Administration, United States Department of Transportation. National Bridge Inventory. http://www.fhwa.dot.gov/bridge/nbi.htm

[7] Hibbitt, Karlsson, Sorenson, Inc. ABAQUS Theory Manual-Version 5.6. HKS; 1996.

[8] Hwang H, Huo JR. Attenuation Relations Of Ground Motions for Rock and Soil Sites in Eastern United States. Soil Dynamics and Earthquake Engineering 1997; 16: 363-372.

[9] Ismail IMK, Mullen CM. Soil Structure Interaction Issues for Three Dimensional Computational Simulations of Nonlinear Seismic Response. EM2000: proceedings of the 14th Engineering Mechanics Conference, 21-24 May 2000, Austin, TX. Reston, VA: ASCE; 2000.

[10] Johnston AC, Schweig ES. The Enigma of the New Madrid earthquakes of 1811–1812. Annual Review of Earth and Planetary Sciences 1996; 24: 339-384.

[11] LeBlanc B, Mullen CL. Characterization of Abutment-Deck Interaction using 3D FEM and Field Vibration Measurements for an Existing Highway Bridge in North Mississippi. In: Uddin, W, Fortes RM, Merighi JV. (eds.) Proceedings of the 2nd International Symposium on Maintenance and Rehabilitation of Pavements and Technological Control, 29 July-1 August 2001, Auburn AL. NCAT; 2001.

[12] Mullen C L, Swann CT. Seismic Response Interaction between Subsurface Geology and Selected Facilities at the University of Mississippi. Engineering Geology 2001; 62(1-3) 223-250.

[13] Mullen C L, Swann CT. The State of Mississippi Standard Mitigation Plan-Earthquake Risk Assessment. Final report to Mississippi Emergency Management Agency-Mitigation Division. Center for Community Earthquake Preparedness, University of Mississippi; 2004.

[14] Mullen CL. Seismic Vulnerability of Existing Highway Bridge Substructures Supporting the I-5 Undercrossing at MS302 (Goodman Road). Final report to Mississippi Department of Transportation-Bridge Division. Center for Community Earthquake Preparedness, University of Mississippi; 2001.

[15] Mullen CL. 3D FEM for Seismic Damage in a Four-Span Interstate Concrete Highway Under-Crossing including Embankment-Structure Interaction. Proceedings of the FHWA National Workshop on Innovative Applications of Finite Element Modeling in Highway Structures, 20-21 August 2003, New York City, NY. UTRC; 2003.

[16] Mullen CL. Seismic Vulnerability of Critical Bridges in North Mississippi. Final report to Mississippi Emergency Management Agency- Mitigation Division. Center for Community Earthquake Preparedness, University of Mississippi; 2011

[17] Swann CT, Mullen CL, Hackett RM, Stewart RK, Lutken CB. Evaluation of Earthquake Effects on Selected Structures and Utilities at the University of Mississippi- A Mitigation Model for Universities. Final report to Mississippi Emergency Management Agency. Department of Civil Engineering, University of Mississippi, and Mississippi Minerals Resources Institute; 1999.

[18] Tuttle MP, Schweig ES. Archeological and Pedological Evidence for Large Prehistoric Earthquakes in the New Madrid Seismic Zone, Central United States. Geology 1995; 23(3) 253-256.

[19] United States Geological Survey. Earthquake Hazard in the Heart of the Homeland, Fact Sheet 2006–3125. http://pubs.usgs.gov/fs/2006/3125/pdf/FS06-3125_508.pdf

Advanced Applications in the Field of Structural Control and Health Monitoring After the 2009 L'Aquila Earthquake

Vincenzo Gattulli

Additional information is available at the end of the chapter

1. Introduction

The earthquake, which has been occurred on 6 April 2009, has been a catastrophic event for both the city and the University of L'Aquila [1]. Nevertheless, the disaster have to be transformed in a tremendous opportunity to revitalize the area, with important benefit for the national and international scientific community to experience the effectiveness of new systems and technologies, and consequently to base, on these results, new developments in several different fields.

The present chapter aims to summarizes the observations made at L'Aquila regarding the dissemination of new technologies belonging to the structural control and health monitoring fields, immediately after the earthquake and in the reconstruction phase [2].

Two synthetic databases are presented and discussed regarding, respectively, the installed seismic protection systems and the structural monitoring experiences, available to the author personal knowledge, and probably mostly incomplete at this moment. Firstly, the large use of new seismic protection systems, using both base isolation and energy dissipation devices, in the new construction and in the retrofitting of existing structures, mainly made in reinforced concrete, is categorized and the main features of the installed systems are synthesized. Secondly, the efforts done in the area of structural monitoring, especially for strongly damaged monumental churches and building, are described and, based on the available information, the characteristics of the used instrumentation, either for permanent or not permanent installation, are classified.

Finally, the results acquired during the development of two different case studies, by a research group of the University of L'Aquila, are presented in detail.

In the first one, the use of energy dissipation devices, such as nonlinear fluid viscous dampers, in a peculiar configuration scheme that make use of the concept of dissipative interconnection in adjacent structures, is illustrated. Indeed during the seismic event of 6th April 2009, the edifices of the Engineering Faculty have suffered particularly for seismic induced large structural displacements and accelerations which have brought them out of order due, mainly, to the failure of non-structural elements [3,4], the breakage of wiring and piping systems and the destruction of furniture and machineries. In particular, among the three recently built buildings of the campus, erected in the early 90's, the so-called "Edifice A" presents the most critical damage scenario, which has been objective of a significant rehabilitating intervention. The critical choice during the design stage and testing are illustrated through several analysis conducted with the aim to construct reliable numerical models reproducing the experienced seismic behaviour and the expected enhancement due to the retrofitting. In particular, the main results of a dynamical testing campaign [5] used to calibrate a series of finite element models, able to reproduce the structural behaviour of the Edifice A, at low oscillation amplitude, are here discussed. Nonlinear static and dynamic structural analysis has been used in the evaluation of the structural performance [4] and of the proposed structural control effectiveness [6]. Device testing [7] and installation procedures have been considered in the overall process to reach high level of confidence in the matching of the rehabilitation goals with the realistically installed seismic protection system.

In the second one, the use of a wireless sensor network (WSN) for permanent structural health monitoring (SHM) of historic buildings in a seismic area is considered, evidencing the conducted specific activities to customize the system for the continuous assessment of the damaged conditions. On the basis of a defined design strategy [8-10], a permanent structural monitoring systems has been installed on the damaged *Basilica of S. Maria di Collemaggio*, at L'Aquila and it is currently working during the whole day. The main findings in the design, delivery, installation and management of the monitoring systems are presented. A series of tests has been conducted for the monitoring systems and the acquired data have been used for structural identification purpose on the basis of clearly stated procedure [11]. Several registrations acquired with the systems during local aftershock or more distant, relatively strong, shocks, as for example the recent Emilia earthquake (20-05-2012), are used to demonstrate the possibility given by the dynamic monitoring to produce valuable information for the structural assessment of historical monuments which can be in strongly damaged condition, such as the case of the Basilica.

2. The use of base-isolation and energy dissipation technologies at L'Aquila

The large number of losses in the property assets caused by the 2009 earthquake, particularly in the case of strategic structures (Hospital, Governance offices, School and University Buildings, infrastructures, Bank Buildings, etc) has demonstrate the large seismic vulnerability of the L'Aquila territory. Probably, the case of the University buildings it is emblematic because these structures were extremely "strategic" from the point of view of the caused disturbance to the local equilibrium reached, before the earthquake, at any level (social, economical, etc). Indeed, the 27000 students attending the classes in the building of the several Faculties constitute a large revitalizing effect for the production realized in the territory of L'Aquila. In contrast the damage suffered by this extremely strategic institution for its territory through the scarce seismic performance of the entire property asset [1] has bad consequence in the reconstruction phase. Notwithstanding the large losses, many projects have been started, immediately after the earthquake, to react immediately to the catastrophic event. Due to a long period of aftershock swarm, still continuing in the area, the main idea, which it was followed, is the realization of safer structures with affordable costs. Therefore, several projects have been realized exploiting the use of passive control for seismic protection, either through the concept of base isolation or by enhancing the dissipation capacity of the structural system. These interventions have been conducted both for buildings devoted to public services and to residential buildings. The realizations using a base isolation system as main seismic protection strategy, available to the author knowledge, are summarized in Table 1 while the structural systems enhanced through dissipative devices are described in Table 2.

Immediately after the earthquake one of the main problems, is to found the right compromise between temporary or definitive construction of houses, which can be used to maintain the population at the site. In the case of L'Aquila a peculiar solution to the problem has been provided directly by the National Government, the Project CASE, consisting in 185 buildings constructed in record time to provide a right accommodation to a large amount of the population through the realization of 4.500 apartments in 185 buildings [12]. Every building has the same structure at the ground floor (columns with seismic isolators and a rigid slab), while the superstructures have been made with different construction solutions and materials.

Among public buildings, the new venue of ANAS, the Italian Infrastructure Public Authority for the management of the road network, has been built in a very short time. It has a circular plant and a base isolation system. Furthermore, it was carried out the demolition and reconstruction of a portion of the Court Law Building, the construction of the new venue of the Faculty of Letters (with the process started in 2006) and the retrofitting of the Faculty of Engineering, project extensively discussed in the following section 4. As important as the public buildings, there were several retrofitting interventions in residential damaged buildings. Among these, quite interesting it is the case of the condominium in via Rauco, being one of the first examples of a peculiar technology application for the uplift of the buildings. During the realization thanks to hydraulic jacks, it was possible to uplift the building of 60 cm and insert seismic isolators at ground floor level. Another example is the case of condominium Habitat, consisting in 10 buildings of different heights connected to one another by 9 bodies scale, arranged to make a semicircular plant all together. The intervention has been characterized by the realization of a single rigid slab to the level of the first deck and the cutting of the columns on the ground floor level, to allow insertion of the devices. In this way it was possible to realize a unique isolation system for all the bodies of the condominium.

The data collected regarding structural control systems, recently, realized in L'Aquila are summarized in Tables 1 and 2, in which is specified, for each construction, the type of intervention, the type and quantity of the devices used and, for some of them, the available specific design characteristics.

Building	Design and Construction Period	Type of Intervention	Construction Material — Substructure	Construction Material — Superstructure	Superstructure Number Floors	Seismic Protection Device	Number of Bearings	Base Isolation Vibration Period (sec)	First Mode Vibration Period (sec)	Bearing δ max (mm)	Bearing Max Vertical Load (kN)
C. A. S. E. Project	2009-2010	Buildings for Homeless (185 buildings)	Steel and reinforced concrete columns, reinforced concrete rigid slab.	Wood, steel, or concrete.	3	Friction Pendulum Bearings	40x28 buildings=1120 Type I (r. c. columns) / 40x91 buildings=3640 Type I (steel columns) / 32x3 buildings=96 Type I (steel columns) / 40x61 buildings=2440 Type II (steel columns) / 32x1 buildings=32 Type II (steel columns)	4	0.5	260	3000 / 3000 / 3000 / 3000
ANAS	2009 - 2010	New construction	Reinforced concrete	Reinforced concrete	3	Elastomeric Bearings - HDRB	60				
Auditorium	2010 - 2012	New construction	Reinforced concrete	Laminated wood	Single building	Elastomeric Bearings	14				
Car Dealership Ford	2011	New construction	Reinforced concrete	Reinforced concrete		Elastomeric Bearings	18				
Residential Building	2012	New construction	Reinforced concrete	Reinforced concrete		Elastomeric Bearings	17				
Condominium Habitat	2011	Retrofitting (19 bodies connected)	Reinforced concrete	Reinforced concrete	3 (edges) and 5 (center)	Friction Pendulum Bearings	277 (4 different sizes)	2.75		300	
Residential Building via Rauco	2011	Retrofitting	Reinforced concrete	Reinforced concrete	6	Friction Pendulum Bearings	32				
Condominium Domus Prima	2011	Retrofitting	Reinforced concrete	Reinforced concrete		Friction Pendulum Bearings	47 (3 different sizes)			390	
Condominium Fortuna 2	2012	Retrofitting	Reinforced concrete	Reinforced concrete		Elastomeric Bearings - HDRB	21				
Condominium Borgo dei Tigli	2012	Retrofitting	Reinforced concrete	Reinforced concrete		Elastomeric Bearings - HDRB	42 (2 different sizes)				
Condominium Apaglia	2012	Retrofitting	Reinforced concrete	Reinforced concrete		Friction Pendulum Bearings	30			350	
Condominium Amiterno	2012	Retrofitting	Reinforced concrete	Reinforced concrete	Single building	Friction Pendulum Bearings	26			300	
Condominium Baratelli	2012	Retrofitting	Reinforced concrete	Reinforced concrete		Friction Pendulum Bearings	66 (3 different sizes)			355	
Condominium Leonardo	2012	Retrofitting	Reinforced concrete	Reinforced concrete		Friction Pendulum Bearings	44			350	
Condominium Acric - Building C2	2012	Retrofitting	Reinforced concrete	Reinforced concrete		Elastomeric Bearings - HDRB	26				
Condominium Andromeda	2012	Retrofitting	Reinforced concrete	Reinforced concrete		Elastomeric Bearings - HDRB	19 (2 different sizes)				
Faculty of Letter	2006 - 2012	Demolition and reconstruction	Reinforced concrete	Reinforced concrete	6 (build. A,B)-7 (build. D)-1 (build. C)	Elastomeric Bearings (HDRB) and sliders	77 + 34	2.6	1	492	14000
Court Law Building - Building B	2011	Demolition and reconstruction	Reinforced concrete	3 Buildings: 1 in steel, 2 in reinforced buildings.	3	Friction Pendulum Bearings	31 Type I, 11 Type II, 15 Type III, 9 Type IV, 6 Type V, 5 Type VI				
3 Buildings via Francia Building B	2011	Demolition and reconstruction	Reinforced concrete	Reinforced concrete	3	Elastomeric Bearings	72 (2 different sizes)				
Building via Caderna	2012	Demolition and reconstruction	Reinforced concrete	Reinforced concrete	3	Elastomeric Bearings	17				
Condominium S. Antonio - Building A	2012	Demolition and reconstruction	Reinforced concrete	Reinforced concrete		Elastomeric Bearings	52 (4 different sizes)				

Table 1. Examples of interventions using a base isolation system in the city of L'Aquila.

Building	Design and Construction Period	Type of Intervention	Construction Material	Number Floors of Superstrucutre	Seismic Protection System	Number of Devices	Device δ max (mm)	Device Max Horizontal Load (kN)
Edifice A Engineering Faculty	2011	Retrofitting	Reinforced concrete	4	Viscous Dampers	18 (Type I)	90	200
						17 (Type II)	60	100
						8 (Type III)	130	400
Condominium Avenia	2012	Retrofitting	Reinforced concrete	4	Elasto Plastic Devices	2 (Type I)		130
						12 (Type II)	15	170
						18 (Type III)		370
Building corso Federico II	2012	Retrofitting	Reinforced concrete	3	Elasto Plastic Devices	14 (Type I)		270
						18 (Type II)		340
						20 (Type III)		480
						10 (Type IV)	15	560
						22 (Type V)		720
						24 (Type VI)		820
						18 (Type VII)		940
						18 (Type VIII)		1170
Condominium via Milonia, 4	2012	Retrofitting	Reinforced concrete	4	Elasto Plastic Devices	4 (Type I)	20	460
						8 (Type II)	20	130
						10 (Type III)	25	130
						2 (Type IV)	25	460
Condominium via Milonia, 2	2012	Retrofitting	Reinforced concrete	4	Elasto Plastic Devices	4 (Type I)	20	460
						8 (Type II)	20	130
						12 (Type III)	25	130
Condominio La Casetta	2012	Retrofitting	Reinforced concrete	5	Elasto Plastic Devices	24	20	150
Building via Rosana - Gioia dei Marsi (AQ)	2012	Retrofitting	Reinforced concrete	4	Elasto Plastic Devices	9	15	560

Table 2. Examples of interventions using passive energy dissipation systems

The data are evidencing the impact of the structural control technology either in the immediate intervention after the earthquake and in the longer reconstruction phase. To the author knowledge, at the city of L'Aquila during the earthquake, base isolation systems or passive energy dissipation devices were not protecting any in-service structure. Only the building of the Faculty of Letter of the University of L'Aquila was under construction, with the isolators on-site but with the superstructure incomplete and the edifice not finished [13]. To have a complete picture, it can be cited that two hysteretic metallic force limiters were installed in the year 2000 at the end of a light truss structure connecting transversally the slender walls of the nave of S. Maria di Collemaggio [14]. The performance of these devices under the earthquake is still under investigation by different research groups, due to the partial collapse occurred in the area of the transept of the Basilica.

Therefore, immediately after the earthquake the base isolated system at L'Aquila, excluding the peculiar project CASE, reaches the number of 20 interventions with a total number of one thousand (1000) installed devices (as reported in Table 1). The data permits to notice that two main classes of seismic bearing insulator have been installed based on viscoelastic behavior (rubber bearing - RB) or friction (sliding pendulum bearing - SPB). The installed devices are almost the same number in each of the two classes (45% RB – 55% SPB). Several data are missed, because are currently not available, as for instance, the average design period of the base isolation systems. Table 2 shows a synthesis of the realized interventions using passive energy dissipation devices. To the author knowledge, three hundred (300) passive devices have been already installed after the earthquake, mostly based on reaching dissipation through the exploitation of confined material in the elasto-plastic regime during the earthquake. Only in the case of the Edifice A of the Engineering Faculty Building forty-three (43) nonlinear viscous fluid dampers of three different types have been installed looking for the increase of dissipation through the relative velocity of adjacent sub-structures.

3. Structural monitoring systems installed at L'Aquila

Before the 2009 L'Aquila earthquake a strong network of seismic accelerometers were functioning close to the epicenter, mostly managed by the Italian Institute of Geophysics and Volcanology (INGV) [15], while very few structure were equipped by a permanent structural monitoring managed by Department of Civil Protection (DPC) [16] (see, also, Table 3). In particular, the response of the Pizzoli Town Hall during the main shock has been recorded and analyzed by DPC, giving special insights on the potentiality of these systems for immediate evaluation of the damaged occurred during an earthquake. The large amount of installed, temporally or permanently, devices of different type (accelerometers, smart wireless devices, displacement and velocity transducers, inclinometers, etc) reach a number of around three hundred (300) evidencing a large impact of this technology in the post-earthquake emergency phase, especially during the earthquake swarms. In particular several monitoring systems have been installed in the emergency phase, during the construction of temporary scaffolding, in order to verify the efficacy of the added structural system especially in the case of monumental building (see for example [17]). Because of this scope, in many cases, the permanent monitoring has worked only for a limited number of months (in the Table 3, the period is not always precisely known to the author and sometimes it should be considered indicative). In other cases, the monitoring system is permanently installed on the structure and it can be used also to determine the change that will occur in the structural behavior during the reconstruction phase [8,9].

In several cases, the structural monitoring system uses only accelerometers, starting from very few measures (three channels in the minor case) to larger number of devices with different characteristics and sensitivity. Instead more complex monitoring systems are used in complex monumental churches and buildings where accelerometers are joined with crackmeters, inclinometers, and temperature measurement devices, etc.

4. Energy dissipation devices installed at university of L'Aquila buildings

Among several interventions, designed with the intent of increasing the dissipative capacity of the structure through seismic protection elements, the case of the Edifice A of the Engineering Campus has been here selected as case study. The peculiarity of this intervention should be searched on the idea of enhancing the control performance through the dissipative connection of adjacent structures. Indeed, the last two decades increasing attention on the mitigation of seismic or wind induced vibrations in adjacent structures through their "smart"

MONITORED STRUCTURE	Building type	Developer/Owner	Prevalent structural material	Monitoring time interval	Overall number of measurment devices	Number of Accelerometers	Number of wireless devices
Pizzoli Town Hall	Building	DPC/Town Council	Masonry	Currently working	17	17 monoaxial	0
Navelli Town Hall	Building	DPC/Town Council	Reinforced concrete	2 months	4	4 triaxial	4
Pianola	Sports Ground Building	DPC/Town Council	Reinforced concrete	2 months	1	1 triaxial	1
Coppito (AQ)	Finance Police School: Sport Palace	DPC/Town Council	Reinforced concrete	2 months	1	1 triaxial	1
Coppito (AQ)	Finance Police School: Auditorium	DPC/Town Council	Reinforced concrete	2 months	1	1 triaxial	1
Reiss-Romoli (AQ)	Building	DPC/Private	Reinforced concrete	2 months	1	1 triaxial	1
School San Demetrio ne Vestini (AQ)	Building	DPC/Town Council	Reinforced concrete	2 months	3	1 triaxial + 2 biaxial	3
Anime Sante	Monumental church	IUAV/Town Council	Masonry	24 months	20+8	16 monoaxial + 4 triaxial	8
Duomo	Monumental church	Private/Town Council	Masonry	daily	4+8	8 monoaxial	4
S. Biagio D'Amiterno Church	Monumental church	UNIPAD/Town Council	Masonry	2 months	6+10	6 monoaxial	0
S. Marco Church	Monumental church	UNIPAD/Town Council	Masonry	2 months	6+10	6 monoaxial	0
S. Agostino Church	Monumental church	UNIPAD/Town Council	Masonry	2 months	16+10	16 monoaxial	0
S. Silvestro Church	Monumental church	UNIPAD/Town Council	Masonry	24 months	11+8	8 monoaxial	0
Palazzo Margherita	Monumental building	UNIVAQ/Town Council	Masonry	daily	8	8 monoaxial	0
Palazzo Camponeschi	Monumental building	UNIVAQ/UNIVAQ	Masonry	12 months	2+2	2 triaxial	2
Scuola De Amicis	Monumental building	UNIVAQ/Town Council	Masonry	Currently working	12+22	12 monoaxial	0
S. Maria di Collemaggio	Monumental church	UNIVAQ/Town Council	Masonry	Currently working	16+11	16 triaxial	27
Forte Spagnolo	Monumetal building	UNIPAD/Town Council	Masonry	5 months	8+6	8 monoaxial	0
New Building ANAS	Public building	ANAS	Reinforced concrete	Currently working	12	6 biaxial + 6 triaxial	0

Table 3. Examples of structural monitoring systems installed at L'Aquila

coupling has been examined. Several studies have been devoted to optimize the dynamic performance of slender structures, such as skyscrapers or tall buildings, introducing dissipation systems acting on the relative motion and aiming to reduce the maximum displacements at the higher floors. Different applications of similar concepts have been applied in the retrofitting of existing adjacent structures. The placements of viscous-type coupling devices into seismic joints have been proposed to dissipate energy and to avoid hammering phenomena [18-21]. In all cases "smart" coupling between adjacent structures has been exploited using passive, semi-active, and active control systems with different features and performances.

Focusing the attention on the passive coupling of adjacent structures, different modelling approaches have been used. The synthetic description of the main problem features through a pair of simple oscillators interconnected by means of a springs and dashpot in series or parallel fashion has been proposed by many authors [22-25]. The use of a simple oscillator pair has been pursued by the research group of L'Aquila both for the proposal of a new design method [26-28] and the use of it at the preliminary stage of the design of the more complex system installed at the Edifice A of the Engineering Faculty [6]. In the following the entire process has been summarized.

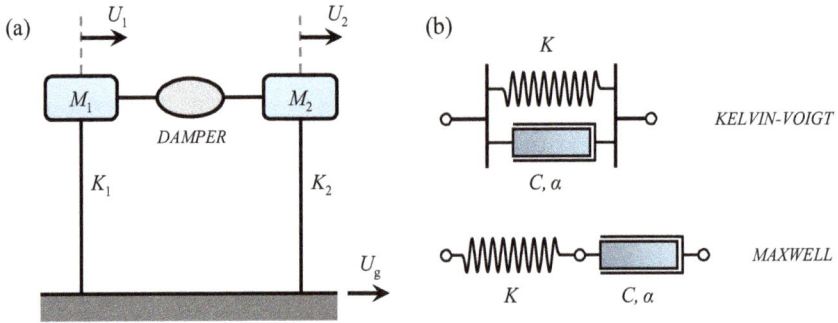

Figure 1. Passive control of adjacent structures: (a) two-dofs model, (b) damper models.

4.1. Simple model of two coupled oscillators for preliminary design

Consider two simple linear oscillators with mass M_j and stiffness K_j, (j=1,2), coupled by a passive damper (Figure 1a). Denoting U_1 and U_2 the relative horizontal displacements and F the mutual force applied by the coupling damper, the dynamic response of the two-degrees-of-freedom (*dofs*) system to a synchronous horizontal ground displacement U_g, is governed by the equations

$$M_1\ddot{U}_1 + K_1 U_1 - F = -M_1\ddot{U}_g$$
$$M_2\ddot{U}_2 + K_2 U_2 + F = -M_2\ddot{U}_g$$

(1)

where dot indicates derivative with respect to time t. Denoting L a convenient reference length, and the following dimensionless variables and parameters can be introduced

$$u_j = \frac{U_j}{L}, \quad u_g = \frac{U_g}{L}, \quad \omega_j^2 = \frac{K_j}{M_j}, \quad \beta = \frac{\omega_2}{\omega_1}, \quad \rho = \frac{M_2}{M_1}, \quad u = \frac{F}{\omega_1^2 M_1 L}, \quad \tau = \omega_1 \tau$$

(2)

where the dimensionless force u is understood as the control variable, and the relevant parameters ρ and β stand for the mass and frequency ratio between the two uncoupled oscillators, respectively. The equations of motion can be rewritten in the synthetic form

$$\mathbf{M}\ddot{\mathbf{u}} + \mathbf{K}\mathbf{u} + su(\mathbf{u}, \dot{\mathbf{u}}) = -\mathbf{M}\mathbf{r}\ddot{u}_g$$

(3)

where **u** is the displacement vector, **M** and **K** are the mass and stiffness matrices, **s** and **r** are
the position vectors of the control and external forces

$$\mathbf{M} = \begin{bmatrix} 1 & 0 \\ 0 & \rho \end{bmatrix}, \quad \mathbf{K} = \begin{bmatrix} 1 & 0 \\ 0 & \rho\beta^2 \end{bmatrix}, \quad \mathbf{u} = \begin{Bmatrix} u_1 \\ u_2 \end{Bmatrix}, \quad \mathbf{r} = \begin{Bmatrix} 1 \\ 1 \end{Bmatrix}, \quad \mathbf{s} = \begin{Bmatrix} -1 \\ 1 \end{Bmatrix} \tag{4}$$

Different rheological models of the coupling damper are introduced to define the constitutive
law $u(u,\ \dot{u})$, relating the control force to the displacement/velocity vector. Adopting a state-
space representation, with the use of the state vector $x = \{u^T,\ \dot{u}^T\}^T$ the equation (3) can be
rewritten as

$$\dot{x} = \mathbf{A}x + \mathbf{b}u + \mathbf{h}\ddot{u}_g \tag{5}$$

where the state matrix A, the allocation control vector b the external input vector h are,
respectively

$$\mathbf{A} = \begin{bmatrix} 0 & \mathbf{I} \\ -\mathbf{M}^{-1}\mathbf{K} & 0 \end{bmatrix}, \quad \mathbf{b} = \begin{Bmatrix} 0 \\ -\mathbf{M}^{-1}\mathbf{s} \end{Bmatrix}, \quad \mathbf{h} = \begin{Bmatrix} 0 \\ -\mathbf{r} \end{Bmatrix} \tag{6}$$

Constitutive models describing with increasing complexity the damper behaviour can be
formulated joining, in different combination schemes, simple elements: a linear spring with
elastic constant K, and a linear dashpot with viscous constant C. Introducing the dimensionless
parameters

$$\eta = \frac{K}{\omega_1^2 M_1}, \quad \gamma = \frac{C}{2\omega_1 M_1} \tag{7}$$

the KV and the Ma model correspond to the alternative parallel or series combination of the
spring and the dashpot, respectively. Consequently, the constitutive law reads

- KV model $u = \eta(u_2 - u_1) + 2\gamma(\dot{u}_2 - \dot{u}_1)$

- Ma model $u = 2\gamma\left(\dot{u}_2 - \dot{u}_1 - \dfrac{\dot{u}}{\eta}\right)$

It is worth noting that the Ma model entails an increment of the model dimension due to the
damper internal dynamics, described by a supplementary half degree-of-freedom. It can be
demonstrated that in the KV case, the design coupling parameters can be chosen according to
the following equations

$$\eta_c = \frac{\rho(1-\beta^2)(1-\rho^2\beta^2)}{(1+\rho)(1+\rho\beta^2)}; \quad \gamma_c = \frac{\rho}{1+\rho}\left(1+\eta_c+\beta^2+\frac{\eta_c}{\rho}-2\left(\beta^2\left(1+\eta_c\right)+\frac{\eta_c}{\rho}\right)^{1/2}\right)^{1/2} \quad (8)$$

in order to assure for the coupled system specific features with respect to the base excitation [28].

Similar characteristics have been found in the *Ma* case for which only numerical analysis have been performed to determine the design coupling parameters η_c, γ_c.

4.2. Seismic protection of Edifice A through nonlinear viscous dampers

During the seismic event of 6th April 2009, the edifices of the Engineering faculty have suffered particularly for seismic induced large structural displacements and accelerations which have brought them out of order due to the failure of non-structural elements [4], the breakage of wiring and piping systems and the destruction of furniture and machineries. In particular, among the three recently-built buildings of the campus, erected in the early 90's, the so-called "Edifice A" presents the most critical damage scenario, which needs a significant rehabilitating intervention.

Edifice A is a four-story building with the resistant structure made of reinforced concrete frames, sitting on a sloping site. Several seismic joints divide the structure into seven independent substructures (Figure 2); some of them are structurally featured by a frame-shear-wall interactive system. In the substructures, the walls are widely used to reinforce and to stiffen the acute corners, the rounded staircases close to the elevator cores and the lower floors. The plan is characterized by asymmetry, with uneven distribution of stiffness and vertical irregularities, and double- or triple-height rooms. The amphitheater facing the main entrance, on the north-west side, is sustained by an independent structure. Concrete slabs are used to realize all the horizontal planes including the roof.

The most evident damages in the Edifice A of Engineering Faculty were found to be localized in the main facade, which has lost large portions of the veneer masonry, made of heavy split-face bricks (Figure 3), laying bare the underlying reinforced concrete structure, remained practically undamaged. All the results collected during the early inspections confirmed that the structure underwent an excessive displacement and acceleration level, surely incompatible with the resistance of many non-structural elements. The massive inward cascade of heavy bricks and sharp glass, fallen down from the facade and the wall of the internal stairs, has realized an unpleasant dangerous scenario [1,4].

Aiming to understand the structural reasons for this inadequate behavior, it should be considered that the design concept follows the idea to have the planar structure sustaining the principal facade rigidly coupled with the three-dimensional frame of the building behind.

Horizontal steel tubes, functioning as interconnecting rods at different floor levels, ensured the coupling between the two substructures (Figure 3c). The bolted anchorages at the rod ends were probably under-dimensioned for the exceptional seismic action, since many of them failed under the combined effects of the unexpected cyclic axial loads and the repeated impacts of the bricks falling down from above. In the progression of the damage the failure of the connection played a great role facilitating the augment of both relative displacements between the two structures (facade and three-dimensional frame) and absolute displacements and acceleration on the facade.

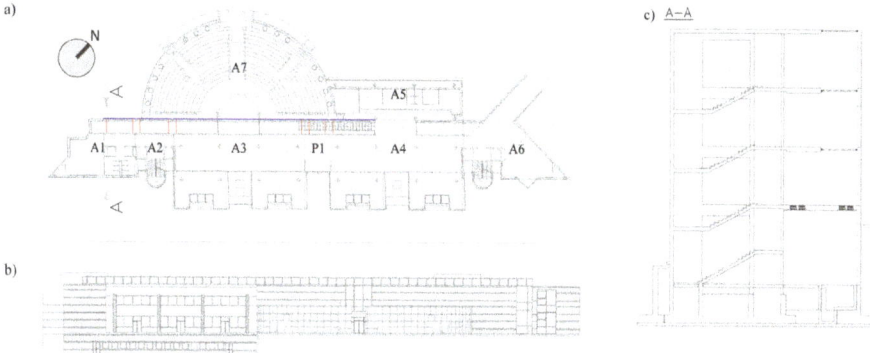

Figure 2. Edifice A: a) plan view at the main entrance level 0, b) facade view c) section A-A

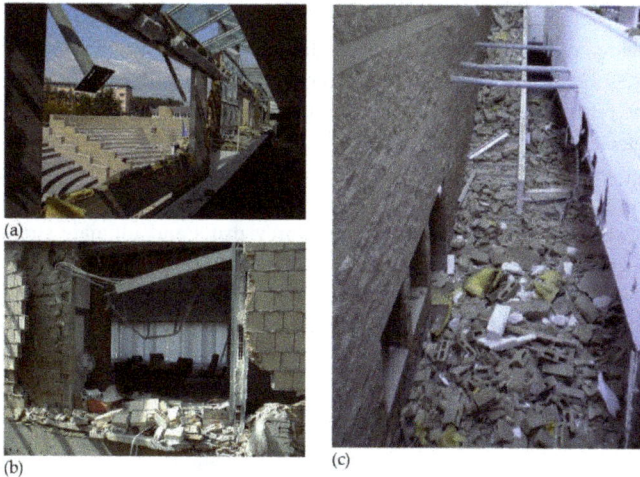

Figure 3. Damages caused by 2009 earthquake to the Edifice A: a) internal view of the main facade, b) internal partitioning walls c) heavy bricks fallen down inside the building from the facade.

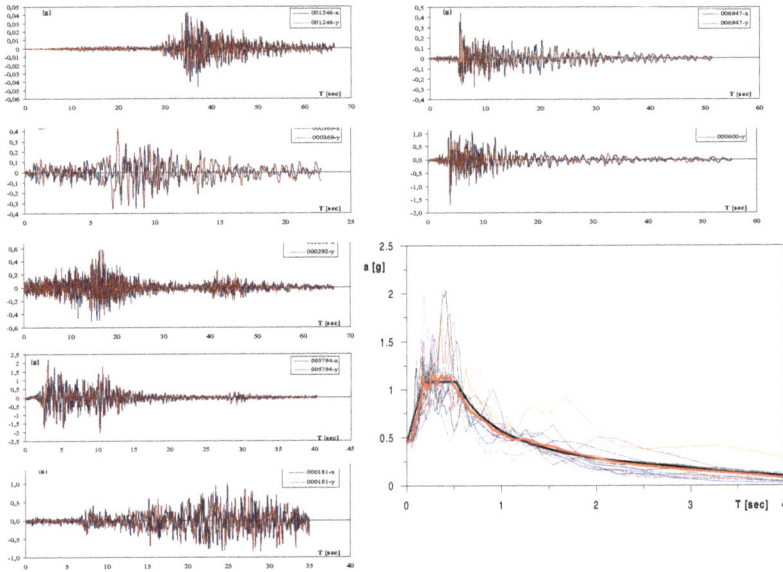

Figure 4. Seven natural earthquake realizations with an average spectrum compatible with the design one used for the evaluation of the seismic protection performances.

Moreover, the overall dynamical phenomenon was probably emphasized by the different mechanical properties of the coupled substructures.

A deep knowledge of the structures has permitted to design an optimized retrofitting intervention, able to satisfy high performance criteria defined in the current Italian National Code [29]. Before the retrofitting interventions, the most vulnerable aspects of the original design have been detected through the comparison with the limitations imposed by the newest national design code. Finite element models for each independent substructure were used, based on the previously obtained information, to verify both the operational limit state and ultimate limit state requests in terms of inter-storey drifts and ultimate strength of each element, respectively. These analyses put into evidence excessive deformation levels of the higher floors, while the other substructures have resulted lower flexible, due to the stiffening presence of fully-height shear walls. The other substructures satisfy the maximum inter-story drift requirements at operational limit state [4]. The effectiveness of the connection between the principal three-dimensional structure and the planar frame sustaining the facade has been recognized as the critical issue to be addressed for the enhancement of the seismic performance.

The limited efficacy of the original metallic tubes, which rigidly couple the facade with the main structure, evidences, also through the occurred damage, large absolute facade displacements and accelerations with high frequency content. This occurrence has suggested considering and comparing different alternatives in reconstructing the damaged coupling elements, exploring new geometric arrangements and technical solutions. After several discussions taking into account comparative criteria (including structural performance, aesthetic outcome and economic aspects), dissipative steel bars, embedding viscous dampers and arranged in a stiff K-shaped configuration, reproducing a planar truss structure, have been selected to restore the facade-structure connection. The leading idea is to realize a dissipative coupling between two adjacent structures with different stiffness, that is the stiff principal three-dimensional structure and the flexible planar frame sustaining the facade. Here, a complete analysis on the benefits reached by the K-shaped dissipative configuration is performed by means of direct time-integration of the nonlinear motion equation numerically obtained through a classical finite element approach in both the case of rigid or dissipative interconnection, for which the nonlinear constitutive relation is fully supported by experimental evidence of the assumed coefficients in the analysis [5]. Seven different acceleration time histories, with different time records (35s-70s) and described through 200 samples per second, have been used to describe the base motion, with spectrum characteristics compatible with the site (Figures 4) [4]. The numerical simulations carried out looking at the complete dynamic structural response (see for example Figure 5) have used as first starting value, the stiffness and viscous coefficient design parameters of the Ma linear model obtained through the method mentioned above for the preliminary design.

However, the final assessment of the viscous coefficient c characterizing mainly the nonlinear viscous dampers (Figure 5d) has been determined from a multistep iterative process, which has allowed the selection of its optimal values [6]. The fractional exponent α has been considered with fixed in the design process (α=0.15), because the manufacturer has assigned it. Selection criteria including both the minimization of the displacements/accelerations at the highest floor, and the reduction of the base section shear stresses have been used.

The analyses show a good performance of the dissipative coupling if both the adjacent structures are subject to significant absolute and relative displacements, as verified in sub-structures A3 and A4. Differently, when the natural frequencies of the coupled structures are appreciably different, as occurred in the stiffer substructures A1, A2 and A6, despite the dissipation is potentially maximized; low displacements are associated to lower dissipated energy.

To reduce the displacements in the longitudinal direction on substructures A3 and A4, a proper coupling to the adjacent A2 and A6 substructures has been designed. The frequency difference in the dominant longitudinal modes of substructures A2 and A3, A4 and A6 has permitted to enhance the efficiency of the dampers in reducing the inter-storey drift in the higher floors. The dampers were installed at the third and fourth floor on the elevator tube-section (A2 and A6), or the frame (A3 and A4). The definition of three synthetic performance indices (J_i), the ratios between the structural performance of each original undamaged and retrofitted substructures, in terms of peak displacements (J_1), accelerations (J_2), and based shear forces (J_3), allow to clearly emphasize the achieved enhancement in the seismic behavior. Moreover, an additional index (J_4) represents the average of previous indices.

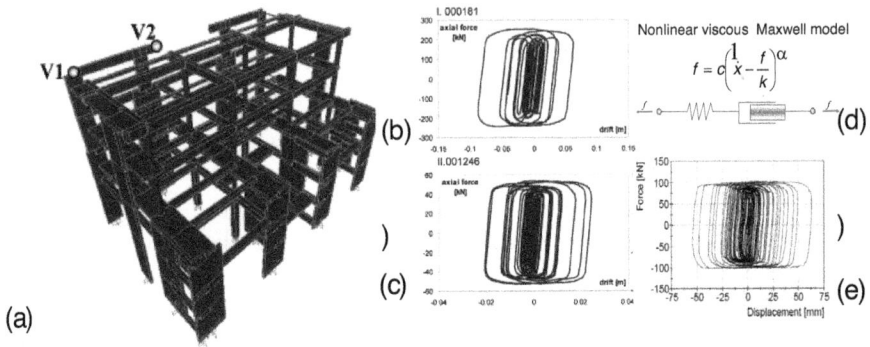

Figure 5. Numerical simulations: a) finite element model of substructure A3, b) c) numerically simulated dissipative cycles for the nonlinear viscous devices, d) nonlinear *Ma* model, e) experimental dissipative cycles.

	J_1	J_2	J_3	J_4
Longitudinal	0.74	0.85	0.73	0.78
Transversal	0.58	0.81	0.96	0.78

Table 4. Performance indices evaluated for the adopted solution.

The rewarding enhancement of seismic structural behavior of substructure A3 and A4, are demonstrated in Table 4, evidencing the effect of viscous coupling in the principal directions, monitored on the top floor. The designed retrofitting reduces substantially the maximum peak displacement (see J_1 in Table 4), particularly in transversal direction, out of plane of the coupled facade frame. The stiffer substructure A2 and substructure A6 contribute, through the viscous coupling, to reduce the maximum displacement in the longitudinal direction. Similar beneficial effects are registered in the peak acceleration reduction, both in transversal and longitudinal direction (see J_2 in Table 4) while the transverse shear force at the base of vertical resistant elements appears not significantly reduced by the viscous coupling (J_3 in Table 4). Figures 5b and c show selected dissipative cycles evaluated during the numerical simulation for a given different base excitation within the seven cases. One of the simulated behavior for a selected device has been also reproduced (Figure 5e) during the campaign tests for the mechanical characterization of the installed devices confirming the expected performances [7].

The use of the performance indexes have permitted to determine an optimized solution which take into account the possibility of having a limited number of different type of dampers, for production reasons. Figures 6a, b and c show the designed constitutive relations for the three selected dampers in the final solution furnished to the manufacturer. Figure 6d shows the results obtained during the test campaign [7].

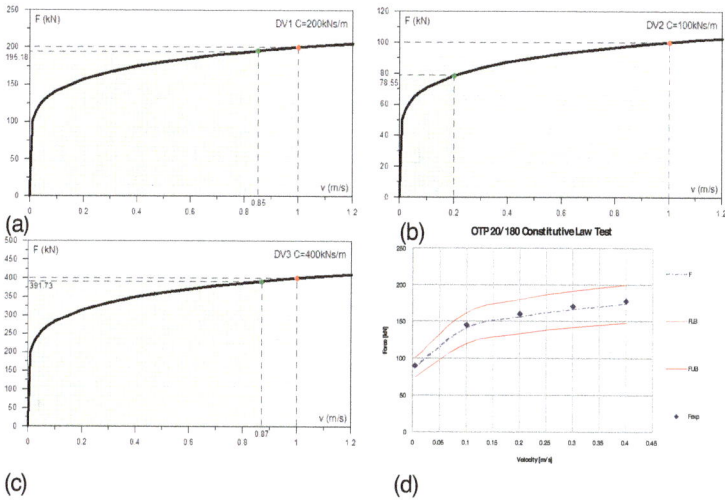

Figure 6. Design force-velocity relations for the three types of nonlinear viscous dampers; a) DV1 b) DV2 c) DV3 d) experimental data from the characterization tests [7].

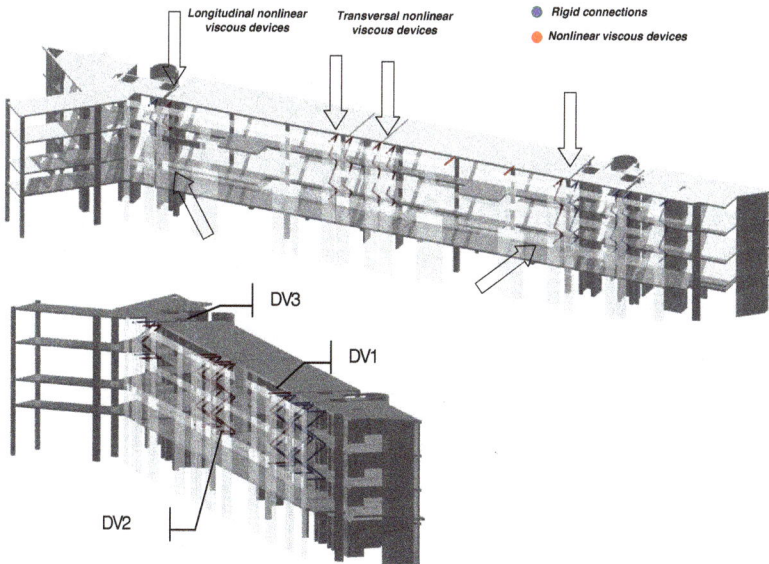

Figure 7. Three-dimensional sketch drawings reporting fluid viscous dampers locations in the structural retrofitting of the Edifice A of the Engineering Faculty of University of L'Aquila.

Figure 7 clarifies the location of the 43 devices: 18 DV1 (Type I); 17 DV2 (Type II) ; 8 DV3 (Type III) (as also reported in Table 2). In particular, in the transversal direction are working 18 DV1 devices in the higher positions (in the A3 substructure: 10 DV1, 4 horizontal and 6 oblique; in the A4 substructure 8 DV1, 2 horizontal and 6 oblique) and 17 DV2 in the lower positions and along the alignment of the slabs P1 (see Figure 2a) (in the A3 substructure: 4 DV2, 2 horizontal and 2 oblique; in the A4 substructure: 1 DV2 oblique; in contrast with the P1 slabs 12 DV2, 4 horizontal and 2 oblique) while 8 DV3 devices are working in the longitudinal direction positioned between A3-A2 and A4-A5.

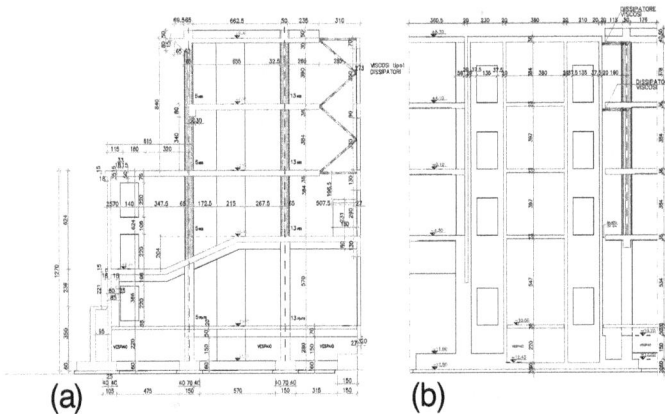

(a)   (b)

Figure 8. Nonlinear viscous damper placement: a) transversal; b) longitudinal.

Figure 8 shows a transversal and a longitudinal section where the protective devices are installed. In particular in Figure 8 it can be noted that a pair of DV3 devices is positioned at each of the two last level working in contrast between the substructures A3 and A2 thanks to the presence of a relevant seismic joint (depth= 20cm).

Figure 9 summarizes some relevant information such as: the large damage scenario appearing in the morning of April 7, 2009 immediately after the earthquake at the main facade of the Edifice A (Figure 9a); the facade completely rebuilt in a picture taken during the reconstruction (September 2011); of the same period two pictures presenting a close view of a DV1 horizontal device in the P1 zone (Figure 9c) and the four alignments of the dissipative trusses that following the perspective belongs the first one to the substructure A3 followed by two alignments in the P1 zone and completed by the last alignment which is the first one for the sub-structure A4 (Figure 9d). It can be noticed that in the last alignment due to the presence to the light stairs coming from the under floor the horizontal device is missed, this occurrence justifies the even total number of installed devices.

Together with the main structural seismic protection, here illustrated, the rehabilitation of the Edifice A has been conducted through the use of several technological applications to avoid

Figure 9. Reconstruction at Edifice A: a) damage scenario involving the facade, b) reconstruction of the facade, c) close view of the installed device, d) dissipative truss structures.

failure at the non-structural elements especially through the connection of both the reconstructed and the remained brick cladding with the reinforced concrete structures to avoid local failure due to the overturning of wall portion. The partition walls inside the building have been completely substituted with plasterboard fixed to aluminum profiles well anchored to the structural elements. Even if in the other two buildings (A and C) it was not necessary the use of seismic devices for structural protection, the approach followed in the work done in the Edifice A through the direct action of a non profit organization, have been extended to the other cases making realizable the return to the campus in the 2013 spring semester.

5. Structural health monitoring research activities at university of L'Aquila

A group of researchers of CERFIS (www.cerfis.it) with complementary skills is conducting a wide plan of activities in the field of dynamic testing under environmental loading and structural health monitoring for a series of buildings, with strategic or historical value, at L'Aquila. In the following a synthetic description of the most challenging findings is reported.

In order to achieve adequate level of confidence on the structural dynamic behaviour of the studied buildings a schedule of consequent activities are currently performed: (*i*) on-site dynamic testing under environmental actions with standard equipments [5,9,11,30]; (*ii*) finite element modelling based on exhaustive survey and material testing; (*iii*) definition of SHM-WSN sensor features; (*iv*) laboratory dynamic testing on 1:3 scaled frame in order to validate procedures and wireless monitoring sensors; (*v*) deployment of structural health monitoring systems with wireless smart sensors; (*vi*) development and installation by remote programming of modal and damage identification procedures taking into account temperature variation effects.

All activities are at different stages of development, therefore in the following a synthetic description for each of them is presented, while the achieved results for the structural health monitoring of the Basilica di Collemaggio are finally reported.

5.1. On-site dynamic testing

The clear comprehension of structural behavior is a consequence of a deep investigation of the different aspects involved. However dynamic testing in operational condition, conducted recording only absolute accelerations at different significant points, can be very helpful [30]. Within the group, the data-recording is generally conducted using a multi-channel acquisition system. Servo-accelerometers (SA107LN-Columbia) have been used in previous experiences [5,30]. The on-site experiences have been recently completed by a comparative studies conducted on real experimental data on the most popular output-only identification procedures for modal model and their use to identify finite element parametrical model [11]. On this basis, the identification of modal parameters from ambient vibration data is currently carried out using two main procedures: Enhanced Frequency Domain Decomposition (EFDD) and Stochastic Subspace Identification (SSI)

The Enhanced frequency domain decomposition is a stochastic technique, operating in the frequency domain, based on the evaluation of the spectral matrix, collecting the frequency-depending power cross-spectral densities of the experimental structure response at different measurement points. The key point of the method is the assumption that, at a certain frequency, only a few significant modes (typically one or two) contribute to determine the spectral matrix.

Instead, the data driven Stochastic Subspace Identification method, representing a time domain technique, allows the modal identification of a structure through the eigenproperties of several stochastic state space models, built to reproduce its experimental response, and characterized by increasing order n. Therefore, the order of the model (or the subspace dimension), which better approximates the experimental response, is a matter of identification too.

5.2. Finite element modeling and updating

The assessment of a representative physical model differs from modal identification in a few conceptual and procedural aspects. Modal models consist of global information, and a few frequencies and mode shapes are expected to capture the dominant structural behaviour. In contradistinction, physical models include local information, such as the stiffness and mass spatial distribution, which in principle should be wholly reconstructed.

The simplifying hypotheses introduced in the modellization phase fix the model dimension, and rigidly determine the inherent structure of the stiffness and mass matrices. Such matrices can be initially evaluated according to nominal, or even estimated values of the mechanical parameters. Forcing the reference model to match the experimental frequencies and modes, the identification process reduces to the calibration, or updating, of the initial parameter values, while the model dimension and the structure of the governing matrices remain unchanged.

Depending on the number, quality, and nature of the available information from the modal identification, different approaches to the physical model updating can be pursued [17]. Generally, the finite element models are used as a reference, taking advantage of the higher flexibility and computational efficiency of the numerical environment to explore different updating schemes [15], corresponding to different sets of free parameters. The data-to-unknowns redundancy is fully exploited, recurring to iterative techniques to minimize purposely-defined objective functions, expressing the error of the updated model in emulating the experimental modal data.

5.3. Definition of SHM-WSN sensor features

Vibration-based SHM requires sensed data that well represents the physical response of the structure both in amplitude and phase. The measurements must have sample resolution to characterize the structural response and must be recorded with a consistent sample rate that is synchronized with other sensed data from the structure. The sensor hardware needs for a sensor board with higher resolution and more accurate sampling rates designed specifically for SHM applications.

The ST Microelectronics LIS344ALH capacitive-type MEMS accelerometer with DC to 1500 Hz measurement range, was chosen for the SHM-A board. This type of accelerometer utilizes the motion of a proof mass to change the distance between internal capacitive plates, resulting in a change of output voltage in response to acceleration. Though MEMS accelerometers are available with lower noise levels, the ST Micro accelerometer offers an excellent price/performance ratio. In addition, it provides three axes of acceleration on a single chip. The specifications for the accelerometer are given in Table 5. The SHM-A sensor board has been designed for monitoring civil infrastructure through the Illinois SHM Project, an interdisciplinary collaborative effort by researchers in civil engineering and computer science at the University of Illinois at Urbana-Champaign [31].

Two hardware configurations of smart sensor nodes are required for the wireless communication and sensing: a gateway node for sending commands and receiving wireless data from network, and the battery powered nodes remote to the base station. To increase the communication range, both nodes are equipped with an antenna, which covers the communication in a range of 30m and a SMA connector to install an external additional antenna. In the CERFIS configuration a watertight partial-gauzy box, allowing an in-the-distance visibility of light sensor to check the efficiency of the remote node, protects the boards. An external cable connecting both the 220V electric web and an energy store box, composed by three rechargeable 1.5V batteries IND alkaline D size with capacity of 20500mAh each, to assure a continuous registration procedure during earthquake events, powers each node. The sensor location, inside historical monuments, does not allow an autonomous powered, as trough the well-known solar panels. An additional USB receptacle is installed to allow the link with a PC. The wireless communication is entrusted to an ADC converter.

Parameter	Value
Axes	3
Measurement range	±2g
Resolution	0.66 V/g
Power supply	2.4 V to 3.6 V
Noise density, x-and y-axes	22 – 28 µg/Hz
Noise density, z-axis	30 – 60 µg/Hz
Temperature range	-40 to 85°C
Supply current	0.85 mA

Table 5. Accelerometer specifications.

5.4. Laboratory dynamic testing and wireless sensor characterization

Preliminary tests are conducted using a modular structural steel frame located at the CERFIS laboratory of University L'Aquila to characterize a SHM-WSN. In particular two different types of test have been performed. In the first series a direct comparison one single wireless sensor (the above described IMOTE 2 type) and one wired accelerometer (SA107LN-Columbia) has been conducted (Figure 10). Within this configuration the frame responses both to a little impulse in longitudinal direction and under environmental noise have been recorded. Others tests have been made using six wireless sensors, two for each slab, placed at diagonally opposite corners. This particular experimental setup has been used to identify the main modal frequencies, shapes and damping. Again both impulsive and ambient tests have been performed. The results are here not reported for sake of brevity. Moreover, in all tests, the wireless sensors, installed in the prototype structure, transfer the collected data to a single wireless node (gateway mode) linked to the acquisition card.

The investigation in the lab environment will be conducted on new sensor configurations fully developed by the CERFIS group. As is well known, one of the major limitations of wireless motes are the limited performances. Therefore, the idea is to use configurable hardware devices (e.g. FPGA) for the creation of hw/sw mixed service based architecture, with processing services directly implemented in hardware. In practice, we want to combine the mote processor with a set of ad-hoc developed co-processors specifically designed for the implementation of various processing modules. We think that this strategy will significantly increase monitoring efficiency, not only allowing a real-time processing, but also enabling the simultaneous support of different analysis techniques addressed to a wide range of application scenarios, from the pure structural health monitoring up to the emergency management, which imply often divergent specific requirements.

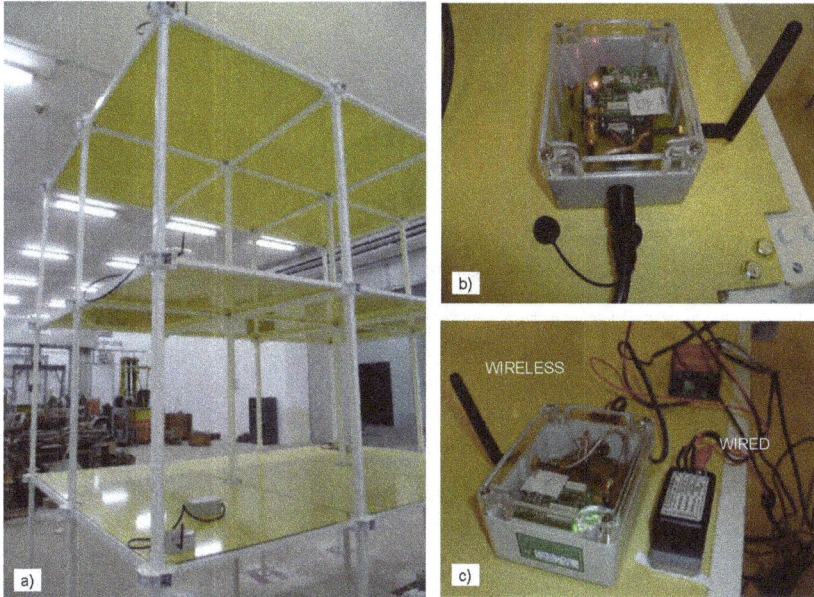

Figure 10. Light model (scale 1:3) of modular steel-made three-dimensional frame: (a) basic configuration, (b) sensor-node of wireless network; (c) comparison with sensor of traditional wired network

5.5. SHM-WSN deployment on strategic and historical structures

Traditionally, a grid of sensor was deployed across a building and the measured data were conveyed via a cable connection to a central processing system (e.g. a personal computer). Recently, Wireless Sensor Networks (WSN) emerged as a possible attractive alternative solution, mainly due to the lower cost, lower size of the systems and ease of setup respect traditional wired systems thanks to the multi-hop connection capabilities which allow the nodes to organize themselves in a network where each node can be source, destination and also a router for the information flowing within the network.

Current wireless monitoring systems are usually based on off the shelf sensor nodes equipped with new generation low cost, small sensors (e.g. MEMS accelerometers). Although these systems are not specifically designed for structural monitoring applications, they can still provide good performances. For example, Illinois Structural Health Monitoring Project (ISHMP) has shown the potential of WSN in several real monitoring scenarios [31]; they used a network of Imote2 motes equipped with a specifically design sensor board (ISM400) and an embedded processing software (ISHMP Toolsuite) based on TinyOS.

Data processing is a key point in the future development of wireless monitoring systems. Many wireless implementations adopt a traditional processing paradigm, with data transmitted from the sensor nodes to a central gateway connected to a PC that performs the entire processing. However, modern sensor nodes are equipped with a microprocessor, allowing them to carry out local processing of data. In other words, data processing can be distributed across the network.

The wireless systems, in fact, have progressed very rapidly in recent years and are now considered the enabling technology for realizing the pervasive ubiquitous computing environment that should support advanced distributed applications in many domains, especially for advanced distributed applications.

Therefore, owing to unprecedented design challenges and potentially large revenues, wireless sensor networks are calling huge interest in both the scientific and the industrial world. Besides a secure optimization of transmission (as shown by ISHMP work, whose software is already partially decentralized), processing de-centralization can bring the advantage of being able to quickly detect local phenomena, even in case of network splitting as a consequence of critical phenomena as an earthquake. This capability can be extremely useful insecurity systems or, generally, in the field of emergency management.

A series of activities are still under development to rethink structural modal analysis techniques, towards the goal of a distributed processing within the network, which could efficiently support real-time monitoring and safety oriented services [10]. Firstly, moving from the achievements and contributions of ISHMP, an iMote2-based monitoring system was developed. Moreover, the ISHMP software tools will be integrated with ad-hoc applications, in order to achieve an efficient distributed processing within our network. Moreover, optimizations of limited energy resources may be achieved through suited techniques of data compression and aggregation, providing reduced energy costs of communications and lower channel capacity for data delivery.

The choice of the ISHMP software tools is not simply determined by the convenience of having a ready-to-use, decentralized-oriented middleware, but has a deeper reason. In fact, given the particular characteristics of the processing, the ISHMP Toolsuite was designed as a service-based software architecture. In other words, the various processing steps are implemented as services, and each application is just a collection of independent modules.

6. The structural health monitoring of the Basilica di Collemaggio

The Basilica S. Maria di Collemaggio is one of the most attractive churches in Centre Italy. It dates from the XV century. The Basilica has a nave and two side aisles. The dimension of the nave is 61m in length and 11.3m in width; its height reaches 18.25m. The two side aisles are 7.8 and 8m in width; two external walls both 12.5m high delimit them. Seven columns, not evenly distanced, on each side separate the nave and two side aisles. The columns are about 5.25m high; a layer of well-laid stone, made of a calcareous material arranged irregularly in a poor quality mortar, encloses their core; the transverse section, approximately circular, is on average 1.00 m in diameter. The thickness of masonry varies from 0.95 m to 1.05 in the external walls; it is 0.9m in the two walls of the nave, over the columns. The four walls are connected on one side to the facade of the Basilica and, on the other side, to the transept. The facade is joined to a thick octagonal tower on the right corner; another masonry building is adjacent to a part of the wall, about 40% of it, behind the tower. The wooden roof is supported from trusses placed in a cross-sectional direction to the walls.

Before the occurring of the 6th April 2009 earthquake a numerical and experimental study has permitted to characterize the dynamic behavior of the Basilica [32-34]. The experimental data were firstly used to identify a modal model and then to determine suitable FE models able to predict and frame the dynamical response of the church. Preliminary numerical analyses were carried out on the basis of several assumptions regarding: (1) mechanical parameters of masonry, (2) timber trusses of the roof, (3) restraints in walls and columns, (4) links among structural components. Afterwards the Basilica was excited at a low level by an instrumented hammer and a mechanical vibration exciter (vibrodyne). Several tests have been carried out, with different positions of the instruments and impact locations, in order to excite and to measure as many modes as possible.

The vibrodyne was located on the top of a lateral wall. The frequency responses were directly measured around the first two modes; these are the most important ones that describe the dynamic response of the church. Experimental data have been used to identify natural

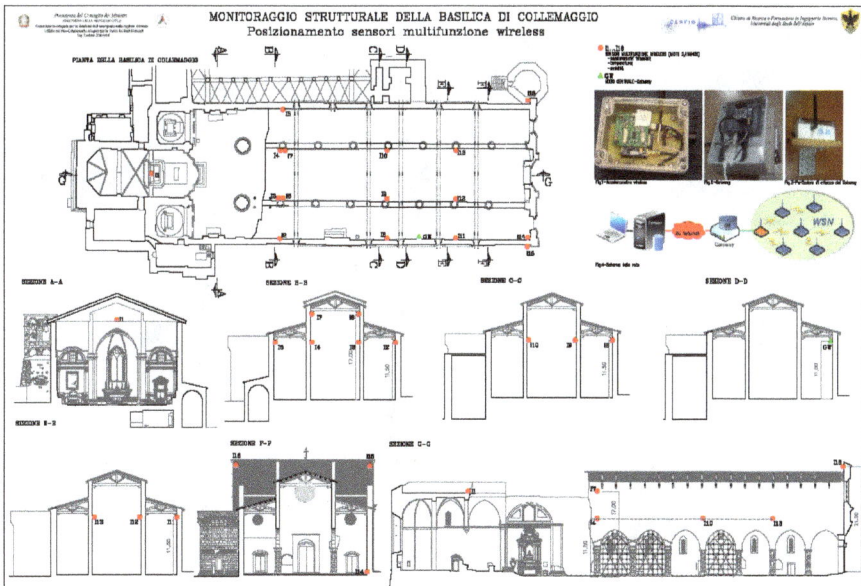

Figure 11. Drawings for the locations of the 16 smart sensors mounting tri-axial MEMS accelerometers, humidity and temperature measuring instruments, installed at the Basilica di S. Maria di Collemaggio, at L'Aquila, Italy.

frequencies, modal displacements and damping factors. The first campaign of tests [32] have permitted to recognise at least four major resonance peaks in the range 0.8÷3.0 Hz.

The first two peaks are around frequencies values, about 1.25 and 1.7 Hz. Other peaks are present over 2 Hz. Two of them, around 2.5 and 2.7 Hz, are well defined in all tests. Secondary peaks, around 2.2, 2.3 and 2.6 Hz, are not always visible in all the responses; they indicate the occurrence of highly coupled modes. These peaks, however, are estimated to be less important: numerical analysis indicates that the participating mass of first two modes is at least 85% of total mass in the transverse direction of the church.

After retrofitting, all peaks are shifted to higher frequencies [33]. The first two are around 1.45 Hz and 2.12 Hz respectively. Other peaks are clearly visible around 2.6 and 2.95 Hz. Secondary peaks, which are not always visible in all the responses, are recognisable even in this case. Higher frequencies are a consequence of the increasing stiffness brought about by retrofitting. It is interesting to observe that now the responses of a pair of accelerometers are basically identical, at least in the range of frequencies examined. This is a clear indication that the retrofitting had improved the link between the longitudinal walls. Other dynamic testing have been performed on the facade [34] which have permitted to evidence that out-of-plane local

Figure 12. Installation phases of the monitoring system at the Basilica di S. Maria di Collemaggio: a) b) sensor positioning on the central walls of the nave, c) sensor positioning beyond the facade, d) f) sensor views, g) phase of on-site testing, h) sensor positioning at the end of the nave walls.

modes of this element are in a frequency range higher than the transversal mode of the nave. Recently, after the earthquake a strong effort has been made to use all the available data from the previous on-site dynamic campaign in order to develop a series of complete finite element models of the Basilica able to reproduce the main modal identified characteristics and the collapse scenario [35]. Starting from these models, a reproduction of the scenario after the collapse has been pursued [8, 36]. A campaign of numerical simulations has been conducted to evaluate the dynamic response of the Basilica together with the temporary retrofitting under small earthquakes characterizing the numerous aftershocks at L'Aquila.

The previous installation for the on-site dynamic testing campaigns together with the observation obtained by the modelling have driven the monitoring installation scheme reported in Figure 11. The wireless network composed by 16 smart sensors (see also Table 3) has been finally installed on June 2011. During the successive months the monitoring system has been enhanced and brought to complete and automatic management to sense seismic induced vibrations. During this path, several test campaigns have been conducted making use of different induced source of vibrations such as hammer, ambient vibrations and free-vibration tests [37]. Finally, in six cases, the seismic induced response of the structures of S. Maria di Collemaggio has been cleared measured, as reported in Table 5. The results of the identification process will be object of further publications.

Number	Earthquakes	Date	Time UTC	Magnitudo MI	Maximum recorded response acceleration (g)
1	Main shock Emilia	20/05/2012	2.03	5,9	0,0054
2	Aftershock Emilia	20/05/2012	13.18	5,1	0,0018
3	Shock Ravenna	06/06/2012	6.08	4,5	0,0014
4	L'Aquila	14/10/2012	16.32	2,8	0,0072
5	L'Aquila	30/10/2012	2.52	3,6	0,0073
6	L'Aquila	16/11/2012	3.37	3,2	0,0082

Table 6. Recorded structural response of S. Maria di Collemaggio.

7. Conclusions

The chapter aims to present the rapid development in the transfer to the real applications of the available technology in the sector of structural control and health monitoring, occurred at L'Aquila immediately after the 2009, L'Aquila earthquake. The benefits in the application of these emerging technologies are still under verification and observation. For the performance evaluation of the installed seismic protections systems, only the occurrence of a relative strong seismic motion, will clearly evidence the benefits introduced in the territory. Differently, the large amount of activities concerning material and in-situ testing together with small or long-term monitoring will surely increment the knowledge regarding the real behaviour of complex masonry or reinforce concrete structures. The amount of obtained data from this large campaign of testing, conducted with different techniques and aims, is in many cases larger than the real possibility of a deep discerning. Indeed a complete extraction of valuable information useful for the understanding of the material and structural behaviour of the large amount of buildings, infrastructure and historical monuments is still undergoing. The presented overview, even if conducted more on an informative level than in a deep scientific manner, remains a valuable starting point for searching innovative procedures and devices in the considered research field. The above references will permit a deeper analysis on specific questions and further publications will make into evidence specific novel findings, developed during the difficult path of doing innovative research in a territory in which a natural disaster has strongly modified the habitual activities conducted before the event.

Acknowledgements

First of all, the author wishes to acknowledge with the numerous young students at any level (5yrs degree, MS and PhD) of the University of L'Aquila and coming from other Universities, which make realizable all the work synthesized in this chapter. Special thank goes to the co-authors of the above referenced publications produced at the University of L'Aquila, from which most of the material here presented is taken and re-arranged. Several research grants have permitted the development of the research such the MIVIS project financed by CARIS-PAQ, the RELUIS projects financed by DPC, the PRIN projects financed by MIUR, the RICOSTRUIRE project financed by the Development Italian Ministry.

Author details

Vincenzo Gattulli

DICEAA, CERFIS, University of L'Aquila, Italy

References

[1] Ceci AM, Contento A, Fanale L, Galeota D, Gattulli V, Lepidi M, Potenza F. Structural performance of the historic and modern buildings of the University of L'Aquila during the seismic events of April 2009. *Engineering Structures* vol. 32(7), pp. 1899-1924, 2010.

[2] Gattulli V. Role and perspective of computational structural analysis for sustainable reconstruction and seismic risk mitigation after an earthquake. in B.H.V. Topping and Y. Tsompanakis, (Editor), *"Civil and Structural Engineering Computational Technology"*, Saxe-Coburg Publications, Stirlingshire, UK, Chapter 1, pp 1-34, 2011. doi: 10.4203/csets.28.1

[3] Beolchini GC, Conflitti G, Contento A, D'Annibale F, Di Egidio A, Di Fabio F, Fanale L, Galeota D, Gattulli V, Lepidi M, Potenza F. Il comportamento degli edifici della Facoltà di Ingegneria dell'Aquila durante la sequenza sismica dell'Aprile 2009. *XIII Convegno di Ingegneria Sismica ANIDIS, Bologna (Italia), Giugno 2009* (in Italian).

[4] Ceci AM, Gattulli V, Potenza F. Serviceability and damage scenario in RC irregular structures: post-earthquake observations and modelling predictions. *ASCE Journal of Performance of Constructed Facilities*, in press, 2013.

[5] Foti D, Mongelli M, Gattulli V, Potenza F, Ceci A. Output-only structural identification of the Engineering Faculty Edifice A at L'Aquila. *Fourth IOMAC Conference, Istanbul, Turkey 2011.*

[6] Gattulli V, Lepidi M, Potenza F, Ceci A. Nonlinear viscous dampers interconnecting adjacent structures for seismic retrofitting. *8th European Conference on Structural Dynamics, Eurodyn11, Leuven*, Belgium, 2011.

[7] Castellano MG, Gattulli V, Borrella R, Infanti S. Experimental characterization of nonlinear fluid viscous dampers according to the New European Standard. *5th European Conference on Structural Control*, EACS12, Genoa, Italy, 2012.

[8] Antonacci E, Gattulli V, Martinelli A, Vestroni F. "Analisi del comportamento dinamico della Basilica di Collemaggio per il progetto di un sistema di monitoraggio". *XIV Convegno di Ingegneria Sismica ANIDIS*, Bari (Italia), Settembre 2011 (in Italian).

[9] Antonacci E, Ceci AM, Colarieti A, Gattulli V, Graziosi F, Lepidi M, Potenza F. Dynamic testing and health monitoring via wireless sensor networks in the post-earthquake assessment of structural condition at L'Aquila. *8th European Conference on Structural Dynamics*, Eurodyn11, Leuven, Belgium, 2011.

[10] Federici F, Graziosi F, Faccio M, Gattulli V, Lepidi M, Potenza F. An integrated approach to the design of Wireless Sensor Networks for structural health monitoring. *International Journal of Distributed Sensor Networks*, Volume 2012 (2012), Article ID 594842, 16 pages doi:10.1155/2012/594842

[11] Antonacci E, De Stefano A, Gattulli V, Lepidi M, Matta E. Comparative study of vibration-based parametric identification techniques for a three-dimensional frame structure. *Journal of Structural Control and Health Monitoring*, Volume 19, Issue 5, pages 579–608, August 2012doi: 10.1002/stc.449, 2012.

[12] Calvi GM. L'Aquila earthquake 2009: reconstruction between temporary and definitive. *2010 NZSEE Conference.*

[13] Salvatori A. Edifici con isolamento sismico alla base: la nuova Facoltà di Lettere, Filosofia e Scienze della Formazione dell'Università dell'Aquila. *XIII Convegno di Ingegneria Sismica ANIDIS, Bologna* (Italia), Giugno 2009 (in Italian).

[14] Cartapati E. L'intervento di miglioramento antisismico delle navate della Basilica di Santa Maria di Collemaggio. *Atti della Giornata di Studio: Patrimonio Storico Architetto-*

nico e Terremoto. Diagnosi ed Interventi di recupero, L'Aquila, November 2000, pp. 155–164 (in Italian).

[15] Maugeri M, Simonelli AL, Ferraro A, Grasso S, Penna A. Recorded ground motion and site effects evaluation for the April 6, 2009 L'Aquila earthquake. *Bull Earth. Eng.* 9, 157–179, 2011.

[16] Spina D, Lamonaca BG, Nicoletti M, Dolce M. Structural monitoring by the Italian Department of Civil Protection and the Case of 2009 Abruzzo Seismic Sequence. *Bull Earth. Eng.* 9, 325–346, 2011.

[17] Russo S. On the monitoring of historic Anime Sante church damaged by earthquake in L'Aquila. *Journal of Structural Control and Health Monitoring*, doi: 10.1002/stc.1531, in press.

[18] Luco JE, Francisco CP, Barros D. Optimal damping between two adjacent elastic structures. *Earthquake Engineering and Structural Dynamics*; 27:649-59, 1998.

[19] Luco JE, F.C.P. De Barros. Control of the seismic response of a composite tall building modelled by two interconnected shear beams. *Earthquake Engineering and Structural Dynamics*; 27:205-33, 1998.

[20] Xu YL, He Q, Ko JM. Dynamic response of damper-connected adjacent buildings under earthquake excitation. *Engineering Structures*; 21:135-48, 1999.

[21] Zhang WS, Xu YL. Dynamic characteristics and seismic response of adjacent buildings linked by discrete dampers. *Earthquake Engineering and Structural Dynamics*; 28:1163-85, 1999.

[22] Aida T, Aso T, Takeshita K, Takiuchi T, Fuji T. Improvement of the structure damping performance by interconnection. *J. of Sound and Vibration*; 242(2):333-53, 2001.

[23] Zhu HP, Wen Y, H. Iemura. A study of interaction control for seismic response of parallel structures. Computers & Structures; 79:231-42, 2001.

[24] Zhu HP, Xu YL. Optimum parameters of Maxwell model-defined dampers used to link adjacent structures. *J. of Sound and Vibration*; 279:253-74, 2005.

[25] Kim K, Rye J, Chung L. Seismic performance of structures connected by viscoelastic dampers. *Engineering Structures*; 28:83-195, 2006.

[26] Gattulli V. On the dissipative coupling of adjacent structures, XX Congresso Nazionale AIMETA'11, Bologna, Italy, 12-15 Settembre, 2011.

[27] Gattulli V, Lepidi M, Potenza F, Ceci AM. Seismic protection and retrofitting through nonlinear fluid viscous damper interconnecting substructures, *EACS 2012, 5th European Conference on Structural Control*, Genoa, Italy – 18-20 June 2012.

[28] Gattulli V, Potenza F, Lepidi M. Damping performance of two oscillators coupled by alternative dissipative connections. *Journal of Sound and Vibration*, in preparation.

[29] NTC 08, Decreto Ministeriale 14/01/2008. Norme tecniche per le costruzioni, Gazzetta
      Ufficiale No. 29 (in Italian) Ministero delle Infrastrutture, 2008.

[30] Lepidi M, Gattulli V, Foti D. Swinging-bell resonances and their cancellation identi-
      fied by dynamical testing in a modern bell tower. *Engineering Structures*, vol. 31, p.
      1486-1500, 2009.

[31] Rice J, Spencer Jr B, Flexible smart sensor framework for autonomous full-scale struc-
      tural health monitoring, NSEL Report Series 018, 2009.

[32] Antonacci E, Beolchini GC, Di Fabio F, Gattulli V. The dynamic behavior of the Basil-
      ica S. Maria of Collemaggio, *Proc. 2nd Intern. Congr. on Studies in ancient Structures*,
      Instanbul, Turkey, 2001.

[33] Antonacci E, Beolchini GC, Di Fabio F, Gattulli V. Retrofitting effects on the dynamic
      behaviour of S. Maria di Collemaggio, *Tenth International Conference on Computational
      Methods and Experimental Measurements*, Alicante, Spain, June 4-6, 2001.

[34] Antonacci E, Beolchini GC, The dynamic behaviour of the facade of the Basilica di S.
      Maria di Collemaggio. *Proc. of Structural analysis of historical constructions, IV Interna-
      tional Seminar*, Padova, Italy, 2004.

[35] Gattulli V, Antonacci E, Vestroni F. Field observations and failure analysis of the Ba-
      silica S. Maria di Collemaggio after the 2009 L'Aquila earthquake. *Engineering Failure
      Analysis* submitted 2012.

[36] Gattulli V, Lampis G, Antonacci E. Sviluppo di modelli strutturali ad elementi finiti
      ed analisi della risposta dinamica della Basilica di S. Maria di Collemaggio. Settem-
      bre 2011 *Pubblicazioni CERFIS*, CEntro di Ricerca e Formazione in Ingegneria Sismica,
      Università degli studi dell'Aquila Rpt. 4/11, 2011 (http://www.cerfis.it/it/download/
      doc_download/82-report-cerfis-n4-2011.html).

[37] Antonacci E, Colarieti A, Federici F, Gattulli V, Graziosi F, Lepidi M, Potenza F. Car-
      atteristiche del sistema di monitoraggio della Basilica di S. Maria di Collemaggio.
      Settembre 2011 Pubblicazioni CERFIS, CEntro di Ricerca e Formazione in Ingegneria
      Sismica, Università degli studi dell'Aquila Rpt. 6/11, 2011 (https://www.box.com/s/
      209e12ac7373ebfd5200)

Chapter 11

Numerical Modelling of the Seismic Behaviour of Gravity-Type Quay Walls

Babak Ebrahimian

Additional information is available at the end of the chapter

1. Introduction

Gravity quay walls are the most common type of construction for docks and harbours because of their durability, ease of construction and capacity to reach deep seabed levels. Gravity quay walls are designed for three main criteria; sliding, overturning and allowable bearing stress under the base of quay wall. Although the design of gravity quay walls has been reasonably well understood for static loads, but analysis under seismic loads is still in being developed. During strong ground shaking, the pore water pressure of cohesionless saturated backfill soils builds up. This pressure increase not only causes the lateral forces on the wall (which leads to wall failure), but also reduces the effective stress of backfill soil which may result in liquefaction. The occurrence of liquefaction in backfill soil was the main reason of damages from past earthquakes to gravity quay walls (e.g., in 1964 at Nigata Port (Hayashi et al. 1966), in 1993 at Kushiro-oki, and in 1994 at Hokkaido Toho-oki (Sasajima et al. 2003)). Moreover, observations of 24 marine structures in 1999 earthquake at Kocaeli, Turkey showed the seaward movement of quay walls due to the liquefaction of backfill soils (Sumer et al. 2002). The same observations were reported in 1999 during the Chi Chi earthquake in Taiwan (Chen and Hwang 1999).

The seismic coefficient method containing Mononobe-Okabe's formula is usually used in the design of gravity-type quay walls to resist earthquake damages but this design method does not take into account the liquefaction of backfill soil and its associated deformations (Sasajima, et al. 2003). Furthermore, conventional design method of quay walls is based on providing capacity to resist a design seismic force, but it does not provide information on the performance of a structure when the limit of the force-balance is exceeded. In this regard, gravity quay walls failures have caused much progress in the development of deformation-based design methods for waterfront structures. Accordingly, much significant experimental and theoretical research works have been done (Sugano et al. 1996; Inagaki et al. 1996; Iai 1998; Iai et al. 1998; Iai and

Sugano 2000; Ichii et al. 2000; Inoue et al. 2003; Nozu et al. 2004; Mostafavi Moghadam et al. 2009 and 2011). A new design methodology, named performance-based design, has born from lessons learned caused by earthquakes in 1990's to overcome the limitations of conventional seismic design (PIANC 2001). In this framework, lateral spreading of the saturated backfill and foundation soils along with the effect of quay wall as the supporting structure (saturated soil-structure interaction) are taken into account as a more logical design.

Predicting the response of a structure retained a liquefiable soil during an earthquake is highly dependent on adequately accounting for the effects of pore water pressure development, stress-strain softening and strength reduction in the soil on the system behavior. Thus, it is required to perform dynamic analyses that account for the saturated soil-structure interaction effects using numerical modeling techniques. Several researchers have reported the use of numerical analysis for predicting the behavior of liquefiable soil measured in laboratory tests or field case histories. Iai et al. (1998) reported that FLIP code can successfully predict the seismic behavior of port structures. Yang and Parra developed CYCLIC code and reported successful predictions of the seismic behavior of gravity quay wall placed on liquefiable sand and calibrated the numerical results with centrifuge tests (Parra 1996; Yang 2000). Both the codes DYSAC2 and DYNAFLOW are reported by Arulanandan (1996) as having adequately predicted the response of a submerged embankment subjected to dynamic loading in a centrifuge test. According to Madabhushi and Zeng (1998), the code SWANDYNE successfully predicted the seismic response of a gravity quay wall rested on liquefiable sand modeled in the centrifuge. Finn reported the successful validation of TARA-3 using centrifuge tests results (Finn et al. 1991). The successful use of FLAC package for prediction of the behavior of caisson retaining walls in liquefiable soils was reported by Dickenson and Yang (1998). They used a nonlinear effective stress analysis method based on the Mohr-Coulomb constitutive model and a pore water pressure increment scheme based on the work of Seed and his co-workers (e.g. Martin et al. 1975; Seed and DeAlba 1986). Likewise, McCullough and Dickenson, who used the same analysis method and soil model in FLAC, reported fairly good agreement between predicted and measured permanent horizontal displacements at top of five anchored sheet pile bulkhead walls in liquefiable soils subjected to different earthquakes in Japan between 1987 and 1993 (McCullough and Dickenson 1998). It should be noted that the assumption of the Yang and McCullough was that the foundation soil is non-liquefiable. A hyperbolic type stress-strain formulation developed by Pyke (1979) along with the pore water pressure build up model proposed by Byrne (1991) was implemented in FLAC code by Cooke (2001). They concluded that Pyke-Byrne model over-predicts the horizontal displacement of gravity quay walls modelled in centrifuge within a factor of approximately two.

Recently, Ebrahimian et al. (2009) have carried out a series of two dimensional fully coupled effective stress dynamic analyses in order to study the deformation of quay walls and the liquefaction potential of backfill soils. Additionally, several 1g shaking table tests have been executed to verify the obtained numerical results. They showed based on the lessons learned obtained from numerical results, a safer design of gravity-type quay walls can be developed. Correspondingly, Mostafavi Moghadam et al. (2009) conducted finite difference effective-stress analyses to investigate the seismic performance of caisson quay walls. Their obtained numerical results in terms of seaward movement, settlement and inclination of wall as well as the total pressure recorded behind the caisson wall were compared with 1g shaking table tests results. It was demonstrated that the numerical results appropriately supports the experimental results obtained by model tests.

In this chapter, firstly some aspects related to calibration of a numerical model are discussed. Then, the numerical results are verified by comparing the calculated values with corresponding ones obtained from 1g shaking table tests. A series of two dimensional fully coupled effective stress dynamic analyses are carried out in order to study the deformation of quay wall, liquefaction potential and failure mechanisms of soil-wall system during seismic loading. The Finn and Byrne model Byrne (1991) is used with some minor modifications to model pore pressure build up during seismic loading. Afterwards, computational parametric studies are performed to investigate the effects of backfill soil properties and input excitation characteristics on the seismic behaviour of gravity-type quay walls including the residual deformation of wall, liquefaction potential and failure modes of soil-wall system.

Several 1g shaking table tests have been executed to verify the obtained numerical results. It is found that the extent of liquefaction and the deformation pattern of soil-wall system that includes seaward displacement, tilting and settlement (as typical failure modes of quay walls due to earthquake) resulted from numerical analyses agree reasonably well with the actual observations in shaking table tests. During seismic excitation, no evidence of liquefaction has detected near the quay wall but liquefaction occurs at some landward distance from the wall. Based on the current results, it seems possible to develop a safer design of gravity-type quay walls by using lessons learned from the present numerical analyses.

2. Description of the numerical method

In this research, a two-dimensional (2D) reference model is developed to simulate the seismic performance of gravity-type quay walls in a rational way. Nonlinear time history dynamic analysis is conducted using computer program FLAC 2D (Itasca 2004). FLAC 2D is an explicit finite difference program for modeling soil-structure interaction analysis under static and seismic loading conditions. Here, numerical approach is based on a continuum finite difference discretization applying Lagrangian approach (Itasca 2004). Every derivative in the set of governing equations is directly replaced by algebraic expression written in terms of field variables (e.g., stress or displacement) at discrete point in space. Regarding dynamic analysis, explicit finite difference scheme is applied to solve the full equation of motion using the lumped grid point masses derived from the real density of surrounding zone. The calculation sequence first invokes the equations of motion for deriving new velocities and displacements from stresses and forces; then, strain rates are derived from velocities, and new stresses from strain rates. Every cycle around the loop corresponds to one time step. Each box updates all grid variables from known values which are fixed over the time step being executed (Figure 1).

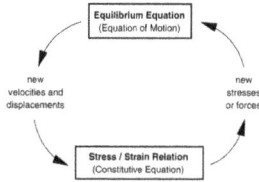

Figure 1. Basic explicit calculation cycle (Itasca 2004)

The equation of motion, in the simplest form, relates the acceleration ($d\dot{u}/dt$) of a mass (m) to the applied force (F) which may vary with time. Newton's law of motion for the mass-spring system is:

$$m\frac{d\dot{u}}{dt} = F \tag{1}$$

In a continuous solid body, Eq. (1) is generalized as follows:

$$\rho\frac{\partial\dot{u}_i}{\partial t} = \frac{\partial\sigma_{ij}}{\partial x_j} + \rho g_i \tag{2}$$

where, ρ = mass density; t = time; x_j = components of coordinate vector; g_i = components of gravitational acceleration (body forces); σ_{ij} = components of stress tensor; i = components in a Cartesian coordinate frame.

For problem analysis, the strain rate tensor and rotation rate tensor, having the velocity gradients, are calculated by the following equations:

$$e_{ij} = \frac{1}{2}\left[\frac{\partial\dot{u}_i}{\partial x_j} + \frac{\partial\dot{u}_j}{\partial x_i}\right] \tag{3}$$

$$\omega_{ij} = \frac{1}{2}\left[\frac{\partial\dot{u}_i}{\partial x_j} - \frac{\partial\dot{u}_j}{\partial x_i}\right] \tag{4}$$

where, e_{ij} = components of strain rate; ω_{ij} = components of rotation rate; $\dot{u}_i$ = components of velocity.

The specific mechanical relationship is used in order to obtain the stress tensor as below:

$$\sigma_{ij} = M\left(\sigma_{ij}, \dot{e}_{ij}, \kappa\right) \tag{5}$$

where, M = specific rule of behaviour; κ = history parameters (based on the specific rules which may or may not exist).

2.1. Modeling procedure

Firstly, a static analysis considering the effect of gravity with the Mohr-Coulomb elastic perfectly plastic constitutive soil model is performed to establish the in-situ stresses before seismic loading. The water within the soil is modeled directly, and is allowed to flow during the static solutions. Static solutions are obtained by including damping terms that gradually remove kinetic energy from the system. Then, boundary pressures equivalent to fluid weight is applied along the bottom of the sea and the front side of the caisson. Once initial stress state is established in the model and the soil model is changed to a pore pressure generation constitutive model, the effective stress dynamic analysis is started. For dynamic analysis, the acceleration record is applied to the nodes along the bottom of numerical grid. During the dynamic solutions excess pore water pressures are allowed to generate and also the dissipation of these pore pressures is modelled. To this end, the Finn and Byrne model (Finn et al. 1977; Byrne 1991) is modified and used to carry out coupled dynamic groundwater flow calculations. This effective stress analysis which take into account the effects of seismically induced pore water pressures is used to investigate the degree and extend of liquefaction that occur in backfill. It should be noted that, all analyses are performed under fully drained condition. One key advantage of the coupled numerical approach is the ability to account for the interde-pendent effects of various mechanisms and phenomena on each other as the numerical computations proceed. For instance, when an effective stress formulation is used, the inclusion of pore water pressure generation in the simulation can impact the strength and stress-strain behavior of the soil during shaking in the same manner as in the field. For the sake of simplicity in defining the performance of a quay wall during an earthquake, the author has focused on the horizontal movement because most of the gravity-type quay walls exhibited a failure pattern predominantly in the form of excessive horizontal movement rather than other damages. The numerical grid for a typical quay wall and foundation soil used in the current study is illustrated in Figure 2. Numerical model represents a backfill of constant depth retained by quay wall rested on a very dense foundation.

2.2. Soil constitutive model

In static analysis, the Mohr-Coulomb constitutive model is used to model the behaviour of sandy soil. The linear behaviour is defined by elastic shear and drained bulk modulus. The shear modulus of sandy soil is calculated with the formula given by Seed and Idriss (1970):

$$G_{\max} = 1000 k_{2\max}\left(\sigma_m'\right)^{0.5} \tag{6}$$

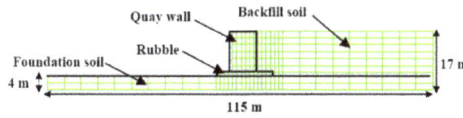

Figure 2. Numerical grid constructed in FLAC

where, G_{max} is the maximum (small strain) shear modulus in pounds per square foot, psf (it is later converted to kPa to be consistent with metric units being used), k_{2max} is the shear modulus number (Seed and Idriss 1970), and σ'_m is the mean effective confining stress in psf. The Poisson's ratio is taken as 0.35.

For pore water generation during dynamic analysis, the updated model proposed by Byrne (1991) is incorporated to account the development of pore water pressure build up as an effect of volumetric strain induced by the cyclic shear strain using the following formulation:

$$\Delta\varepsilon_v = C_1 \exp\left(C_2\, \varepsilon_v/\gamma\right) \tag{7}$$

where, $\Delta\varepsilon_v$ is the volumetric strain increment that occurs over the current cycle, ε_v is the accumulated volumetric strain for previous cycle, γ is the shear strain amplitude for the current cycle, and C_1 and C_2 are constants dependent on the volumetric strain behavior of sand. According to Byrne (1991), the constant C_1 in equation 7 controls the amount of volumetric strain increment and C_2 controls the shape of volumetric strain curve. These constants are estimated using:

$$C_1 = 7600\left(Dr\right)^{-2.5} \tag{8}$$

$$C_2 = 0.4/C_1 \tag{9}$$

where, Dr is the relative density of soil in percent. To provide constitutive model that can better fit the curves of shear modulus reduction and damping ratio derived from the experimental tests data, two different modifications to soil model are implemented as a part of this research to assess the potential for predicting liquefaction phenomenon and associated deformations. To represent the nonlinear stress-strain behavior of soil more accurately that follows the actual stress-strain path during cyclic loading, the masing behavior is implemented into FLAC which works with Byrne model by a FISH subroutine as a first modification.

Since, there is a need to accept directly the same degradation curves derived from the test data in fully nonlinear method to model the correct physics, the second modification is related to incorporate such cyclic data into a hysteretic damping model for FLAC.

Modulus degradation curves imply a nonlinear stress-strain curve. An incremental constitutive relation can be derived from the degradation curve, described by $\tau/\gamma = M_s$, where τ is the normalized shear stress, γ is the shear strain and M_s is the normalized secant modulus. The normalized tangent modulus, M_t, is descried as

$$M_t = \frac{d\tau}{d\gamma} = M_s + \gamma \cdot \frac{dM_s}{d\gamma} \qquad (10)$$

The incremental shear modulus in a nonlinear simulation is then given by GM_t, where G is the small-strain shear modulus of the material.

2.3. Full non-linear dynamic analysis

Equivalent linear analysis is the common method used for evaluating the seismic behaviour of earth structures. In this approach, first, the responses are linearly analyzed using the initial values of damping ratio and shear modulus. Then, the new values of damping ratio and shear modulus are estimated, using maximum value of shear strain and laboratory curves. These values are used for redoing the analysis. This procedure is repeated several times until the material properties show no variation. Therefore, no non-linear effect is directly captured by this method as it assumes linearity during the solution process. Strain-dependent modulus and damping functions are considered roughly in order to approximate some effects of non-linearity (damping and material softening).

In the non-linear analysis, employed in this study, the non-linear stress-strain relationship is directly followed by each zone. Damping ratio and shear modulus of the materials are calculated automatically at different strain levels. The real behaviour of soil, under cyclic loading, is non-linear and hysteretic. Such behaviour can be simulated by Masing model (Masing 1926), which can model the dynamic behaviour of soil. In this model, the shear behaviour of soil may be explained by a backbone curve as:

$$F_{bb}(\gamma) = \frac{G_{max}\gamma}{1 + (G_{max}/\tau_{max})|\gamma|} \qquad (11)$$

where, $F_{bb(\gamma)}$ = backbone or skeleton function; γ= shear strain amplitude; G_{max} = initial shear modulus; τ_{max} = maximum shear stress amplitude.

Stress-strain curve follows the backbone curve in the first loading, as shown in Figure 3(a); however, for explaining the unload-reload process, the above equation should be modified. If load reversal occurs at the point (τ_r, γ_r), stress-strain curve follows the path given by the below formula:

$$\frac{\tau - \tau_r}{2} = F_{bb}\left[\frac{\gamma - \gamma_r}{2}\right] \qquad (12)$$

In other words, the shapes of unload-reload curves are similar to that of backbone curve (with the origin shifted to the loading reversal point) except they are enlarged by a factor of 2, as shown in Figure 3(b). The Equations (9) and (10) describe the Masing behaviour (Masing 1926).

Masing rules seem not to be enough for precise explanation of soil response under general cyclic loading. Finn et al. (1977) developed modified rules to describe the irregular loading. They suggested that unloading and reloading curves follow the concerning two rules. If the new unloading or reloading curve exceeds the last maximum strain and cut the backbone curve, it will follow the backbone curve up to meeting the next returning point, as shown in Figure 3(c). If a new unloading or reloading curve passes through the previous one, it will follow the former stress-strain curve, as shown in Figure 3(d). According to this model, the tangent shear modulus can be defined at the points on the backbone and new reloading-unloading curves by the Formulas (11) and (12), respectively, as:

$$G_t = G_{max} \left/ \left[1 + \frac{G_{max} |\gamma|}{\tau_{max}} \right]^2 \right.$$

(13)

$$G_t = G_{max} \left/ \left[1 + \frac{G_{max}}{2\tau_{max}} |\gamma - \gamma_r| \right]^2 \right.$$

(14)

Based on the results, obtained in this research, the shear stress decreases as the number of load cycles increases; it means that shear stress-strain curves are more inclined. In this study, Masing rules are implemented into FLAC via a series of FISH functions in order to simulate the non-linear stress-strain relationships.

2.4. Material properties and input seismic loading

For the gravity quay wall, the concrete caisson is modeled by linear elastic elements. The granular backfill is modeled as a purely frictional, elastic–plastic soil with a Mohr–Coulomb failure criterion. Strength properties for sandy soil at relative densities of 25 and 85 percent are obtained based on correlations in literature. Maximum shear moduli (i.e. modulus at a strain level of approximately 0.0001%) for Sandy soil are determined from formula developed by Seed and Idriss (1970), given in Equation (6). Hydraulic conductivities are derived from constant head permeability tests. Values for the constants C_1 and C_2 used in the volumetric strain equation (Equation (7)) are obtained from Equations 8 and 9. The material properties of soil layers are listed in Table 1.

After static equilibrium is achieved (end of static construction stage), the full width of the foundation is subjected to the variable-amplitude harmonic ground motion record illustrated in Figure 4. The mathematical expression for input acceleration is given by:

$$\ddot{U}(t) = \sqrt{\beta e^{-\alpha t} t^\eta} \sin(2\pi f t)$$

(15)

where, $\alpha=3.3$, $\beta=1.3$ and $\eta=10$ are constant coefficients, f is the base acceleration frequency and t is the time.

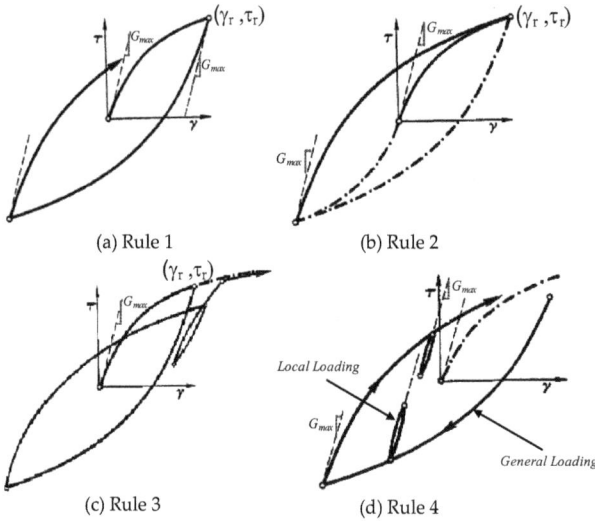

(a) Rule 1   (b) Rule 2

(c) Rule 3   (d) Rule 4

Figure 3. General patterns of loading, unloading and reloading paths in Masing model

	γ_d	G	K	Friction	C_1	C_2
	(kN/m³)	(MPa)	(MPa)	(degree)		
Caisson quay wall	2400	83.33	111.11	-	-	-
Backfill soil	1520	35	75.83	25	2.432	0.164
Foundation soil	1600	130	281.66	38	0.114	3.508
Rubble foundation	1800	155	206.66	40	-	-

Table 1. Soil properties

2.5. Structural properties

The structural element consists of a gravity quay wall supported on very dense foundation soil. The structural element is modeled within FLAC as elastic element. The soil-structure interaction between soil and wall including normal and shear springs. The FLAC manual (Itasca 2004) recommends as a rule of thumb that k_s and k_n be set to ten times the equivalent stiffness of the stiffest neighboring zone (see equation (16) below).

$$k_n = k_s = 10 \times \max\left[\frac{\left(K + \frac{4}{3}G\right)}{\Delta Z_{\min}}\right] \tag{16}$$

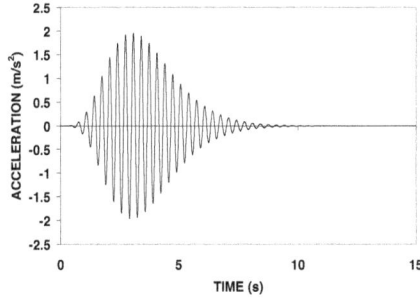

Figure 4. Seismic excitation applied to the bottom of numerical model

where K and G are the bulk and shear moduli; and $\Delta z_{\min}$ is the smallest width of an adjoining zone. The max [ ] notation indicates the maximum value over all zones adjacent to the interface.

In the case of a rough wall, modelling the interface between the soil and the wall is invariably an integral part of the analysis. In the case of soil–structure interaction, the interface is considered stiff compared to the surrounding soil, but it can slip and open in response to the loading. Joints with zero thickness are more suitable for simulating the frictional behaviour at the interface between the wall and the soil. The interface model is described by Coulomb law (Itasca 2004) to simulate the soil/wall contact. The interface element properties are summarized in Table 2.

Surface	k_n	k_s	Friction
	(Pa/m)	*(Pa/m)*	*(degree)*
Quay wall with rubble	1e^{11}	1e^{11}	20
Quay wall with backfill soil	1e^{11}	1e^{11}	12

Table 2. Interface element properties

2.6. Fluid properties

The modeled fluid is representative of seawater with a unit weight of 1027 kg/m³ and a bulk modulus of 2 GPa. An estimated pore pressure distribution is initialized in the model prior to the initial static stress state solution.

2.7. Hydrodynamic effects

Based on Westergaard's approach (1931), effects of hydrodynamic pressure can be approximately considered as added masses acting with the quay wall. The added mass increases parabolically with depth and is defined by

$$m(y) = \frac{7}{8} M_W \sqrt{h.y} \tag{17}$$

Where m(y) is the variation of mass with depth y. M_w is the mass density of water and h is the overall depth of the water in front of the quay wall. The added mass can be reasonably modeled as a simple-supported beam element which possess the corresponding mass m(y) and is free to move in the horizontal direction.

The water in front of the wall is modeled indirectly by including the resulting water pressures along the boundaries. This allows a simple modeling of the water, but does not account for the dynamic interaction of the wall and water.

2.8. Boundary conditions

Many geotechnical problems can be idealized by assuming that the regions remote from the area of interest extend to infinity. As the capability of computer is limited, the unbounded theoretical models have to be truncated to a manageable size by using artificial boundaries. In practice, the numerical model for the quay wall system should be extended to a sufficient depth below the ground level and to a sufficient width to consider local site effects and soil-structure interaction. During the static analysis, the bottom boundary is fixed in the both horizontal and vertical directions and the lateral boundaries are just fixed in the horizontal direction. In dynamic problems fixed boundary condition will cause the reflection of outward propagating waves back into the model.

Thus, absorbing boundaries proposed by Lysmer and Kuhlemeyer (1969) are applied to the lateral boundaries in the model during dynamic analyses. It is based on the use of independent dashpots in the normal and shear directions at the lateral boundaries.

2.9. Damping

Material damping in a soil is generally caused by its viscous properties, friction and the development of plasticity. Indeed, the role of the damping in the numerical models is to reproduce in magnitude and form the energy losses in the natural system when subject to a dynamic load. The dynamic damping in the model is provided by the Rayleigh damping option provided in FLAC. A damping percentage of 5 percent is used which is a typical value for geologic materials (Itasca 2004). The damping frequency is chosen by examining the undamped behavior of the numerical model. A damped frequency of 1 Hz is used for the present model. In each dynamic analysis, 5 percent Rayleigh damping is included for the soil elements in addition to the hysteretic damping already incorporated in the nonlinear stress-strain model.

2.10. Element size

To avoid the numerical distortion of the propagating wave in dynamic analysis the spatial element size, ΔL, must be smaller than approximately one-tenth to one-eighth of the wavelength associated with the highest frequency component of the input wave (Kuhlemeyer and Lysmer 1973):

$$\Delta L = \lambda/9 \tag{18}$$

In general, the cut-off frequency for geotechnical earthquake engineering problems should be no less than 10 Hz (ASCE 2000). Considering above criteria, element size is defined small enough to allow seismic wave propagation throughout the analysis. A finer mesh is used in sensitive areas such as below and near quay wall. A coarser mesh has been chosen for other areas in order to save computer analysis time.

2.11. Time step

To complete the numerical solution, it is necessary to integrate the governing equations in time in an incremental manner. The time step of the solution should be sufficiently small to accurately define the applied dynamic loads and to ensure stability and convergence of the solution. In the current FLAC model, the time step is approximately 10^{-6} second.

The convergence criterion for FLAC is the ratio defined to be the maximum unbalanced force magnitude for all the gridpoints in the model divided by the average applied force magnitude for all the gridpoints.

3. Verification

To validate the implementation of the masing rule and hysteresis damping in FLAC program, the simulation of one-zone sample with modified Byrne model is done by using the unit cell as shown in Figure 5 incorporated with the implemented rules.

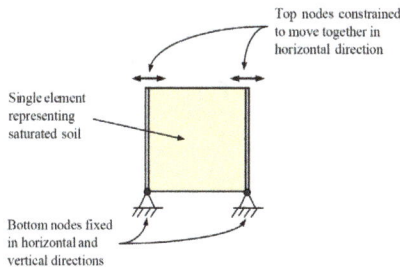

Figure 5. One-zone model in FLAC for simulating cyclic simple shear test

The one-zone sample is modeled with FLAC that this consist of a sandy soil which is given a periodic motion at its base. Vertical loading is by gravity only. Equilibrium stresses and pore pressures are installed in the soil, and pore pressure and effective stress (mean total stress minus the pore pressure) are established within the soil. The modified Byrne model is applied for the soil. The results based on Equation (7) are shown in Figures 6 and 7.

Figure 6 indicates the pore pressure build up in a single zone. It can be seen that the effective stress reaches zero after about 20 cycles of shaking (2 seconds, at 5 Hz). At this point, lique-faction can be said to occur. This test is strain-controlled in the shear direction. The stress/strain loops for the one-zone sample for several cycles are shown in Figure 7. It can be observed that shear modulus decreases with increasing shear strain. The hysteretic model seems to handle multiple nested loops in a reasonable manner. There is clearly energy dissipation and shear stiffness degradation during seismic loading. Due to the satisfactory modeling of the validation case, the numerical model is used to perform parametric studies on caisson quay wall, as described in the pervious sections.

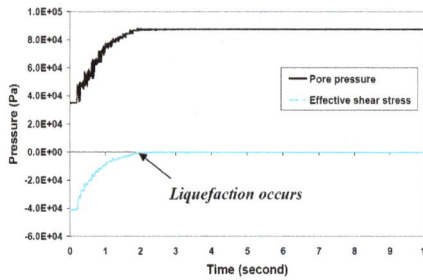

Figure 6. Pore pressure generation and effective stress time histories during seismic loading

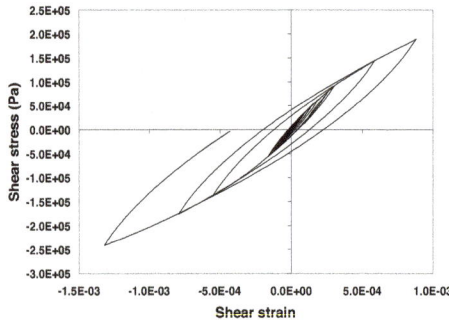

Figure 7. Hyteresis loop in a one-zone sample element

A series of 1g shaking table tests have been executed in order to verify the obtained numerical results. It is attempt to create almost similar conditions between laboratory model test and numerical model. The liquefiable soil is modeled by loose sand and non-liquefiable soil is modeled by very dense sand. The seismic excitation is shown in Figure 8. The numerical results are presented and compared to those of corresponding shaking table test. Figure 9(a) shows the permanent deformation pattern of the numerical model after dynamic excitation. The nodal displacement vectors are presented in Figure 9(b). As may be expected, more ground surface settlement is observed in the backfill near the wall than at far field. A rigid body rotation of the wall (tilt) to the seaward direction is also clearly seen. The deformation pattern of model test at the end of seismic loading is presented in Figure 10. The trend of deformation behind quay wall and movement of the wall are in fairly good agreement with numerical results. Comparisons between the calculated and measured results (Calculated: numerical and Measured: experimental results) are made in Figure 11.

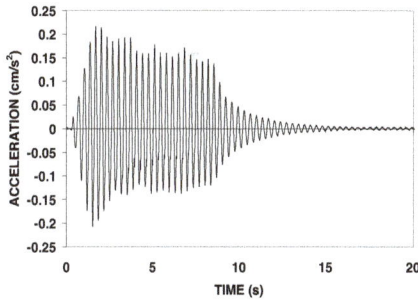

Figure 8. Input base excitation

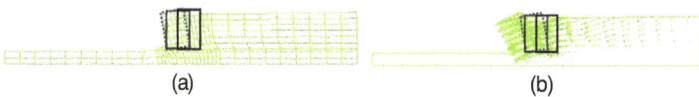

Figure 9. Computed post-earthquake (a) deformed shape, (b) nodal displacement vectors for the quay wal after the seismic loading

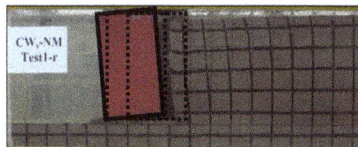

Figure 10. Measured post-earthquake deformed shape after the seismic loading

One might notice that the calculated results are rather close to measured values. This clearly demonstrates that the current numerical procedure captures very well the seismic behavior of the gravity-type quay wall and surrounding soils.

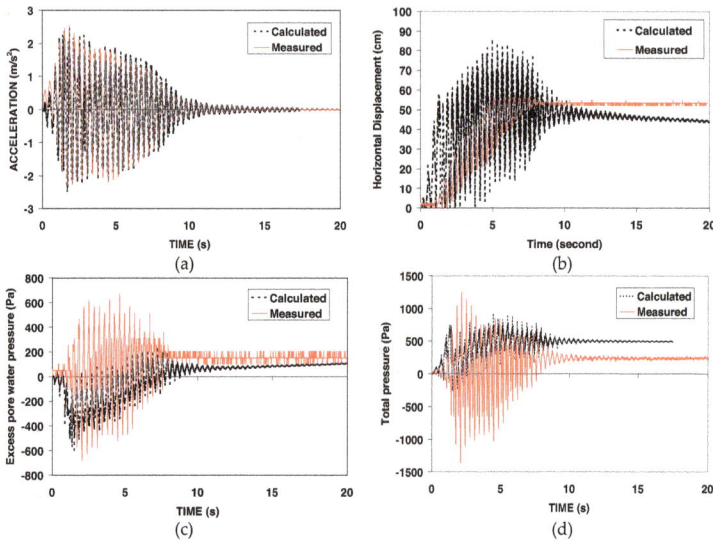

Figure 11. Recorded versus computed (a) acceleration, (b) horizontal displacement at top of the quay wall, and (c) excess pore water pressure, (d) total pressure behind the quay wall

4. Numerical results and discussion

Results of nonlinear effective-stress dynamic analyses are presented in this section to investigate the effects of soil properties and input excitation characteristics on liquefaction potential, deformation of quay wall and failure mechanisms of soil-wall system during seismic loading. For gravity quay walls on firm foundations, typical failure modes during earthquakes are seaward displacements and tilting. Therefore, the horizontal displacement of quay wall head is selected as a key parameter to judge about the stability of quay wall.

4.1. Influence of relative density of backfill soil

Three additional sets of soil properties with different relative densities are selected for backfill material (Dr = 15%, 25% and 40%). Figure 12(a) depicts the computed lateral displacement of the quay wall's head. The horizontal deformation for all relative densities is greater than allowable value proposed by PIANC (2001) and the quay wall system goes toward failure. After the earthquake, the system reaches to equilibrium. The final permanent deformations and rate of increase for Dr=15% is much more than the others. Figure 12(b) shows cumulative vertical displacement (settlement) of the quay wall's head for the various backfill soil relative densities. As expected, the higher the relative density, the less the accumulated vertical permanent deformation. Figure 12(c) shows the excess pore water pressure ratio at far field in depth of 13 m for three different materials. It is clearly seen that the free field backfill is quickly liquefied. Figure 12(d) depicts the excess pore water pressure ratio time histories for various backfill soil relative densities behind the quay wall in depth of 13 m. It is found that the liquefaction does not occur in backfill soil behind the quay wall. This can be attributed to the movement of quay wall due to seismic loading which directly influences the pore pressure build up.

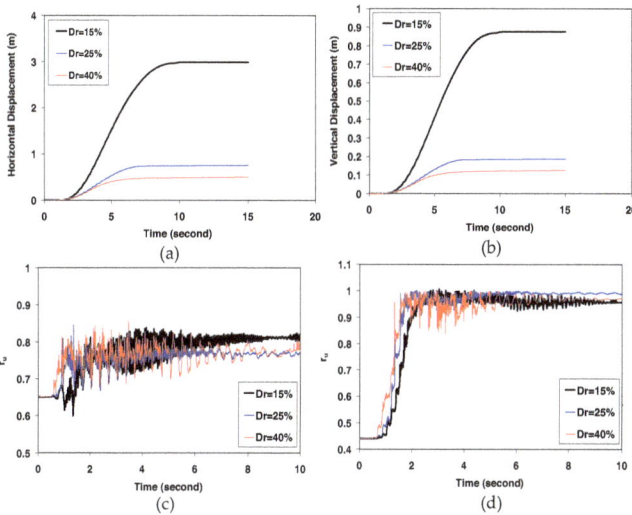

Figure 12. Computed values of: (a) horizontal deformation, and (b) vertical deformation of quay wall's head, and (c) excess pore water pressure ratio at far field, and (d) excess pore water pressure ratio behind the quay wall for various relative densities of backfill soil

Figure 13 shows the deformed shape after the seismic excitation for three sets of relative densities. As may be expected, more ground surface settlement is observed in the backfill near the wall than at the far field. One may observe that there is a significant movement of quay wall for Dr=15% rather than Dr=25% and 40%. Note also that the lateral spreading of soil is clearly visible near the area influenced by the quay wall especially for Dr=15%. In addition, differential settlements between the quay wall and the apron as well as the deformation of the foundation rubble beneath the quay wall are also observed in Figure 13. This is consistent with the common mode of deformation for gravity quay walls. The analysis indicates translation and rotation mode (rocking) of the wall.

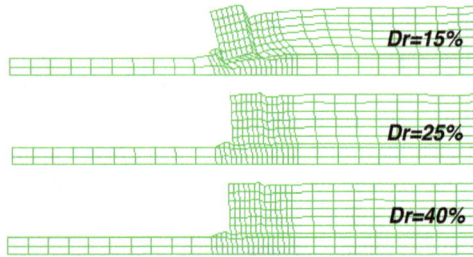

Figure 13. Deformed shape of the quay wall system for various relative densities of backfill soil

4.2. Influence of shear stiffness of backfill soil

Three different shear modulus values are adopted (G=20, 35 and 50 MPa) to investigate the effect of shear stiffness of the backfill soil. The other parameters are the same for all the analyses. Figures 14(a) and (b) show the permanent horizontal and vertical displacements time histories of the quay wall's head during seismic loading. It is observed that quay wall has significant movement toward the seaside. Both horizontal and vertical deformations are greater than the allowable limits which have been mentioned in PIANC (2001).

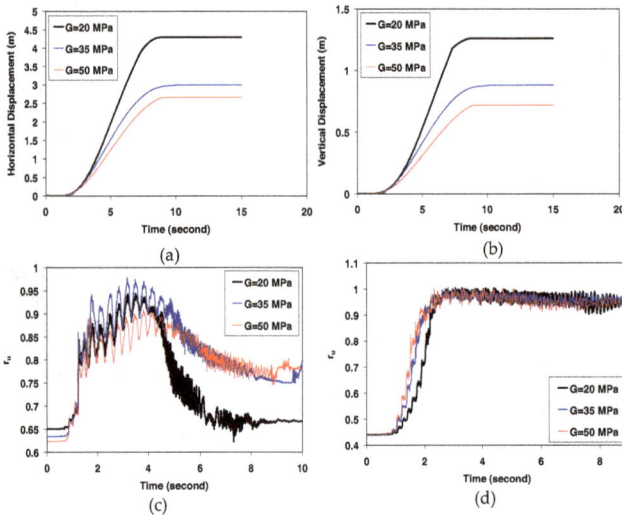

Figure 14. Computed values of: (a) horizontal deformation, and (b) vertical deformation of quay wall's head, and (c) excess pore water pressure ratio at far field, and (d) excess pore water pressure ratio behind the quay wall for various shear moduli of backfill soil

Figures 14(c) and (d) show the excess pore water pressure ratio time histories for the far field location and the area behind quay wall. The Figure 14(d) exhibits a significant increase in pore pressure which leads to liquefaction (r_u=1) for three types of backfill soil. But for the soil behind quay wall and adjacent to it, at first, the excess pore water pressure ratio increases quickly and after 4 seconds, the excess pore water pressure is dissipated rapidly and consequently, r_u decreases with time. This can be pertained to the quay wall movement during seismic loading which influences the excess pore water pressure. In addition, significant reduction is observed after 4.5 s for r_u at behind quay wall.

In Figure 15, lateral spreading is clearly obvious behind the quay wall for all the analyses. Deferential settlements between the quay wall and the apron are also visible. The major failure pattern is tilting and rotation of the quay wall toward the seaside which is consistent with the actual failure mode of quay wall movement in literature.

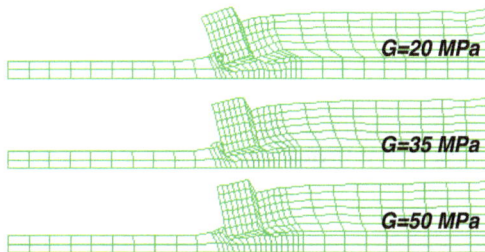

Figure 15. Deformed shape of the quay wall system for various shear moduli of backfill soil

4.3. Influence of friction angle of backfill soil

Figure 16(a) shows the calculated lateral deformation of quay wall's head for three different values of backfill soil friction angle (ϕ=25 , 30 and 35 degree). As may be expected, the higher the friction angle, the less the accumulated permanent deformation. The same trend as horizontal displacement is observed for the vertical displacement of quay wall's head. Both horizontal and vertical displacements generally increase with decreasing the friction angle of backfill soil.

Figures 16(c) and (d) show the time histories of excess pore water pressure ratio at the far field and the area behind quay wall, respectively. It is seen that the excess pore water pressure ratio at the far field reaches its maximum value (r_u=1) at around 2 s. At this time, liquefaction occurs. But for the region behind quay wall, liquefaction does not occur because the volume of backfill soil near the wall tends to increase during the outward movement of the wall. As previous results, Figure 17 demonstrates that the failure mode is rotation and the wall tends to rotate at the bottom.

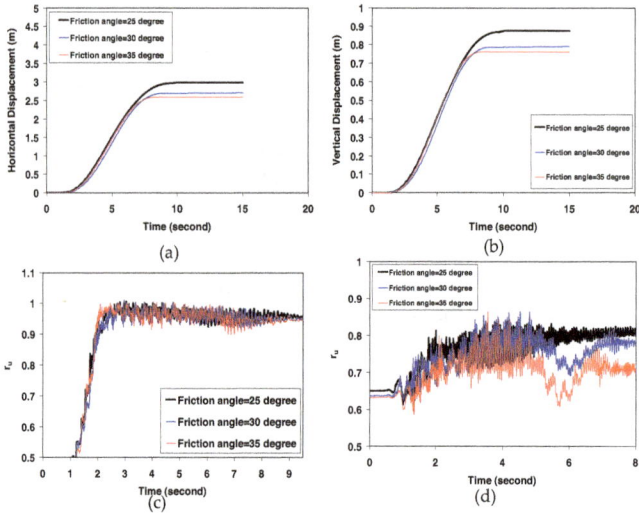

Figure 16. Computed values of: (a) horizontal deformation, and (b) vertical deformation of quay wall's head, (c) excess pore water pressure ratio at far field, and (d) excess pore water pressure ratio behind the quay wall for various friction angles of backfill soil

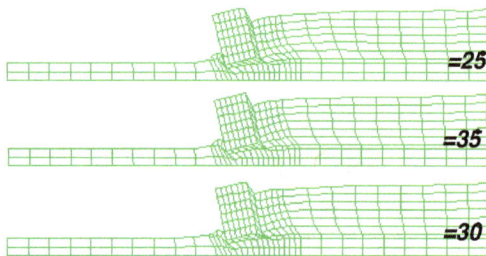

Figure 17. Deformed shape of the quay wall system for various friction angles of backfill soil

4.4. Influence of maximum amplitude of seismic loading

Three different maximum amplitudes are selected for the input excitation applied to the base of the model (a_{max}=0.15g, 0.2g and .25g) to consider the effect of input excitation intensity. All other parameters are the same for all the analyses. Figures 18(a) and (b) show the horizontal and vertical displacements time histories at the top of quay wall subjected to seismic loading with different maximum amplitudes. As expected, by increasing the maximum amplitude of seismic loading, the deformation of quay wall increases, both laterally and vertically. It is noticed that the values of displacements are much more than the allowable values recommended by PIANC (2001). It means that in such cases, the quay wall completely fails during loading.

Figures 18(c) and (d) depict r_u time histories at the far field and the area behind quay wall. As it is seen, the backfill soil liquefy at far field for all the maximum amplitude but at first liquefaction occurs for a_{max}=0.25g. For the area behind quay wall, r_u is less than 0.85 during seismic loading. Therefore, it can be concluded that liquefaction does not occur nearby (behind) the quay wall. The difference in pore pressure build up pattern between the far (free) field and near-wall field is mainly due to the fact that near the wall, soil experiences significant compression and extension alternatively during shaking (due to wall oscillation). In the free field, soil mainly experiences shear during shaking, allowing for high r_u and leading eventually to liquefaction.

Figure 19 indicates translation and rotation (rocking) of the quay wall. Lateral spreading and ground failure behind the quay wall are clearly observed. As may be expected, more ground surface settlement is noticed in the backfill near the wall than at the far field. A large tilting of the wall to the seaward is obviously observed.

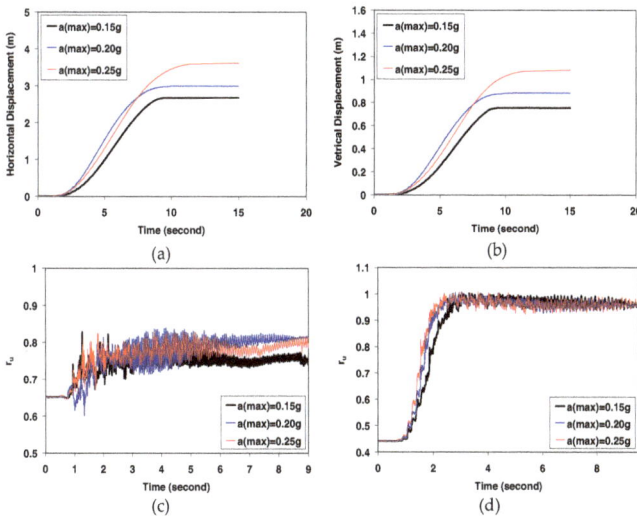

Figure 18. Computed values of: (a) horizontal deformation, and (b) vertical deformation of quay wall's head, (c) excess pore water pressure ratio at far field, and (d) excess pore water pressure ratio behind the quay wall for various maximum amplitudes of the seismic loading

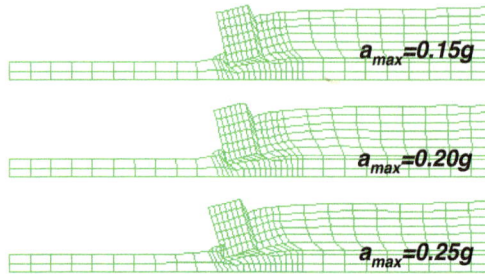

Figure 19. Deformed shape of the quay wall system for various maximum amplitudes of the seismic loading

4.5. Influence of frequency of seismic loading

Figures 20(a) and (b) indicate the horizontal and vertical displacements time histories of the quay wall's head during seismic loading. As may be expected, by increasing the frequency of seismic loading, the displacements increase excessively and the quay wall system entirely fails. The values of displacements are so much higher than allowable values proposed by PIANC (2001).

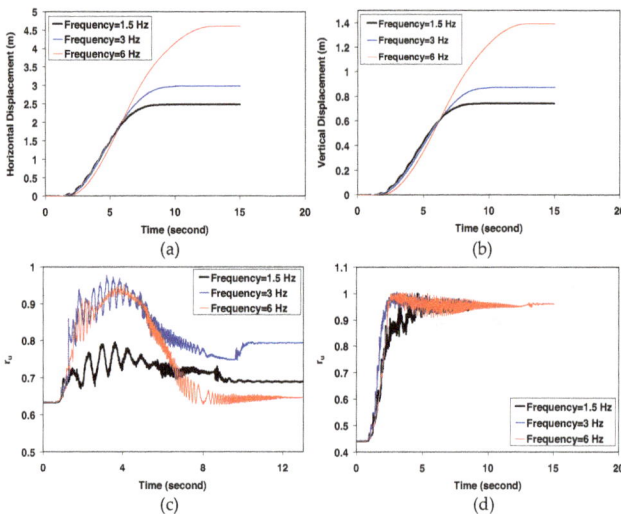

Figure 20. Computed values of: (a) horizontal deformation, and (b) vertical deformation of quay wall's head, and (c) excess pore water pressure ratio at far field, and (d) excess pore water pressure ratio behind the quay wall for various frequencies of the seismic loading

As pervious, liquefaction occurs at the far field which is not affected by the quay wall movement (Figure 20(c)) but for the area behind quay wall, liquefaction does not occur due to seaward movement of the quay wall (Figure 20(c)). As it is clear in Figure 20(d) that the excess pore pressure ratio increases till 4 seconds but after that dissipation is observed for all frequencies. The rate of dissipation for frequency of 6 Hz is higher than the others. Figure 21 depicts the deformed shape of quay wall system after seismic loading. As it is seen, the lateral spreading behind quay wall completely observed for all frequencies but it is much more severe for F=6 Hz which the quay wall has been failed entirely and the area behind quay wall has subsided excessively. The failure mode of the quay wall is translation and rotation. When the frequency of seismic loading increases, the quay wall rotates more.

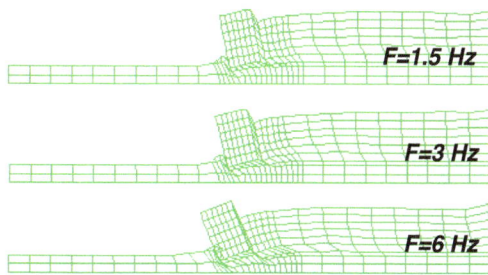

Figure 21. Deformed shape of the quay wall system for various frequencies of the seismic loading

5. Conclusions

In this study, 2D nonlinear effective stress dynamic analyses have been carried out to investigate the seismic behavior of gravity-type quay walls. A reference model has been constructed and then subjected to seismic loading. The Finn and Byrne model has been adopted with some slightly modifications which take into account the pore water pressure generation and liquefaction process under dynamic loading. The numerical model has been validated by simulating 1g shaking table test. It is shown that the obtained numerical results agree reasonably with actual observation in the shaking table test. The seismic response of the gravity wall itself has well captured by the numerical analyses with satisfactory predictions of acceleration, displacement, total pressure and pore water pressure time histories. Additional computational parametric studies have been conducted by varying backfill soil relative density, shear modulus, friction angle and maximum amplitude and frequency of input excitation to study the extent of liquefaction and deformation mechanism of quay wall system. It is concluded that soil properties and input motion characteristics are among the most influential factors in dictating seismic performance of the quay wall system. The results show that the backfill in the soil-wall interaction zone and the foundation soils beneath the quay wall experiences less excess pore water pressure even liquefaction occurs in the far field during shaking. The alternative pumping and suction process in excess pore water pressure which are caused by wall's vibrations increase the level of damage because large amounts of backfill are forcedly leaked into the sea. The lack of backfill liquefaction near the wall is attributed to the lateral displacement of the wall. In the other words, excess pore water pressure does not attain 100% liquefaction behind the quay wall contrary to the far field. The current study states that the numerical simulation incorporated with the special numerical techniques is capable of modeling the seismic response of gravity-type quay walls.

Author details

Babak Ebrahimian*

School of Civil Engineering, Faculty of Engineering, University of Tehran, Tehran, Iran

References

[1] Arulanandan, K. (1996). Application of numerical procedures in geotechnical earthquake engineering, *Proceeding of Application of Numerical Procedures in Geotechnical Earthquake Engineering*, National Science Foundation Workshop/ Conference, October 28-30.

[2] American Society of Civil Engineers, (2000). ASCE 4-98 Seismic Analysis of Safety-related Nuclear Structures and Commentary. *ASCE*, Virginia, USA.

[3] Byrne, P. M. (1991). A cyclic shear-volume coupling and pore pressure model for sand, *Proceeding of Second International Conference on Recent Advances in Geotechnical Earthquake Engineering and Soil Dynamics*, Vol. 1, University of Missouri, Rolla, Missouri, pp. 47-55.

[4] Chen, C., Hwang, G. (1969). Preliminary analysis for quay wall movement in Taichung harbour during the September 21, 1999, Chi-Chi earthquake, *Earthquake Engineering and Engineering Seismology*, Vol. 2, pp. 43-54.

[5] Cooke, H. G. (2000). Ground improvement for liquefaction mitigation at existing highway bridges, *Ph.D. dissertation*, Department of Civil and Environmental Engineering, Polytechnic Institute and State University, Virginia.

[6] Dickenson, S. E., Yang, D. S. (1998). Seismically-induced deformations of caisson retaining walls in improved soils, *Proceeding of Geotechnical Earthquake Engineering and Soil Dynamics III*, Vol. II, Geotech. Special Pub. No. 75., ASCE, Reston, VA, pp. 1071-1082.

[7] Ebrahimian, B., Mostafavi Moghadam, A. A., Ghalandarzadeh, A. (2009). Numerical modeling of the seismic behavior of gravity type quay walls, *Proceeding of Perform-

ance-based Design in Earthquake Geotechnical Engineering—Kokusho, Tsukamoto , Yoshimine (eds), Taylor & Francis Group, London, ISBN 978-0-415-55614-9, Tokyo, June 15–18.

[8] Finn, W.D.L., Lee, K.W., Martin, G.R. (1977). An effective stress model for liquefaction. *Journal of Geotechnical Engineering Division ASCE*, Vol. 103, No. 6, pp. 517-553.

[9] Finn, W. D. L. (1988). Dynamic analysis in geotechnical engineering, *Proceeding of Earthquake Engineering and Soil Dynamics II - Recent Advances in Ground Motion Evaluation*, ASCE Geotechnical Engineering Division,Park City, Utah, June, pp. 523-591.

[10] Finn, W. D. L. (1991). Estimating how embankment dams behave during earthquakes, *Water Power and Dam Construction,* London, April, pp. 17-22.

[11] Hayashi, S., Kubo, K., Nakase, A. (1966). Damage to harbour structures by the Nigata earthquake, *Journal of Soils and Foundations,* Vol. 6, No. 1, pp. 89-111.

[12] Iai, S. (1998). Seismic analysis and performance of retaining structures, *Proceedings of the 1998 Conference on Geotechnical Earthquake Engineering and Soil Dynamics III*, Part 2 (of 2), ASCE, pp. 1020-1044.

[13] Iai, S., Ichii, K., Liu, H., Morita, T. (1998). Effective stress analysis of port structures, *Special issue of soils and foundations*, pp. 97-114.

[14] Iai, S., Sugano, T. (2000). Shaking table testing on seismic performance of gravity quay walls, *12 WCEE.*

[15] Ichii, K., Iai, S., Morita, T. (2000). Performance of the quay wall with high seismic resistance, *Journal of Computing in Civil Engineering*, Vol. 17, No. 2, pp. 163-174.

[16] Inagaki, H., Iai, S., Sugano, T., Yamazaki, H., Inatomi, T. (1996). Performance of cassion type quay walls at Kope Port, *Special issue of soils and foundations*, pp. 119-136.

[17] Inoue, K., Miura, K., Otsuka, N., Yoshida, N., Sasajima, T. (2003). Numerical analysis of the earth pressure during earthquake on the gravity type quay wall, *Proceedings of the Thirteenth (2003) International Offshore and Polar Engineering Conference*, International Society of Offshore and Polar Engineers, Honolulu, United States., pp. 2095-2099.

[18] Itasca, (2004). FLAC User's Guide, Version 4.0, *Itasca Consulting Group. Inc,* Minnesota, USA.

[19] Kuhlemeyer, R.L., Lysmer, J. (1973). Finite element method accuracy for wave propagation problems, *Journal of Soil Mechanics and Foundations ASCE*, Vol. 99, No. SM4, pp. 421-427.

[20] Lysmer, J., Kuhlmeyer, R.L. (1969). Finite element method for infinite media, *Journal of Engineering Mechanics ASCE*, Vol. 95, No. EM4, pp. 859-877.

[21] Madabhushi, S. P. G., Zeng, X. (1998). Seismic response of gravity quay walls. II: numerical modeling, *Journal of Geotechnical and Geoenvironmental Engineering ASCE*, Vol. 124, No. 5, pp. 418-427.

[22] Martin, G. R., Finn, W. D. L., Seed, H. B. (1975). Fundamentals of liquefaction under cyclic loading, *Journal of Geotechnical Division ASCE*, Vol. 101(GT5), pp. 423-438.

[23] Masing, G. (1926). Eigenspannungen und Verfestigung Beim Messing, *Proceedings of 2nd International Congress on Applied Mechanics*, Zurich.

[24] McCullough, N. J., Dickenson, S. E. (1998). Estimation of seismically induced lateral deformations for anchored sheetpile bulkheads. *Proceeding of Geotechnical Earthquake Engineering and Soil Dynamics III*, Seattle, USA.

[25] Mostafavi Moghadam, A. A., Ghalandarzadeh, A., Towhata, I., Moradi, M., Ebrahimian, B., Haji Alikhani, P. (2009). Studying the effects of deformable panels on seismic displacement of gravity quay walls, *Ocean Engineering*, Vol. 36, pp. 1129–1148.

[26] Mostafavi Moghadam, A. A., Ghalandarzadeh, A., Moradi, M., Towhata, I., Haji Alikhani, P. (2011). Displacement reducer fuses for improving seismic performance of caisson quay walls, *Bulletin of Earthquake Engineering*, Vol. 9, pp. 1259–1288.

[27] Nozu, A., Ichii, K., Sugano, T. (2004). Seismic design of port structures, *Journal of Japan association for earthquake engineering*, Vol. 4, pp. 195-208.

[28] Parra, E. (1996). Numerical modeling of liquefaction and lateral ground deformation including cyclic mobility and dilation response in soil systems, *Ph.D. Dissertation*, Department of Civil Engineering, RPI, Troy, NY.

[29] PIANC, 2001. Seismic Design Guidelines for Port Structures, A.A. Balkema, Rotterdam.

[30] Pyke, R. (1979). Nonlinear soil models for irregular cyclic loadings, *Journal of Geotechnical Division ASCE*, Vol. 105(GT6), pp. 715-726.

[31] Sasajima, T., Sakikawa, M., Miura, K., Otsuka, N. (2003). In-situ observation system for seismic behavior of gravity type quay wall, *Proceedings of the Thirteenth (2003) International Offshore and Polar Engineering Conference*, International Society of Offshore and Polar Engineers, Honolulu, United States, pp. 2087-209.

[32] Seed, H. B., Idriss, I. M. (1970). Soil moduli and damping factores for dynamic response analyses, *Report EERC 70-10, Earthquake Engineering Research Center*, University of California, Berkeley, CA.

[33] Seed, H. B., DeAlba, P. (1986). Use of SPT and CPT tests for evaluating the liquefaction resistance of soils, *Proceedings of Insitu 1986, ASCE*.

[34] Sugano, T., Morita, T., Mito, M., Sasaki, T., Inagaki, H. (1996). Case studies of cassion type quay wall damage by 1995 Hyogoken-Nanbu eartjquake, *Eleventh world conference on earthquake engineering*, Elsevier Science Ltd.

[35] Sumer, B.M., Kaya, A., Hansen, N. E. O. (2002). Impact of liquefaction on coastal structures in the 1999 Kocaeli, Turkey earthquake, *Proceedings of the Twelfth (2002) International Offshore and Polar Engineering Conference*, International Society of Offshore and Polar Engineers, Kitakyushu, Japan, 12, pp. 504-511.

[36] Westergaard, H. (1931). Water pressure on dams during earthquakes, *Transactions of ASCE*, pp. 418-433.

[37] Yang, Z. (2000). Numerical modeling of earthquake site response including dilation and liquefaction, *Ph.D. Dissertation*, Department Of Civil Engineering and Engineering Mechanics, Columbia University, New York, NY

Seismic Evaluation of Low Rise RC Framed Building Designed According to Venezuelan Codes

Juan Carlos Vielma, Alex H. Barbat,
Ronald Ugel and Reyes Indira Herrera

Additional information is available at the end of the chapter

1. Introduction

Along its history, Venezuela has been severely affected by destructive earthquakes [1]. Approximately 80% of the population lives in seismically active areas, where have occurred destructive earthquakes even in recent times [2]; The seismic hazard, inadequate design and construction of buildings as well as the damage occurred from previous earthquakes, demonstrate a high vulnerability in existing buildings. Then it is essential to continuously make progress and research in the field of earthquake engineering and upgrade the seismic design codes. Seismic upgrade requires the evaluation or predictions of the expected damage to structures at the time of an earthquake of a certain severity occur. From this prediction it can be defined solutions for the reduction of structural vulnerability [3].

The damage occurred in buildings after an earthquake indicates the need for reliable methodologies for the evaluation of seismic behavior of the existing buildings. According to current technical and scientific advances, seismic evaluation of reinforced concrete (RC) structures can be done by two different approaches: empirical methods and mechanical methods [4]. The current tendency of earthquake engineering in the evaluation of structural behavior is the application of simplified mechanical methods based on performance, involving the capacity spectrum [5], because there are developed refined models and detailed analysis.

This study used a mechanical method that involves non-linear analysis with deterministic and probabilistic approaches, as well as procedures of analysis based on Limits States defined by displacements [6], in order to evaluate the behavior of a low rise RC building with plan irregularity, designed according to Venezuelan codes [7]-[9] and subjected to seismic action effect. Through the use of mathematical models and computational tools, seismic behavior of the building is obtained in a suitable way. Among these tools any procedure was chosen: the quadrants method, which leads to the rapid assessment of the seismic capacity of a structure through its non-linear response [10]. Results of the research shown that the current design of this kind of structures is not safe when they are under the maximum seismic actions prescribed by codes, then it is necessary to review the design procedures in order to find more realistic designs that fulfill the goals of the performance-based design.

2. Case studied

A two story RC framed building was studied, (Figure 1a), which contains internal staircase and 220 m² total plan area. This structure represents a common typology used for residential buildings in Venezuela, for prone seismic zones. The structure was designed and detailed for a high ductility value (response reduction factor of 6).

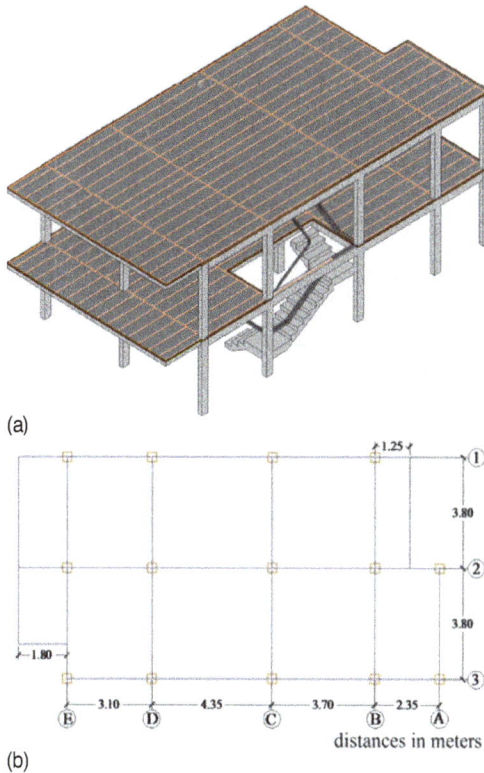

(a)

(b)

Figure 1. Low rise RC building(a) 3D view (b) Plan view

The building was modeled according its original design, called *original building (OB)*, with plan asymmetry (Figure 1b) and one way 25 cm depth slabs in *X* direction. A second model was designed adjusted to seismic performance requirements formulated by Herrera *et al.* [12], called *resizing building (RB)*, which presents equal geometrics and mechanics character- istics than *OB* model, but considering the "strong column-weak beam" condition. It was also used the displacement-based seismic design procedure of *Priestley et al.* [13] in order to de- sign of a third model, called *displacement-based design building (DBDB)*. These three models differ only in the dimensions of its structural elements (Table 1).

Building	Axis X beams (cm)	Axis Z beams (cm)	First level columns (cm)	Second level columns (cm)
OB	20x35	20x35	20x30	20x30
RB	20x45	20x35	30x30	30x30
DBDB	20x40	20x40	35x35	30x30

Table 1. Geometric characteristics of elements from each modeled building

3. Assessment method

The Quadrants Method is based on the results of the non-linear static analysis (Pushover analysis). This analysis results are plotted in a displacement vs. base shear format, this generate the capacity curve which represents the overall capacity of the whole structure against lateral forces. In order to evaluate the capacity curve two of the main structural parameters are taken into account. The first one is the design elastic shear, obtained from the elastic analysis of the structure using the elastic design spectrum. The second parameter is the threshold that defines the Repairable Limit State, obtained from [14] for RC framed buildings with similar charac- teristics to the studied ones. The thresholds have been computed from characteristic values of three levels of damage proposed in [15] and are showed in Table 2. Both values are used to define two axes over the capacity curve, the elastic base shear defines an horizontal axis and the damage threshold defines a vertical axis, then Capacity Curve is divided in four spaces or quadrants, see Figure 2.

Limit State	Damage type	Seismic hazard	Probability event	Inter-storey drift in (%)
Service	Non-structural	Frequent	50% in 50 years	$0,2<\delta<0,5$
Damage control	Moderate structural	Occasional	10% in 50 years	$0,5<\delta<1,5$
Collapse prevention	Severe structural	Rare	2% in 50 years	$1,5<\delta<3,0$

Table 2. Inter-storey drifts adopted for the damage thresholds determination

The performance point is a common procedure accepted among the scientific community to evaluate the seismic performance of a structure under a specific demand. It is usually obtained from the idealized shape of the Capacity Curve as is shown in the Figure 2 [16]. The Quadrants Method also uses this parameter in order to define the roof displacement of the case studied, defined according to the N2 method [5]. If the performance point is under the axis defined by the elastic base shear (Quadrants III or IV), the design does not meet the basic objective of the seismic design because the building does not have enough lateral strength. If the performance point is on the right side of the vertical axis (Quadrant I) means that the building has adequate stiffness, otherwise (Quadrant II) it means that the stiffness is very low and the displacements can be longer than the displacements that can produce advanced structural damage, techni-cally or economically irreparable. These lateral displacements are usually computed from the dynamic response of the structure submitted to a strong motion with a return period of 475 years, or an occurrence probability of 10% in 50 years [17].

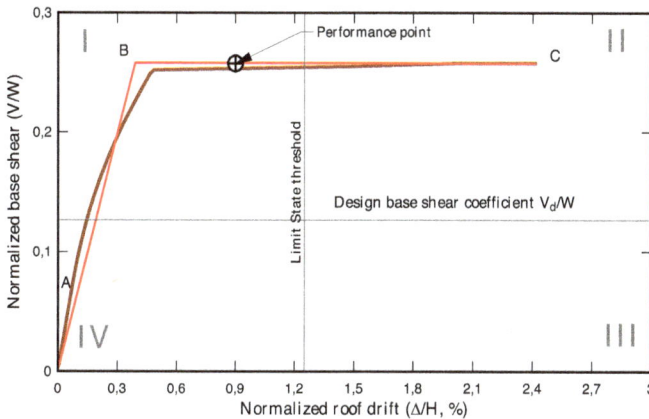

Figure 2. Capacity curve and the axis that define the Quadrants Method

The Quadrants Method can provide an objective criterion in order to upgrade the seismic capacity of a structure. If the performance point is on the Quadrant I, the structure has enough lateral strength and stiffness, so does not need to be reinforced. If the structure is on the Quadrant II, it is necessary to provide additional stiffness by using conventional procedures like RC or steel jacketing. If the performance point is on Quadrant III, the structure requires a more radical intervention, adding stiffness and lateral strength. In this case it is possible to combine some traditional reinforcement techniques with new ones like FRP jacketing. In this case the columns are the subject of the main intervention. Finally, if the performance point is on the Quadrant IV, the structure does not has enough lateral strength and then the reinforce-ment technique must be FRP jacketing.

4. Nonlinear analysis

The structures are modeled by incorporating the structural response when it incurs in the material and geometrical non-linear range, produced by high deformations caused by accidental excitations (earthquakes) [11]. The analyses were performed using ZEUS-NL software [18], which allows to model complex structures with "n" number of finite elements, thus to know the elements in the building which are most vulnerable to damage. Each building is modeled in two dimensions, spitting each frame to get a more detailed response for the seismic behavior of each frame; a 3D dynamic analysis was applied to the ER model.

The static Pushover analysis is performed once the frames have been subjected to action of gravity loads, based on the pseudo-static application of lateral forces equivalent to displacements of seismic action [5]. The pattern of lateral seismic loads consist in increasing loads with height (triangular distribution) applied in a monotonic way until the structure reaches its maximum capacity [20].

This procedure applies a solution of equilibrium equations in an incremental iterative process form. In small increments of linear loads, equilibrium is expressed as:

$$K_t \Delta_x + R_t = \Delta F \tag{1}$$

Where Kt is the tangent stiffness matrix, Rt is the restorative forces at the beginning of the increased load. These restorative forces are calculated from:

$$R_t = \Sigma K_t, K \Delta_u \tag{2}$$

While this procedure is applied, the strength of the structure is evaluated from it is balance internal conditions, updating at each step the tangent stiffness matrix. Unbalanced loads are applied again until it can satisfy a convergence criterion. Then, a new load increase is applied. The increases are applied until a predetermined displacement is reached or until the solution diverges.

From the capacity curve provided in this analysis, it is determined the structural ductility (μ) by the quotient between the ultimate displacement and cadence point displacement, as shown in the following expression:

$$\mu = \Delta_u / \Delta_y \tag{3}$$

Where Δ_u is Ultimate displacement and Δ_y is the global yield displacement. Both values are computed from the idealized capacity curve of the structure.

By the other hand, the dynamic analysis is an analysis method that can be used to estimate structural capacity under seismic loads. It provides continuous response of the structural system from elastic range until it reaches collapse. In this method the structure is subjected to one or more seismic records scaled to intensity levels that increase progressively. The maximum values of response are plotted against the intensity of seismic signal [21-22]. The procedure to perform the dynamic analysis from the seismic signal is:

- To define a seismic signal compatible with the design scenario;

- To define the scaled earthquake intensity a monotonic way;

- To define the extent of damage or damage Limit States;

- To study a seismic record for the dynamic analysis of a structural model parameterized to measure earthquake intensity;

The non-linear dynamic analyses provide a set of curves which are a graphical representation of the evolution of the drifts respect time. Results let to compute the damage lumped in specific elements of the structure, but these results are beyond the objective of this Chapter.

Analysis earthquake	Limit State	Return period (years)	Occurrence probability in 50 years	Interstorey drift δ (%)
Frequent	Serviceability	95	50 %	δ < 0,5
Rare	Reparable damage	475	10 %	δ < 1,5
Very Rare	Collapse Prevention	2475	2 %	δ < 3,0

Table 3. Limit States and seismic hazard level

For the dynamic analysis the structures were subjected to seismic action (see Table 2) defined by accelerograms built on the basis of a likely value of maximum acceleration of the soil and the hazard level associated with the location of the structure and other seismic characteristic design parameters [16]. These accelerograms called "synthetic accelerograms" are generated through the implementation of a set of earthquakes with wide frequency content, using the PACED program [17], based on the Venezuelan code's elastic design spectrum. For the dynamic analyses of the three buildings (*OB, RB, DBDB*), it were used 3 synthetic accelerograms with duration of 60sec.

Non-linear dynamic analysis was applied to all buildings in order to verify if the performance evaluated by the Quadrants Method is reliable in order to evaluate the fulfilment of the thresholds defined in the precedent section. For this purpose they has been computed three synthetic elastic design spectrum-compatible accelerograms by means of the PACED program [23]. In Figure 3 are shown the Venezuelan rigid-soil elastic design spectrum with the response spectra obtained from the synthetic accelerograms.

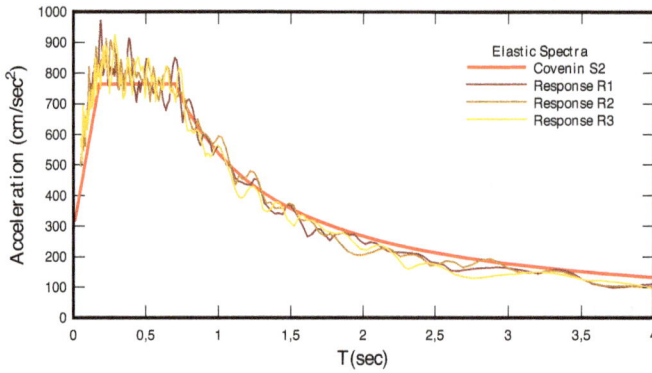

Figure 3. Elastic response spectra from elastic design spectrum-compatible accelerograms

These three earthquakes were applied to all frames from the three buildings evaluated, in order to obtain maximum displacement that can be reached by each one. In the software used [18], it was required the implementation of dynamic loads in direction X and the assignation of a control node located in the gravity center of the roof level.

The 3D non-linear dynamic analysis is based on the procedure explained in [20]. The RB building is analyzed, defining its geometry, materials and sections, serviceability loads in Y direction in all beams-columns joints, and dynamic loads on outer nodes with directions and combinations shown in Table 4. One direction ribbed slabs were modeled as rigid diaphragms in its plane by using additional elements with no flexural capacity (Figure 4).

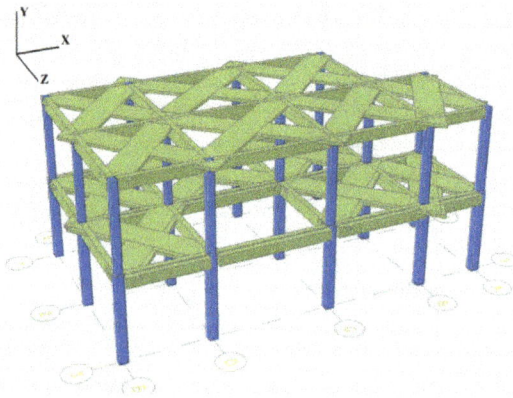

Figure 4. Rigid diaphragms in 3D *RB* framed building

Once built the model, there were applied all the accelerograms with the combinations shown in Table 3, for the interstorey drifts and maximum torsional moments on supports. These combinations are based on the Venezuelan seismic code [7] and following established by [24] about the seismic response of asymmetric structural systems in the inelastic range.

Nº	Seismic combination
1	100 % (X)
2	100 % (Z)
3	100% (X) y 30% (Z)
4	100% (Z) y 30% (X)

Table 4. Applied seismic combinations.

5. Results

From classic elastic analysis, the verification of interstorey drifts of the *OB* building, shows that they exceed the limit established in [7], while in the *RB* model it was obtained that it meets the code's parameters, which limits the inter storey drift to 0,018. By the other hand, in the *DBDB* building were not performed drifts verifications, since it was designed based on the method performed in [13], where the generated seismic forces are originally limited to not exceed the limit value of drift specified in the applied code.

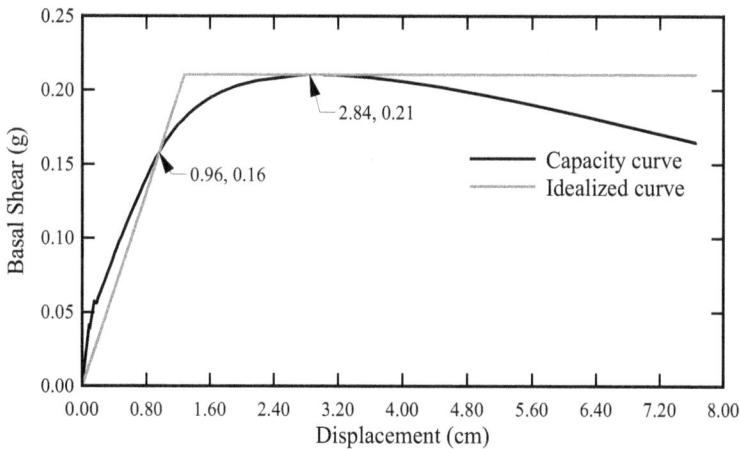

Figure 5. Normalized and idealized capacity curves. Frame C. *OB* building

To determine the values of structural ductility it was necessary to plot the idealized curve in function of the capacity curve obtained from non-linear pseudo-static analysis (pushover analysis), in order to know the point at which the structure begins to yield. Figure 5 shows an example of the normalized capacity curve with the idealized (bi-linear) curve of Frame C of *OB*. Structural ductility for each evaluated building values are presented in Table 5. From this Table it is evident that the original building, designed according to current Venezuelan codes has ductility values lower than the redesigned and the displacement-based buildings.

	Frame	Building		
		EO	ER	DBDB
	A	5,56	5,52	4,77
	B	2,22	6,04	5,38
Structural Ductility	C	2,17	4,69	5,25
	D	2,21	5,54	5,59
	E	2,23	7,07	6,06
	1	2,66	5,29	6,69
	2	2,20	4,17	5,92
	3	2,83	5,95	6,24

Table 5. Structural ductility results

From the obtained capacity curves there were computed the Performance point (Pp) of every frame of each evaluated building. Table 5 presents the values of Pp of all the frames of evaluated buildings. Figure 6 shows the determination of the Pp of Frame C from *OB* building using the N2 procedure proposed by Fajfar [5].

FRAME	Pp (cm)		
	EO	ER	DBDB
A	5,94	2,42	2,52
B	13,89	9,47	7,43
C	15,22	9,50	9,38
D	14,01	9,50	7,57
E	13,45	9,55	6,60
1	12,62	9,35	6,07
2	15,74	11,48	9,29
3	10,92	7,57	4,23

Table 6. Performance points (Pp) of studied buildings frames

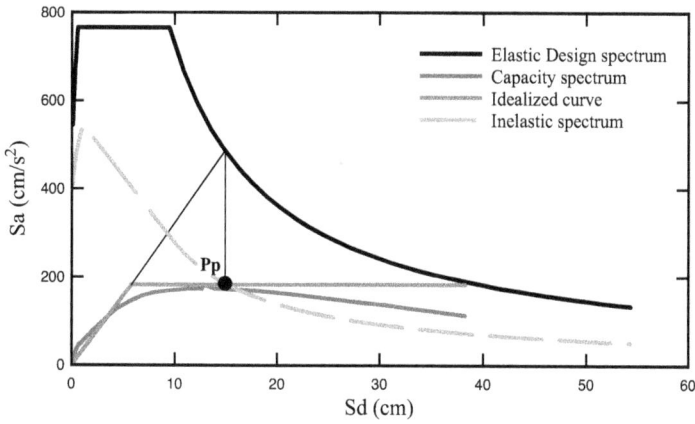

Figure 6. Performance point of Frame C. *OB* Building, determined by N2 procedure

From dynamic analyses, there were determined global and interstorey drifts of each frame from all three models studied. Both types of drifts were calculated on the basis of the application of synthetic accelerograms with different intensities, representing the lateral forces applied to frames in order to generate their respective maximum displacements. Figures 7 to 9 show the evolution of the global (Δ/H) drifts expressed as a percentage, respect to time (sec) of the frame C from *OB*, *RB* and *DBDB* models for a peak ground acceleration of 0,3 g, respectively.

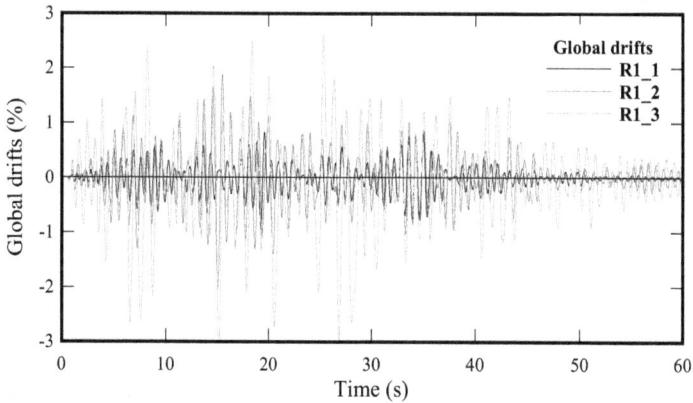

Figure 7. Global drifts. R1 earthquake. Frame C. *OB* building.

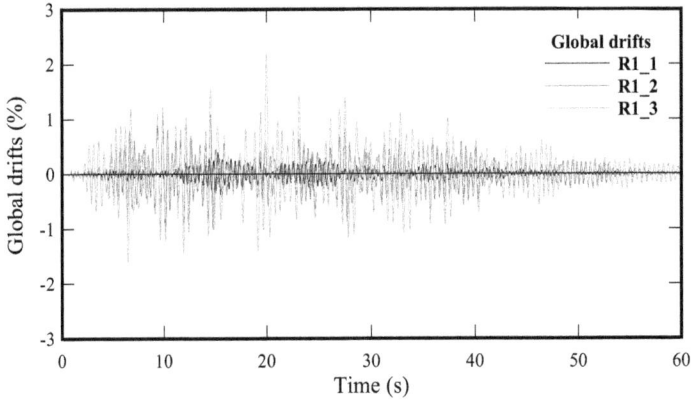

Figure 8. Global drifts. R1 earthquake. Frame C. *RB* building

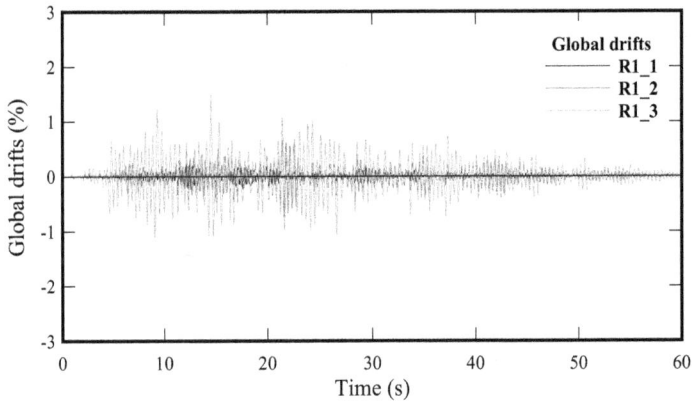

Figure 9. Global drifts. R1 earthquake. Frame C. *DBDB* building

Figures 10 to 12 show the results for interstorey drifts of frame C from *OB, RB* and *DBDB* buildings, taking into account the R1_3 earthquake with duration of 60 seconds. Similarly, interstorey drifts for applied earthquakes, R1, R2 and R3 with its three intensities, were obtained. It were verified for each Limit State considered in this study. Table 6 reflects the values of interstorey drifts of buildings in study for earthquake R1, taking into account the three levels of hazard, 0,5%, 1,5% and 3%, for the Limits States considered.

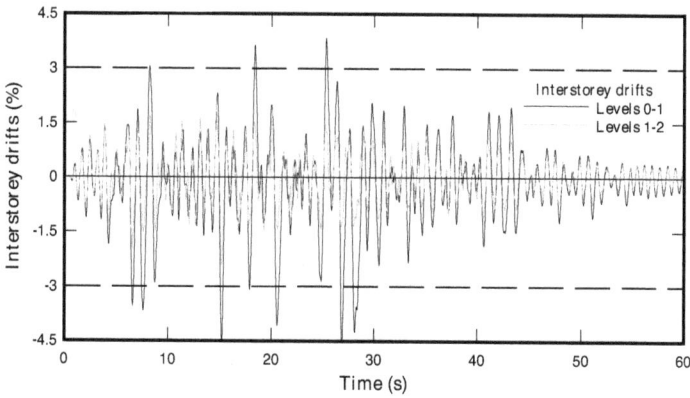

Figure 10. Interstorey drifts. R1_3 earthquake. Frame C. *OB* building

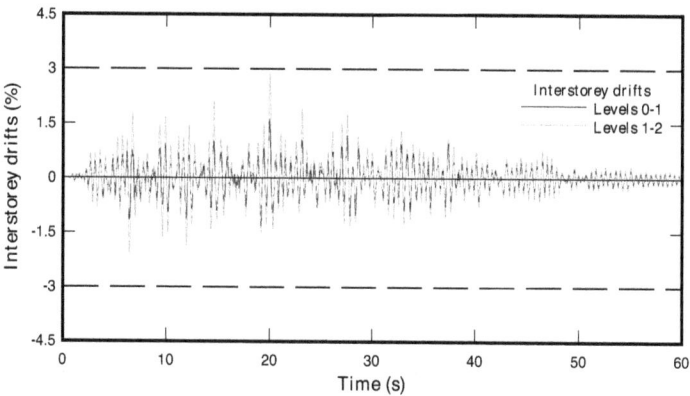

Figure 11. Interstorey drifts. R1_3 earthquake. Frame C. *RB* building

3D Nonlinear dynamic analysis

Interstorey drifts in frames of *RB*, were obtained by applying the R1_3 earthquake for the combinations 1 and 2 (Table 7). According to results obtained, interstorey drifts in 2D and 3D modeled buildings differ greatly from each other, for this reason it is important to take into account the 3D analysis in order to evaluate the drifts of buildings, because irregularities can produce lateral displacements that does not match with the obtained in 2D analysis.

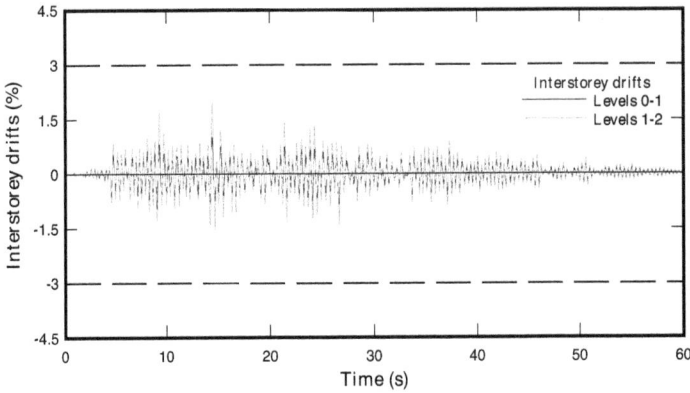

Figure 12. Interstorey drifts. R1_3 earthquake. Frame C. *DBDB* building

FRAME	LIMITS STATES								
	OB			RB			DBDB		
	SLS	RDLS	PCLS	SLS	RDLS	PCLS	SLS	RDLS	PCLS
A	-	-	-	-	OK	OK	OK	OK	OK
B	-	-	-	OK	OK	OK	OK	OK	OK
C	-	-	-	OK	OK	OK	OK	OK	OK
D	-	-	-	-	OK	OK	OK	OK	OK
E	-	-	-	-	OK	OK	OK	OK	OK
1	-	-	-	-	OK	OK	OK	OK	OK
2	-	-	-	-	OK	OK	OK	OK	OK
3	-	-	-	-	OK	OK	OK	OK	OK

SLS: Serviceability Limit State, **RDLS** Reparable damage Limit State, **PCLS**: Prevention of Collapse Limit State; -: No meet the norm,+: Checks the Venezuelan seismic code

Table 7. Interstorey drifts verification. R1 earthquake. *OB*, *RB* and *DBDB* building

In the 2D model greater drifts were obtained, while in 3D model the drifts were reduced by the contribution of the diaphragms. Also it were determined the maximum torsional moments in each column before the implementation of R1_3 earthquake in all supports for the four combinations described in Table 8. In Figure 13 have been plotting torsional moments in function of time for the four combinations, where nodes appointed by n111 until the n513 are corresponding to supports, while Figure 14 shows the maximum torsional moment range for each column from three-dimensional analysis. It is evident that for the accelerograms used, the maximum torsional moments occurs in the extreme columns and in the columns located in the intersection of the structure. This is an important feature that confirms the negative effect of the irregularity combined with the seismic action. The torsional moments for the other seismic combinations used in this study were obtained using the same procedure.

Seismic combination	Node-column Description	Max. Torsional Moment. (Nxm)
1	Corner column. n513	64225
2	Corner column. n512	76000
3	Corner column. n513	41000
4	Corner column. n512	65000

Table 8. Maximum torsional moments for seismic combinations.

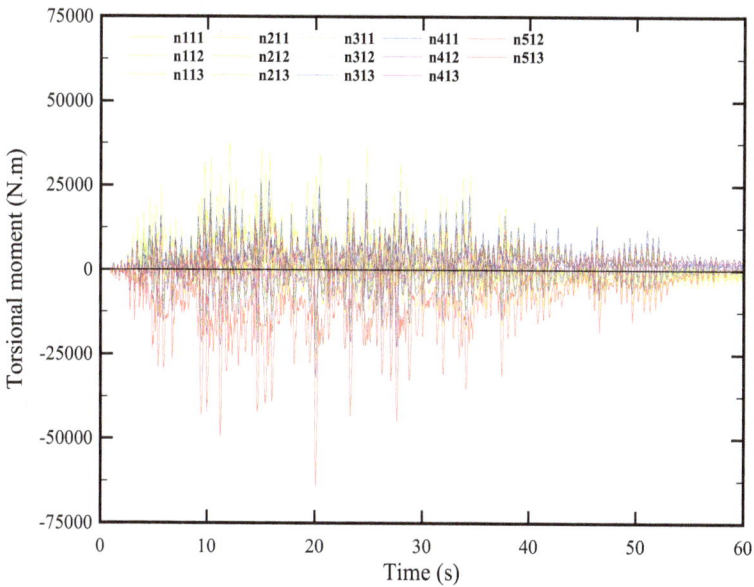

Figure 13. Torsional moments for earthquake 100% (X.

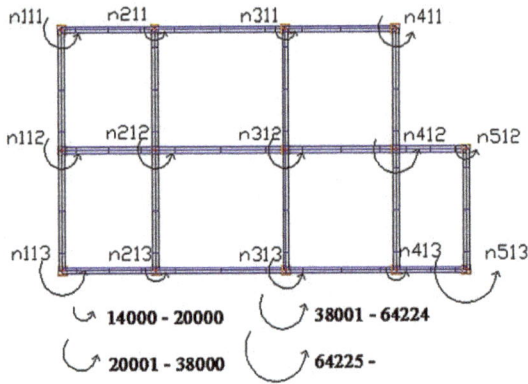

Figure 14. Torsional moments for earthquake 100% (X). Plant detail.

6. Conclusions

In order to know the seismic response of the studied building it were used analytical methods considering the seismic hazard level and structural regularity criteria. The elastic analysis applied to the *OB* building identified elastic displacements greater than maximum value of interstorey allowed by Venezuelan seismic code. From the resizing model *RB* it was obtained interstorey drifts that satisfied the maximum value established in the code. Thus, the sections of the structural elements of *OB* are insufficient to properly control the damage caused by seismic forces.

From dynamic analysis there were computed the global and interstorey drifts for all three evaluated models determining the dynamic response of these structures and controlling the damage level reached in them. With the global drifts, it was evaluated the threshold of the collapse Limit State, which corresponds to the maximum value of 2.5%. *RB* and *DBDB* buildings reached drifts values below this limit, proving good seismic performance on both buildings; *OB* presented drifts values which exceeded this limit. In the verification of interstorey drifts it was generally noted that interstorey drifts of *OB* building were longer than the considered by hazard levels, while the two resized buildings reached values within the thresholds established for each Limit State.

Three-dimensional dynamic analysis applied to *RB* building allowed determine that interstorey drifts values were under the threshold of the Limit States considered. On the other hand, in order to know the maximum torsional moments for each column in this model, there were applied four seismic combinations where it was noted that there was greater torsion in the case of the component of the earthquake in Z-direction. Based on these results it was demonstrated the structural asymmetry of the assessed building since the center of mass does not coincide with the center of rigidity, determining that the greatest torsional moments are on outer columns and inner corners.

Interstorey drifts of *RB* building obtained from 2D and 3D nonlinear dynamic analysis, it was noted that 2D model provided greater drifts values than the 3D model drifts. This is a logical and expected result since the 3D dynamic analysis considers the rigid diaphragm, which introduces restrictions to the number of degrees of freedom in the structure.

Inelastic static analysis is more reliable than linear methods in the prediction of the parameters of response of buildings, although this method has no response on the effects of higher modes of vibration. A more reliable and sophisticated method is the 2D no linear dynamic analysis, where it can be better determined the likely behavior of the building in response to the earthquake. However, the uncertainties associated with the definition of accelerograms used in these analysis and properties of coplanar structural models can be reduced with the implementation of the dynamic 3D analysis because there are considered factors associated with structural redundancy and are used more actual values in terms of rigidity of resistant structural lines.

The Quadrants Method presented in this paper is suitable for rapid and reliable evaluation of structures, with a low calculation effort. The cases studied demonstrated that the method can provide a reliable criterion to predict if any structure would have an inadequate seismic performance based on the results of static non-linear analysis. Results obtained from dynamic non-linear analysis confirmed the results obtained from the application of the Quadrants Method.

Despite the plan irregularity of the studied building, the Quadrants Method was suitably in order to predict that its lateral stiffness was not enough. Dynamics analysis confirmed this feature, then the cross sections of this building was resizing and details of the confinement were improved in order to meet the regulations of the current version of the Venezuelan seismic code. The seismic performance of the new designed building was tested with the Quadrants Method and dynamic analysis, showing that the resizing structure met all the Limit States used in this research.

Acknowledgements

Authors wish to acknowledge to the Scientific and humanistic Council of Lisandro Alvarado University for the financial support of the research in the field of non-regular structures. Also the authors would highlight the role of the International Center of Numerical Methods for Engineering in the collaborative research.

Author details

Juan Carlos Vielma[1*], Alex H. Barbat[2], Ronald Ugel[1] and Reyes Indira Herrera[1]

*Address all correspondence to: jcvielma@ucla.edu.ve

1 Structural Engineering Department, School of Civil Engineering, Lisandro Alvarado University, Venezuela

2 Department of Strength of Materials and Structural Analysis in Engineering, Technical University of Catalonia, Barcelona, Spain

References

[1] Grases J., Altez R. and Lugo M. Destructives Earthquakes Catalogue. Venezuela 1530/1998. Central University of Venezuela. Natural Sciiences, Physics and Mathematics Academy. Engineering School. Caracas, Venezuela, 1999.

[2] Pérez O. and Mendoza J. Seismicity and tectonics in Venezuela and surroundings areas. Earth Physics,1998, 10, pp 87-110.

[3] Barbat A., Mena U. and Yépez F. Probabilistic evaluation of seismic risk in urban zones. International magazine for numerical methods for Calculus and engineering projects. 1998, 14, 2, 247-268.

[4] Calvi, G., Pinho, R., Magenes, G., Bommer, J., Restrepo, L and, Crowley, H. Development of Seismic Vulnerability Assessment Methodologies over the Past 30 Years, ISET Journal of Earthquake Technology, 2006, Paper No. 472, 43, 3, 75-104

[5] Fajfar P. Nonlinear analysis method for performance based seismic design. Earthquake Spectra, EERI, United States of America, 2000, 16, 3, 573-591.

[6] Vielma, J. C., Barbat, A. H. and Oller, S. Framed structures earthquake resistant design. International Center for Numerical Methods in Engineering (CIMNE) Monograph. Earthquake Engineering Mongraphs. Barcelona, Spain, 2011.

[7] Covenin 1756:01. Earthquake-resistant Design code. Part 1. Fondo Norma. Caracas, Venezuela, 2001.

[8] Covenin 1753:06. Design and construction of buildings with structural concrete. Fondo Norma. Caracas, Venezuela, 2006.

[9] Covenin 2002:88. Minimum design loads and criteria for buildings code. Fondo Norma. Caracas, Venezuela, 1988.

[10] Vielma, J. C., Barbat, A. and Martínez, Y. The Quadrants Method: A procedure to evaluate the seismic performance of existing buildings. 15 World Conference on Earthquake Engineering. Lisbon, Portugal, 2012.

[11] Herrera, R., Vielma, J. C., Ugel, R., and Barbat, A. Optimal design and earthquake resistant design evaluation of low rise framed rc building. Natural Science, Earthquake Special Issue. August 2012. California, USA. Doi:10.4236/ns.

[12] Vielma, J. C., Barbat, A. H. and Oller, S. Seismic safety of RC framed buildings designed according modern codes. Journal of Civil Engineering and Architecture. 2011, 5:7, 567-575. David Publishing Company. Chicago, United States of America.

[13] Priestley M., Calvi G. and Kowalski M. Displacement-based seismic design of structures. IUSS Press. Pavia. Italia, 2007.

[14] Vielma, J. C., Barbat, A. H. and Oller, S. Nonlinear structural analysis. Application to evaluating the seismic safety. Nova Science Publishers. New York, 2009.

[15] Di Sarno, L. and Elnashai, A. Fundamentals of Earthquake Enginering. John Wiley and Sons. Chichester, United Kingdom, 2008.

[16] Park, R. State-of-the-art report: ductility evaluation from laboratory and analytical testing. Proceedings 9th WCEE, IAEE, Tokyo-Kyoto, Japan, Vol VIII: 605-616, 1988.

[17] Vielma, J. C., Barbat, A. H. and Oller, S. Seismic sizing of RC buildings according to energy-based amplification factors (In Spanish). Hormigón y acero. 2011. 63:263, 83-96.

[18] Elnashai A., Papanikolau V. and Lee, D. ZEUS-NL, A system for Inelastic Analysis of Structures. User Manual. Mid-America Earthquake Center report no. MAE, Illinois University. Urban, Champagne, Illinois, 2011.

[19] Papanikolaou, V. and Elnashai, A. Evaluation of conventional and adaptive pushover analysis I: methodology, Journal of Earthquake Engineering, 2005, 9, 6, 923-941

[20] Mwafy, A. and Elnashai, A. Static pushover versus dynamic collapse analysis of RC buildings, Engineering Structures, 2001, 23, 407–424

[21] Vamvatsikos D. and C. Allin Cornell (2002). Incremental dynamic analysis, Earthquake Engineering and Structural Dynamics, 2002, 31, 491–514.

[22] Kappos, A. and Stefanidou, S. A deformation-based seismic design method for 3D R/C irregular buildings using inelastic dynamic analysis. Bulletin of Earthquake Engineering, Springer, Netherlands, 2010, 8, 4, 875-895

[23] UCLA-CIMNE. Compatible accelerograms with elastic design spectrums generation software. (PACED). International Center for Numerical Methods in Engineering. Universidad Centroccidental Lisandro Alvarado. Venezuela, 2009.

[24] Fajfar, P., Marusic, and D., Perus, I. Torsional effects in the pushover-based seismic analysis of buildings. Journal of Earthquake Engineering, 2005, 9, 6, 831-854.

Permissions

The contributors of this book come from diverse backgrounds, making this book a truly international effort. This book will bring forth new frontiers with its revolutionizing research information and detailed analysis of the nascent developments around the world.

We would like to thank Sebastiano D'Amico, for lending his expertise to make the book truly unique. He has played a crucial role in the development of this book. Without his invaluable contribution this book wouldn't have been possible. He has made vital efforts to compile up to date information on the varied aspects of this subject to make this book a valuable addition to the collection of many professionals and students.

This book was conceptualized with the vision of imparting up-to-date information and advanced data in this field. To ensure the same, a matchless editorial board was set up. Every individual on the board went through rigorous rounds of assessment to prove their worth. After which they invested a large part of their time researching and compiling the most relevant data for our readers. Conferences and sessions were held from time to time between the editorial board and the contributing authors to present the data in the most comprehensible form. The editorial team has worked tirelessly to provide valuable and valid information to help people across the globe.

Every chapter published in this book has been scrutinized by our experts. Their significance has been extensively debated. The topics covered herein carry significant findings which will fuel the growth of the discipline. They may even be implemented as practical applications or may be referred to as a beginning point for another development. Chapters in this book were first published by InTech; hereby published with permission under the Creative Commons Attribution License or equivalent.

The editorial board has been involved in producing this book since its inception. They have spent rigorous hours researching and exploring the diverse topics which have resulted in the successful publishing of this book. They have passed on their knowledge of decades through this book. To expedite this challenging task, the publisher supported the team at every step. A small team of assistant editors was also appointed to further simplify the editing procedure and attain best results for the readers.

Our editorial team has been hand-picked from every corner of the world. Their multi-ethnicity adds dynamic inputs to the discussions which result in innovative outcomes. These outcomes are then further discussed with the researchers and contributors who give their valuable feedback and opinion regarding the same. The feedback is then collaborated with the researches and they are edited in a comprehensive manner to aid the understanding of the subject.

Apart from the editorial board, the designing team has also invested a significant amount of their time in understanding the subject and creating the most relevant covers. They scrutinized every image to scout for the most suitable representation of the subject and create an appropriate cover for the book.

The publishing team has been involved in this book since its early stages. They were actively engaged in every process, be it collecting the data, connecting with the contributors or procuring relevant information. The team has been an ardent support to the editorial, designing and production team. Their endless efforts to recruit the best for this project, has resulted in the accomplishment of this book. They are a veteran in the field of academics and their pool of knowledge is as vast as their experience in printing. Their expertise and guidance has proved useful at every step. Their uncompromising quality standards have made this book an exceptional effort. Their encouragement from time to time has been an inspiration for everyone.

The publisher and the editorial board hope that this book will prove to be a valuable piece of knowledge for researchers, students, practitioners and scholars across the globe.

List of Contributors

Alejandro Gaytán
Universidad de Guadalajara, Departamento de Ciencias Computacionales, Centro Universitario de Ciencias Exactas e Ingeniería, Mexico

Carlos I. Huerta Lopez
University of Puerto Rico at Mayaguez Civil Engineering and Surveying Department, Puerto Rico Strong Motion Program Home Institution: Research Center and Higher Education at Ensenada, CICESE Seismology Department, Earth Sciences Division, Puerto Rico

Jorge Aguirre Gonzales and Miguel A. Jaimes
Instituto de Ingeniería, Universidad Nacional Autónoma de México, Circuito Interior, Ciudad Universitaria, Mexico

Sayed Hemeda
Conservation Department, Faculty of Archaeology, Cairo University, Egypt

M.R. Costanzo, C. Nunziata and V. Gambale
Dipartimento di Scienze della Terra, dell' Ambiente e delle Risorse, Univ. Napoli Federico II, Italy

Babak Ebrahimian
School of Civil Engineering, Faculty of Engineering, University of Tehran, Tehran, Iran

F. Panzera and G. Lombardo
Dipartiemnto di Scienze Biologiche, Geologiche ed Ambientali, Universita' di Catania, Italy

S. D'Amico and P. Galea
Department of Physics, University of Malta, Malta

Won Sang Lee and Joohan Lee
Korea Polar Research Institute, Republic of Korea

Sinae Han
Korea Polar Research Institute, Republic of Korea
Kangwon National University, Republic of Korea

Alessandro Contento, Daniele Zulli and Angelo Di Egidio
Dipartimento di Ingegneria Civile, Edile-Architettura, Ambientale University of L'Aquila, Italy

Vitaly Yurtaev and Reza Shafiei
Moscow State University of Railways, Department Of Civil Engineering, (MIIT), Moscow, Russia
Washington State University, Pullman, WA, USA

C. Mullen
Department of Civil Engineering, University of Mississippi-Oxford, USA

Vincenzo Gattulli
DICEAA, CERFIS, University of L'Aquila, Italy

Babak Ebrahimian
School of Civil Engineering, Faculty of Engineering, University of Tehran, Tehran, Iran

Juan Carlos Vielma, Ronald Ugel and Reyes Indira Herrera
Structural Engineering Department, School of Civil Engineering, Lisandro Alvarado University, Venezuela

Alex H. Barbat
Department of Strength of Materials and Structural Analysis in Engineering, Technical University of Catalonia, Barcelona, Spain